Intellectual Property for Electronic Systems: An Essential Introduction

Executive Editors

Brian Bailey
Proprietor
Brian Bailey Consulting

Kathy Werner
IP Strategy and Business Manager
Freescale Semiconductor
President
VSI Alliance

IEC
Chicago, Illinois

ISBN: 978-1-931695-57-2

Intellectual property for electronic systems : an essential introduction / executive editors, Brian Bailey, Kathy Werner.
p. cm.
Includes bibliographical references.
ISBN 978-1-931695-57-2
1. Integrated circuits--Design and construction. 2. Semiconductors--Design and construction. 3. Intellectual property--Handbooks, manuals, etc. I. Bailey, Brian, 1959- II. Werner, Kathy.
TK7874.I47156 2007
621.381068--dc22
2007040901

International Engineering Consortium
300 West Adams Street, Suite 1210
Chicago, Illinois 60606-5114, USA
+1-312-559-4100 voice • +1-312-559-4111 fax
publications@iec.org • *www.iec.org*

About the Executive Editors

Brian Bailey is an independent consultant who helps electronic design automation (EDA) and system design companies with technical, marketing, and managerial issues related to verification and electronic system level (ESL). He was with Mentor Graphics for 12 years, where he last was chief technologist for verification.

He is the co-editor of *Taxonomies for the Development and Verification of Digital Systems* (Springer, 2005), the executive editor and author for *The Functional Verification of Electronic Systems: An Overview from Various Points of View* (IEC Press, 2005), and co-author of *ESL Design and Verification: A Prescription for Electronic System Level Methodology* (Morgan Kaufmann/Elsevier, 2007). He has published many technical papers, has given keynote speeches at conferences, performed seminars around the world, and has been both a contributor and moderator of panels at many major conferences.

Brian established the functional verification track at the DesignCon conferences, which has quickly grown to be one of the major tracks of the conference. He also serves on the technical program committees of many major conferences, including the Design Automation Conference (DAC). He chairs the interfaces standards group and the coverage terminology sub-group within Accellera and has also chaired other standards groups in Accellera and VSIA. Brian graduated from Brunel University in England with a first-class honors degree in electrical and electronic engineering.

Kathy Werner is the IP strategy and business manager at Freescale Semiconductor and oversees IP coordination, standardization, and quality, and Freescale's interaction with indsutry organizations for systems-on-chip (SoC) and IP issues. She is the president of the VSI Alliance, the leading industry organization for IP standards, and she facilitated the successful transfer of VSI's work to other industry organizations. Kathy chaired the VSI Alliance quality IP (QIP) pillar for six years during which the first QIP metric was introduced. She is working with other industry organizations to extend this metric and further adopt VSI standards. Kathy is also the chair for the DesignCon IP business and engineering impacts track.

About the Publisher

The International Engineering Consortium (IEC) is a nonprofit organization dedicated to catalyzing technology and business progress worldwide in a range of high technology industries and their university communities. Since 1944, the IEC has provided high-quality educational opportunities for industry professionals, academics, and students. In conjunction with industry-leading companies, the IEC has developed an extensive, free on-line educational program. The IEC conducts industry-university programs that have substantial impact on curricula. It also conducts research and develops publications, conferences, and technological exhibits that address major opportunities and challenges of the information age. More than 70 leading high-technology universities are IEC affiliates, and the IEC handles the affairs of the Electrical and Computer Engineering Department Heads Association and Eta Kappa Nu, the honor society for electrical and computer engineers. The IEC also manages the activities of the Enterprise Communications Consortium.

IEC Corporate Members / University Program Sponsors

The IEC's University Program, which provides grants for full-time faculty members and their students to attend IEC Forums, is made possible through the generous contributions of its Corporate Members. For more information on Corporate Membership or the University Program, please call +1-312-559-4625 or send an e-mail to *cmp@iec.org*.

IEC Corporate Members

These are some of the universities that participate in the University Grant Program:

The University of Arizona
Arizona State University
Auburn University
University of California at Berkeley
University of California, Davis
University of California, Santa Barbara
Carnegie Mellon University
Case Western Reserve University
Clemson University
University of Colorado at Boulder
Columbia University
Cornell University
Drexel University
École Nationale Supérieure des Télécommunications de Bretagne
École Nationale Supérieure des Télécommunications de Paris
École Supérieure d'Électricité
University of Edinburgh
University of Florida
Georgia Institute of Technology
University of Glasgow
Howard University
Illinois Institute of Technology
University of Illinois at Chicago
University of Illinois at Urbana/Champaign
Imperial College of Science, Technology and Medicine
Institut National Polytechnique de Grenoble
Instituto Tecnológico y de Estudios Superiores de Monterrey
Iowa State University
KAIST
The University of Kansas
University of Kentucky
Lehigh University
University College London
Marquette University
University of Maryland at College Park
Massachusetts Institute of Technology
University of Massachusetts
McGill University
Michigan State University
The University of Michigan
University of Minnesota
Mississippi State University
The University of Mississippi
University of Missouri-Columbia
University of Missouri-Rolla
Technische Universität München
Universidad Nacional Autónoma de México
North Carolina State University at Raleigh
Northwestern University
University of Notre Dame
The Ohio State University
Oklahoma State University
The University of Oklahoma
Oregon State University
Université d'Ottawa
The Pennsylvania State University
University of Pennsylvania
University of Pittsburgh
Polytechnic University
Purdue University
The Queen's University of Belfast
Rensselaer Polytechnic Institute
University of Southampton
University of Southern California
Stanford University
Syracuse University
University of Tennessee, Knoxville
Texas A&M University
The University of Texas at Austin
University of Toronto
VA Polytechnic Institute and State University
University of Virginia
University of Washington
University of Wisconsin-Madison
Worcester Polytechnic Institute

Preface

Brian Bailey, Proprietor, Brian Bailey Consulting

Kathy Werner, IP Strategy and Business Manager, Freescale Semiconductor; President, VSI Alliance

The electronic design automation (EDA) industry solely exists for one reason—to increase the level of productivity in the creation of electronic systems. This productivity increase can be brought about through better tools, languages, or methodologies and can be either direct or indirect in their benefits. For example, a synthesis tool allows more gates to be designed per day, not only because of the efficiencies that the tool provides, but also by enabling the usage of higher abstraction languages that allow more complex things to be described with fewer errors. The first gain is the direct impact of the tool, whereas the gain from the higher level of abstraction is an indirect gain. This does not imply that one is more valuable than the other, yet no one has individually measured the productivity impact of each of these.

A secondary effect of this productivity increase is that larger and more complex systems can be created; as a result, this fuels the entire electronic industry as consumers appear to have a never-ending thirst for new, more capable gadgets that make their lives either more fun or more efficient. The increase in abstraction allows engineers to comprehend and reason about larger systems because many of the lower levels of details have been removed. Thus, it would appear that the human mind has the capability to deal with a relatively fixed number of concepts as a time, and so the more abstract we can make those concepts, the larger the total system complex can be, once it has been implemented.

At the same time, the technology associated with fabricating electronic systems—be they in complete systems, boards, or integrated circuits—has been through the same kind of continuous improvement, and the advancements of silicon chip fabrication technology has been well-documented and discussed for the last 40 years or more.

A Tale of Two Moores

Few engineers have not come across the name of Moore and Moore's Law, that is to say, Gordon Moore, one of the co-founders of Intel in 1968. His observation that became known as Moore's law predates this when in 1964, as the head of research and development at Fairchild Semiconductor, he observed that the number of transistors that could be put on a single chip would double every year. This was quite a bold call given that he had only a little over four years of data to base this on. For a decade, his observation held truth, but then the pace slowed to a rate closer to every two years. And in 1975, he officially changed the growth rate, but the industry prefers to remember it as 18 months.

Year of introduction	# of Transistors	Processor
1971	2,250	4004
1972	2,500	8008
1974	5,000	8080
1978	29,000	8086
1982	120,000	286
1985	275,000	386 processor
1989	1,180,000	486 DX processor
1993	3,100,000	Pentium® processor
1997	7,500,000	Pentium II processor
1999	24,000,000	Pentium III processor
2000	42,000,000	Pentium 4 processor

Table 1: *Moore's Law at Intel*

After it became popularized, Moore's Law became the driver for the industry. Every few years, the physicists and fabrication technology people herald the end of Moore's law, only to find a new material or method which allowed what had been an unbreakable barrier before to be brushed aside with the new view being that we could go on almost forever. Nowadays, there is no sign to an end to the complexity of systems, even though we are seeing perhaps the end of the line for the processor, namely, the "grand-daddy" of the industry. It has become too difficult and expensive, mainly in terms of cost and power, to make a single processor run faster. At the same time, the size reductions in traditional complimentary metal oxide semiconductor (CMOS) fabrication technology may be closer to the end than the beginning, but even a dead-end there will not stop systems from becoming bigger and more complex.

There is a second Moore, Geoffrey Moore, who may not be as familiar to engineers as Gordon Moore, but who will probably be more familiar to their managers. He is the author of a number of business books, and the first one which was published in 1991 is called *Crossing the Chasm* [1]. Later in his career, he refined and extended the lifestyle concepts of technology products with such releases as *Living on the Fault Line* [2] and *Dealing with Darwin* [3].

In 2004, Geoffrey Moore published an article in the *Harvard Business Journal* [4]. In that article he wrote:

> "It's important to recognize that differentiation-creation innovation and productivity-creating deconstruction must be conducted in tandem. If you try the former without the latter, the inertia demon defeats you. If you try the latter without the former, you do nothing to overcome the forces of commoditization..."

This statement takes a little deciphering. He affirms that there are two types of innovation, namely technological and process. (1) Technological innovations allow you to create bigger and better things and to differentiate yourself from the competition and (2) process innovations enable you to do things more efficiently. We need both types of innovation.

If we concentrate only on being able to create things cheaper and designing them faster, then quite quickly the things that we are designing will become commodities. This has been frequently occurring in the manufacturing centers of the world as cheaper, or more productive, labor forces have become available, and the transportation necessary to move the finished goods has improved. This is also what mostly happens in the memory industry. For example, as soon as the technology has become known to manufacture a bigger memory, then the next few years are all about increasing the yield to bring down costs. The same applies to the design of logic functions. For example, as previously stated, the introduction of logic synthesis made it much more efficient to create certain kinds of circuits, but if we used synthesis just as a productivity tool to design those same circuits faster, then there would be little need for people to use synthesis at all since it is unlikely that the new designs and consumer devices would be any better than the previous generation.

We thus need the second kind of innovation—technological innovation—which allows us to distinguish one product from another. This is what most engineers consider the most important part of their jobs. However, when the next design starts, as engineers know, it will be bigger and more complex, but there will be no extra people on the team and the schedule will not get any longer. If you do every design with the same tools and the same level of productivity, then you will very quickly not be able to handle these larger designs. Can you imagine attempting to create a multi-function, multi-media phone starting from scratch at the gate level? It would just not be economically possible even if you did manage to succeed technically. So, even though engineers may like to concentrate on the technological innovation, they must also improve their efficiency with each new design.

Bailey's Law

When we combine the teachings from these two Moore's, Gordon and Geoffrey, we can make another observation about our industry which I will call Bailey's Law. I first offered this observation in a book entitled *The Functional Verification of Electronic Systems* [5] which states that the productivity of a design team must double every 12 months. This is the rate of productivity increase necessary to keep up with the growth of system complexity as outlined by Gordon Moore. This productivity increase is at a rate higher than Moore's law because systems are not just about the electronic hardware. They are also about the software component which is increasing at an even faster rate, along with the degree of solution connectivity and the difficulties associated with functional verification which grow much faster than design size. Systems are no longer working in isolation, but they can talk to many other systems spread over a wide geographic distance in order for them to perform their intended function.

In 2001, the International Technology Roadmap for Semiconductors (ITRS) introduced a cost model that was then presented in a paper by Kahng and Smith [6]. This cost model showed the escalating costs associated with the design of chips and the need for process improvements if the costs were not to get out of hand. The biggest single factor in those costs is manpower and thus indirectly the need for improvements in designer efficiency. *Figure 1* shows this cost model. The top line (register-transfer level [RTL] methodology only) is the escalating cost that would be incurred if not for process innovations and the adoption of new tools. While some people disagree about those estimated costs, it is within the right order of magnitude. The lower line shows the impact that a number of discrete improvements have made in the overall costs. Again, some people may argue about the actual process innovation steps and the impact they had, but the result is still within the right ballpark. If we draw a straight line through those points, it would be flat, showing that the cost reductions and total engineering efficiency is keeping pace with the increase in design sizes, and thus one can see the doubling of efficiency every twelve months.

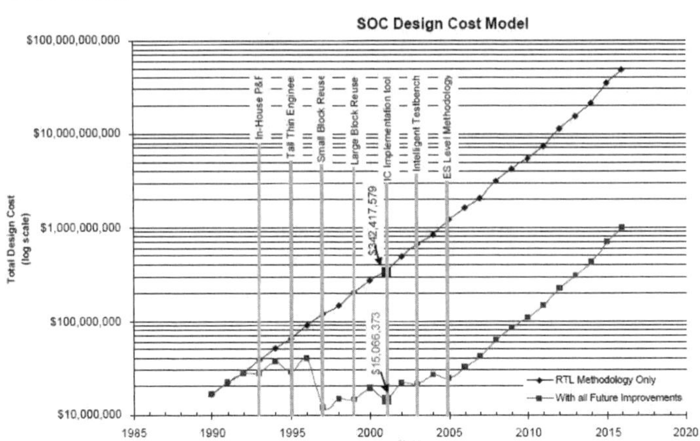

***Figure 1:** Cost Model*

Two of the named improvements are small and large block reuse, and in more recent times, a third would have to be added called platform-level reuse. In addition, some of the gains from the intelligent testbench have come from the reuse in the verification world. This clearly demonstrates the impact and value that reuse has had on this industry so far, but getting there was not easy and significant efforts are still required to get the full impact of these process innovations.

History Repeats Itself

I remember a remark that was made to me many years ago which was that, in this industry, we have to resolve the same problems every twenty years or so. By the time the problem comes around again, it has either been transposed into a slightly different form, or the people involved with it are no longer around, and the solutions have been forgotten. An example of this is how we migrate to a new level of abstraction, which is nowadays a big EDA issue. Few people were around during the last such transition when the industry migrated from gate level schematics to language-based RTL. That was not only a change in abstraction, but also in the way in which systems were described. People tend to remember it the way they want it remembered and the tales of all of those companies that failed to make the change, be they on the EDA side or the designer side which get lost in time. This should come as no surprise because the side that wins the war writes the history books in general. However, just looking at the past for direction is not always the right thing to do. In the case of our example, RTL simulation took hold long before synthesis came along, but did not become a significant industry until after logic synthesis established it in the flow, that is, even after that gate level simulation persisted in the flow for many years. Today, we are seeing the migration to the electronic system (ES) level, which presents a very confusing picture to many people—partly because they expect it to happen in the same way that the transition to RTL was imagined to have happened.

Looking back over history, the reuse problems of today were also present and solved in the past. Any design engineer with 20 years of experience will probably still be able to tell you half of the devices in the 7400 series. These were the reusable building blocks of the board era, and Texas Instruments set the standards that almost everyone else followed. They defined the logic family, the interconnect, and the form of the data sheets. Other industry standards were adopted for the physical interconnect such as package sizes, pin spacing, among others. Parallel device families existed for faster devices, more powerful drivers, low power, etc., that were identical in every other way. While some of these packages were primitive, especially in terms of today's levels of reuse, others were quite extensive. An entire set of tools grew up around these standards and this made for plenty of choice and competition within the tool industry. The key to its success was due to constraints on the interfaces, consistency between parts and manufacturers, and consistent, if not complete, documentation.

Re-use and Standards

The success of the intellectual property (IP) industry and standards go hand-in-hand. Almost every aspect of the design flow contributes to the ability to have a thriving IP marketplace. For example, a piece of IP needs to fit into a design flow. This means that two languages, Verilog and VHDL, along with certain synthesis tools and flows have to be supported, and certain libraries have to be targeted. But, these are just about the flow. What about the design? If we assume for a moment that the IP is a peripheral that will attach to a bus, then which bus fabrics are to be supported? What about on-chip debug and connecting this to other debug blocks on the chip? How about physical test of the chip? The next set of issues relates to verification, namely how the IP was tested and how is it going to be tested once it has been integrated into the system. Further, which verification languages are to be supported and which verification methodology if any has been used. Then, there are the less tangible aspects about a block, as if it was designed well, how often has it been used in real chips, and how many people are using it.

Without standards, it would not only become impossible to create a piece of IP that could target enough customers to be economically and technically feasible, but it would also be difficult to tell a potential customer what they could expect to receive if they purchased your product. In the early days of this virtual IP market, there were a lot of disappointed customers and a lot of frustration felt on both sides of the fence. This gave the early pioneers in this area a bad taste of re-use and scared many people away from it for quite some time. It was not uncommon for people to say that it took them longer to fix third-party IP than it would have taken to design it themselves. Of course, there was another side to this as well which is that few people would leave a piece of IP un-altered and thought that they could make a few improvements before installing it into their product. This, of course, destroyed any verification that had been done on the IP and since verification often takes more time than the design itself, it is not surprising that people arrived at this conclusion.

Almost ten years ago, organizations such as the Virtual Socket Interface (VSI) Alliance acknowledged the problems at hand and affirmed that the only way to fix them was through standards. As much as anything else in the early days, it was an education issue about what IP vendors and users should expect to provide and the ways in which they should conduct business. The problem is that standards are very difficult to define ahead of the proper formulation of the problem. The early IP providers and users were experimenting with business models, delivery vehicles, and many of the other issues that are now being addressed by the industry.

It is a rare situation when a good standard comes from a committee. When everyone in the room is trying to learn the best way to do something or trying to think of the ultimate solution, the solutions, at best, tend to be overcomplicated or miss obvious issues that would have come about when looking at an implementation. There is also the issue of politics in which every party to the standards effort wants to have their contribution or special case included. Again, this leads to bloated and inefficient standards in most cases.

The best standards within the EDA industry are those that have been developed by one or few people working in close co-operation that have come up with a solution that has been proven to work in practice. This provides a firm foundation onto which additional functionality can be added, if necessary, but makes it more difficult to be side-tracked into feature explosion. This is now the model that many standards bodies such as Accellera encourage. After an issue has been established and fairly well-defined, a call for donations is made. Several competing solutions may then be offered and the best one selected as the basis for the standardization work, often with features from the others being included.

The Emergence of This Book

With all of the importance and attention on the IP industry, it would be natural to think that plenty of books have been written on the subject, but in fact I do not know of any that cover all aspects of the IP flow, or provide the kind of insights necessary for someone thinking of being a provider or user of IP for the first time to be able to learn about all of the issues they need to be aware of. The only book worth mentioning is the *Reuse Methodology Manual* [7], now in its third edition, which when first published was almost a proposal of how IP should be designed and verified, and a lot has changed since then. The book has evolved to keep pace with the changes over the years. The later editions have also included some examples of how companies put the methodology into practice.

The International Engineering Consortium and DesignCon

Education is an important part of every engineer's career. This industry moves so fast and new developments turn up every day that require us all to learn new languages, tools, techniques, and methodologies. If we are to keep up with these changes, then we risk falling behind the rest of industry. But, who are the educators? While the EDA industry does a good job of providing training for languages and tools, they are doing this not only for the benefit of the industry, but also to increase sales of the tools they have available and that favor the feature sets that they can provide. We must thus have additional ways to obtain this education. One of those additional sources is through the conferences that publish papers about the new advances made in academia and research departments, such as the Design Automation Conference (DAC) and Design, Automation, and Test in Europe (DATE). Another source is conferences such as DesignCon, which is organized by the International Engineering Consortium (IEC), a non-profit educational group serving the industry. This conference does not publish research, but instead concentrates on what is real and available now, presented by engineers in a format that allows them to discuss a subject in great detail

A number of years back, the IEC and DesignCon, agreed to create some additional tracks that were outside of their normal focus of high-speed design issues, including more general design and verification issues for system-on-a-chip (SoC). These tracks enabled area practitioners to inform the rest of the community about their work and research, along with their success and failures. In the few years that these new tracks have existed, they have attracted many of the most highly talented people in the field, both as attendees and as presenters, along with high quality papers, and its stature as a premiere event for design and verification continues to grow. Two years ago, the IEC asked me to put together a book that leveraged the papers that had been submitted to the functional verification track. This book, *The Functional Verification of Electronic Systems: An Overview from Various Points of View* was a big success and a first for the IEC. It not only drew upon the published papers from the conference, but also invited many industry leaders to create additional chapters for the book that would make it a much more complete text about verification than would have been possible by just republishing papers. At the same time, many of the papers were significantly updated to include more current information. These changes made it much more valuable than just a compendium of papers that many other conferences publish.

More recently, the IEC approached me again to create a second book for them. I reviewed all of the papers from the last four years from all of the design and verification tracks and noticed the large quantity of papers on intellectual property. This was thus a book subject that would provide significant value and a subject in which engineers believed they had a lot of good information to share with other companies. When the subject was decided upon, the planning for the book then started. As with the previous book, this book draws from some of those previously published papers, and it also pulls together many invited chapters that deal with important information. Those chapters were written by some of the most important and influential people in the field.

Together, these contributions will provide the reader with some insights into the future of the industry in terms of new tools, new languages, and new methodologies, and provide direct examples of how certain problems are being overcome.

References

[1] Geoffrey A. Moore, *Crossing the Chasm: Marketing and Selling High-Tech Products to Mainstream Customer*, Harper Business Press, 1991.

[2] Geoffrey A. Moore, *Living on the Fault Line: Managing for Shareholder Value in any Economy*, Harper Business Press, 2000.

[3] Geoffrey A. Moore, *Dealing with Darwin: How Great Companies Innovate at Every Phase of Their Evolution*, Portfolio, 2005.

[4] Geoffrey A. Moore, *Darwin & the Devil: Innovating within Established Enterprises*, Harvard Business Review, July-August 2004, pp. 86-92.

[5] Brian Bailey (Exec. Ed.), *The Functional Verification of Electronic Systems. An Overview from Various Points of View*, Professional Education International, 2005.

[6] Andrew B. Kahng and Gary Smith, *A New Design Cost Model for the 2001 ITRS*, Proceedings of the International Symposium on Quality Electronic Design, 2002

[7] Keating, Michael; Bricaud, Pierre (Eds.), *Reuse Methodology Manual for System-on-a-Chip Designs: Third Edition*, Springer, 2002.

[8] Brian Bailey, Grant Martin, and Andrew Piziali, *ESL Design and Verification: A Methodology for Electronic System Level Methodology*, Morgan Kaufmann/Elsevier, 2007

Acknowledgements

I thank the IEC for giving me the opportunity to make this book available, and I hope that it can help everyone in the never-ending quest to improve development productivity. I also thank all of the book's contributors for their time and energy and especially Kathy Werner for acting as my co-executive editor to bring together the people necessary to make this book happen.

Brian Bailey
August 2007

Most things in life come about through team work, and this book was certainly no exception to that. When Brian first approached me about the possibilities for this book, I was both flattered and a little nervous as I had never done anything like that in the past. However, while Brian is an experienced writer, I knew the IP industry and the people that would be able to bring this project to life—so together, we made a good team. I would like to thank all of the contributors who dedicated many hours to complete this work and to all the people that have built VSIA, and more importantly, enabled the industry to progress and mature. Organization should be judged not only by the standards or documents that they produce, but also by the impact they have in providing the right kind of forum that attracts the brightest people in an industry to come together, discuss, and solve the problems that inhibit an industry's growth. Without this pioneering spirit, the industry would not have advanced as much as it has over the past ten years. The decision was recently made to close down VSIA and transfer the work to other organizations as part of an industry standards consolidation. As the past president of VSIA, I hope that I can continue to foster the spirit of co-operation that made VSIA so successful.

Kathy Werner
August 2007

Table of Contents

About the Executive Editors iii
About the Publisher........ iv
Acronym Guide........ 479

Preface vii
Acknowledgements xix

Part I: Overview

Chapter 1: Introduction 3
Brian Bailey, Proprietor, Brian Bailey Consulting
Kathy Werner, IP Strategic Business Manager, Freescale Semiconductor; President, VSI Alliance
1.1: Introduction 3
1.2: Historical Overview 4
1.3: Business Models 7
1.4: Who Should Read This Book? 8
1.5: The Book's Structure and How to Read It 9
1.6: Chapter Listing........ 10

Chapter 2: The Business of SIP........ 17
Jack Horgan, Managing Editor, EDACafe
2.1: Introduction 17
2.2: Value Proposition........ 20
2.3: Market Size 23
2.4: How to Find IP Vendors 28
2.5: Publicly-Traded Companies........ 31
2.6: The Big Three EDA Companies 48
2.7: Privately-Held Companies 50
2.8: Non-SIP Companies........ 50
2.9: Entering the IP Market........ 51

Chapter 3: Building a Legally Sound Intellectual Property Portfolio 53
Dennis Fernández, Managing Partner, Fernández & Associates, LLC
Behfar Bastani, United States Patent Agent
Susan Fintz, Attorney, Wilson, Sossini, Goodrich, and Rosati
3.1: Introduction 53
3.2: Forms of Protection 54
3.3: Design States 56
3.4: License Terms, Risks, and How to Mitigate 60

Part II: The IP Provider

Chaper 4: Analog, Star, and Commodity IP Blocks 71
Henry Chang, Vice President and Co-Founder, Designer's Guide Consulting
Larry Cooke, Engineering Consultant
Grant Martin, Chief Scientist, Tensilica, Inc.; Senior Member, IEEE
4.1: Introduction 71
4.2: Analog IP 73
4.3: Star IP 81
4.4: Commodity IP 92
4.5: Emerging Issues 101

Chapter 5: IP Protection 107
Saverio Fazzari, Associate, Booz Allen Hamilton
5.1: Introduction 107
5.2: Why Protect IP? 108
5.3: History of IP Protection 110
5.4: EDA and IP Protection 113
5.5: Different Technical IP Protection Methods 114
5.6: Legal IP Protection Issues 116
5.7: Different IP Types 117
5.8: Challenges for Successful IP Protection 119
5.9: Future of IP Protection 122

Chapter 6: Software 123
Richard Vlamynck, Verification Engineer, Mentor Graphics Corp.
6.1: Introduction 123
6.2: Software Considerations for Hardware IP 123
6.3: The Fundamental Problems of Software Drivers for Hardware IP 125

6.4: Binaries versus Source Code 133
6.5: Software Verification and Debug 134
6.6: Conclusion 138

Chapter 7: Verification IP 141
Sean Smith, Chief Verification Architect, Denali Software
7.1: What is Verification IP? 141
7.2: Benefits of IP 143
7.3: Make versus Buy 146
7.4: Evolution of Verification IP 149
7.5: Architecture of Modern VIP 155
7.6: Future VIP Directions 158
7.7: Checklists for Evaluating VIP 161
7.8: Summary 166

Chapter 8: Verification of IP 169
Leena Singh, Senior Staff Manager and Engineer, Qualcomm, Inc.
8.1: Introduction 169
8.2: Challenges in Verifying IP and Industry Trends 171
8.3: Creating a Verification Environment 179
8.4: Example Implementation 184
8.5: Following Standards 193

Chapter 9: IP Quality 197
Kathy Werner, IP Strategic Business Manager, Freescale Semiconductor; President, VSI Alliance
9.1: Introduction 197
9.2: Internal IP Reuse Quality 198
9.3: Third-Party IP Quality 207
9.4: Quality Specific to IP Types 212
9.5: Future Quality Considerations 216

Part III: The IP User

Chapter 10: IP Selection 221
Adam Traidman, President, Chip Estimate Corporation
Pallab Chatterjee, Chief Technical Officer, SiliconMap, LLC
10.1: Introduction 221
10.2: The Process of IP Selection 222
10.3: IP Sourcing 223

10.4: Selection Criteria ... 230
10.5: IP Evaluation ... 239

Chapter 11: Block-Level and Platform-Level Reuse ... 245
Jeff Freeman, Member, Technical Staff, Transportation and Product Standard Group, Freescale Semiconductor
11.1: Introduction ... 245
11.2: Architecture ... 245
11.3: Block-Level Reuse ... 247
11.4: Platform-Level Reuse ... 248
11.5: Verification Reuse ... 251
11.6: Documentation Reuse ... 254
11.7: Points of View ... 254
11.8: Conclusion ... 255

Chapter 12: Integration, Test, and Debug ... 257
Neil Stollon, Principal Engineer, HDL Dynamics
12.1: Introduction ... 257
12.2: Integration Issues ... 258
12.3: Digital IP Integration ... 265
12.4: Analog IP Integration ... 275
12.5: Integration of Soft versus Hard IP ... 275
12.6: Test Issues ... 277
12.7: Design for Debug Issues ... 283
12.8: Other Related Issues ... 295
12.9: Conclusion ... 298

Part IV: Case Studies

Chapter 13: Back-End Methodology and Techniques for a Multiprotocol Mixed Signal IP Design ... 303
Ken Umino, Consultant, Synopsys Professional Services
Jason Upton, Synopsys Professional Services
Ross Segelken, Design Engineer, nVidia
John T. Stonick, Application Engineer, Synopsys Professional Services
13.1: Introduction ... 303
13.2: Basic Architecture Description ... 304
13.3: Sub-Block and Lane Configuration Description ... 305
13.4: Library Preparation Flow ... 309
13.5: Constraints Strategy ... 311

13.6: Physical Design Strategy 314
13.7: Final Remarks 320

Chapter 14: eVC for Functional Simulation of Voltage/Temperature Variation and Noise on an I/O Bus 321
Joseph D. Mendenhall, IBM
Robert B. Likovich, Jr., IBM
Chad E. Winemiller, IBM
14.1: Introduction 321
14.2: Problem Description 321
14.3: Selection of Verification Language 324
14.4: Elastic Interface eVC Capabilities 325
14.5: eVC Parameters 331
14.6: Resolution of eVC Parameters 336
14.7: Phase Completion Mechanism 341
14.8: Conclusion 342

Chapter 15: Toward Automation of IP (Reuse) Quality Measurement 345
Michael Rohleder, Principal Staff Engineer, Freescale New Product Development Center
Ahmed Dabbagh, Design-for-Reuse Team Leader, Freescale Semiconductor
Joachim Fader, Principal Staff Engineer, Freescale New Product Development Center
15.1: Introduction 345
15.2: Common Design Environment 347
15.3: Certification 348
15.4: Initial Analysis 349
15.5: Related Work 351
15.6: Simplifying Certification 352
15.7: Rule Pruning 353
15.8: Meta Pruning 355
15.9: Golden Rules 357
15.10: Automation 358
15.11: Integration and Data Integrity 362
15.12: Effort for Implementation versus Update 364
15.13: Some First Results 367
15.14: Incremental Certification 371
15.15: Rule Characterization and Classification 372

15.16: IP Reuse Quality Requires Verification IP Reuse Quality ... 376
15.17: Structural and Interface Requirements for VIP ... 378
15.18: Verification Flow Automation ... 383
15.19: Summary ... 386

Chapter 16: Strategies of IP Evaluation, Selection, and Management ... 389
Mobashar Yazdani, ASIC Program Manager and Technologist, Hewlett-Packard
Mike Stahl, Alliance Manager and Design Technology Scientist, Hewlett-Packard
16.1: Introduction ... 389
16.2: The Challenge ... 390
16.3: The Next Seven Years ... 392
16.4: Design or Buy ... 393
16.5: IP Life Cycle ... 393
16.6: IP Categorization ... 398
16.7: Point IP or Portfolio Assessment ... 399
16.8: IP Vendor/Supplier Evaluation Areas ... 403
16.9: IP Acquisition ... 408
16.10: Transferring IP ... 410
16.11: Summary ... 412

Chapter 17: Considerations for ASIC Physical Design ... 415
Ramesh Banda, Distinguished Member of Technical Staff, LSI Logic Corporation
17.1: Introduction ... 415
17.2: Front-End Design and Test Flow Diagram ... 416
17.3: Tools and Associated Versions Audits ... 418
17.4: RTL Deliverables Audits ... 421
17.5: RTL Audits ... 423
17.6: DFT/Verification Audits ... 429
17.7: Synthesis, Timing, and Constraints Audits ... 431
17.8: Front-End Flow Audits Checklist ... 435
17.9: Unique Features in Design Flow ... 437
17.10: Conclusion ... 438

Chapter 18: Case Study, Multiformat Video Decoder IP 439
Design Based on Configurable and Extensible Processor IP
Gulbin A. Ezer, Hardware Design Manager, Tensilica, Inc.
Dennis Moolenaar, Applications Engineer, Tensilica, Inc.
18.1: Introduction 439
18.2: Block–Based Video Decompression Algorithms 440
18.3: Architectural Evaluation Phase and IP Selection 441
18.4: Design Phase 446
18.5: MP Partition of Decoder Applications and 460
Validation on ESL Models
18.6: System Emulation on FPGA Platform 461
18.7: Physical Design of Full System 462
18.8: Packing Video Decoder IP 463
18.9: Conclusion 465

Author Biographies 469

Part I:
Overview

Chapter 1

Introduction

Brian Bailey, Proprietor, Brian Bailey Consulting

Kathy Werner, IP Strategy and Business Manager, Freescale Semiconductor; President, VSI Alliance.

1.1 Introduction

It is quite possible that the size of the intellectual property (IP) market already exceeds the size of the electronic design automation (EDA) industry, although nobody actually keeps complete and accurate figures. This lack of firm numbers is due to the fact that, for many companies, IP is an integral part of a more complex business. For example, is the TI DaVinci platform IP or silicon? While the end goal is to sell silicon, what value should be attributed to the IP itself? In the recent book about the electronic system level of abstraction [1], an argument is made that tools at this level of abstraction may actually contribute very little to the total sales value, but that most of the value should be associated with IP, and the tools are necessary to configure and integrate them into systems. However, due to past changes in the industry, we have seen the pendulum swing between IP and tools several times already. There was industry disagreement when Gartner and the EDA Consortium (EDAC) started to combine both IP and EDA revenues. Some argued that this was a way to hide how much trouble the EDA industry was in. This is clearly not the case since EDA revenue, excluding IP and services, was up 14.4 percent in 2006. However, IP alone was up 21.1 percent showing the growing importance of this category.

IP today is not just about hardware blocks, but contains many discrete pieces including software drivers, integration aids, test strategies, and verification models. It is often the case that hardware models must be written to target both application-specific integrated circuits (ASICs) and field-programmable gate arrays (FPGAs). This all increases the complexity behind the creation, selection, and integration of IP. However, choosing an IP vendor is not just about features and cost, but it is rather about the willingness of provider and user to partner in product creation. Depending on the newness of the IP, this partnership may have to be very close, whereas with commodity IP, the deal is closer to a pure sale. Whichever end of the scale the partnership falls, there is a high degree of trust that must exist between both parties, and trust only comes about by knowledge and by understanding the issues that each side of the partnership is dealing with. This is one of the prime motivations for this book and why the book has been divided into several sections, each focusing on a particular side of the issues. These partnerships, for the latest technology are often three-way, which involve an IP developer, user, and foundry. This is especially true when an IP block is being put into silicon for the first time.

1.2 Historical Overview

It is unknown when exactly reuse started for the chip industry, however, early attempts were not reuse by today's definitions, but rather salvage. Pieces that had been used in previous designs were adapted for use in new designs, and this worked when it was the same designer or design team that was conducting the reuse. During the early to mid-1990s, the industry was beginning to raise questions on the new geometries available at that time, namely 0.35 and 0.25 micron. So, what was going to be done with all of the whitespace of the chip area that needed to be filled up?

Part of the industries solution to this issue was the creation of the Virtual Socket Interface Alliance (VSIA), in 1996, that was chartered with defining a common interface through which all IP could be connected and standards that would ensure a consistent form and quality of all IP delivered. VSIA was probably the most successful organization that brought every player in the market together in record time. Its membership, which consequently soared, contained all IP providers, EDA companies, design houses, and fabrication houses.

Initially, VSIA produced large quantities of documents describing an on-chip bus principle and plug-and-play documents. However, in reality, the problem was a lot more complex than originally conceived and with the on-chip bus, every IP vendor wanted their bus and their capabilities included. It took several years before the On-Chip Bus Development Working Group (OCB DWG) managed to come up with something for which all vendors agreed, but during that period, other bus standards were initially becoming accepted by the industry. In addition, a bus that was designed by committee, as often happens, finished up being too general purpose and too inefficient for most applications. Luckily, the industry has converged on just a few busses with the logical choice often being application area dependent.

Synopsys was one of the major and early developers of general purpose IP, and it was also pro-active in the development of quality standards. Each block of their IP was shipped along with a quality checklist called OpenMore. Primarily a marketing effort, its value became very apparent to the industry as it provided potential customers with a list of things to demand from potential suppliers. Synopsys, and their OpenMore collaborator Mentor Graphics, eventually donated this to VSIA as the basis for their quality IP (QIP) standard.

Quality standards are used to gauge the trustworthiness of an IP supplier, which is unnecessary with the providers of star IP, such as ARM or IBM. For these providers, their reputation is more important than a metric, but it does provide a way for new start-ups to provide a certain level of confidence in their product quality and the processes used to develop them.

There is always a balance that must be reached for an IP provider. The IP has to be flexible enough to be used in multiple applications so that it can reach a large enough potential market base, but also constrained enough to make its own verification manageable. This is perhaps one of the most common complaints heard about IP, even when a certain block has been successfully taken to silicon multiple times. A new customer will pick up the block and in their first test find a major flaw in the IP because it was used in a different configuration or different mode of operation than any of the previous customers. For that IP user, then, they lose confidence in the IP supplier and if the design is not a recent one for the IP supplier, they may also have lost much of the talent used to create it, and thus long delays and continued quality problems persist.

A significant number of problems also remain for the industry and the success of IP reuse in general. Assuming a block has been properly verified, how do you verify its usage within a larger context without indirectly having to verify its internal functionality as well? When VSIA was first formed, one of their slogans was "reuse without redesign." we have always maintained that this is not enough and what is required is "reuse without re-verification." For processors, this problem has been partly solved, but there is great industry reluctance to extend this solution to other parts of the chip. This involves the use of abstract models rather than the implementation models.

When verifying a processor–based design, no company would insert a gate or register transfer level (RTL) model of the processor. These are implementation models and, while they may be used for limited back-end verification runs, they are not used to establish the basic functionality of the chip. Instead, instruction set simulators (ISSs) are used which are abstract models of the processor that have the same behavior, but do not attempt to replicate the internal device structure. These models can run many orders of magnitude faster than the implementation models, and processor vendors continue to push the performance of these models ever higher. Today, it is not unusual to have processor models running at 10 to 100 millions of instructions per second. Processor vendors spend a lot of time ensuring the quality of these models and that they properly reflect the implementation of the processor in silicon. However, high performance models are seldom used for any other parts of the design and, as a result, complete chip simulations run very slow and limit the amount of verification that can be performed. This is a problem that the industry must actually remediate soon.

Finding IP has become a lot easier in recent years through such catalog sites as design and reuse and chip estimate. These are sites that not only catalog the available IP, but also have extensive and often intelligent search engines that enable a user to locate the available cores that fit any number of specific criteria that they may have. Chapter 10 will further discuss this process and show some examples from the chip estimate Website.

1.3 Business Models

The IP industry does not have a single, all encompassing business model, and this is partly dependent on the type of IP acquired and the IP supplier's registration. Licensing models in use are royalty–based, per library, per design, per instance, or a combination of these.

Smaller IP suppliers often prefer to have a license fee but no royalty since this provides them with the money up front. However, this does not provide them the incentive to continue working with a client when problems are found. As with all generalizations, there are exceptions, and some large IP suppliers also prefer this licensing mechanism since it is very easy to administrate. Further, it does not require that the IP user keep records that could be examined in an audit to establish exactly how many units have been shipped for royalty payments. A common extension is for an IP company to also offer consulting services such that a design could be adapted to their customers' specific needs. One should note that in these cases, significant total cost savings over an in-house design is quite unlikely, but the value comes from having an expert who has worked on a particular design possibly several times in the past. It thus becomes vital to make sure that the expert is still with the company and how much of his time will be made available to you.

Some IP vendors have also combined multiple blocks into a library and have charged a single fee based on gaining access to that library. There are also differences in that some vendors levy a single fee per chip, while others require additional revenues if multiple copies of a block exist in the same chip. When up-front costs are involved, the user has to trust the IP supplier.

At the other end of the scale, star IP vendors charge both license fees and royalties. The advantage of a royalty–based revenue stream is that the IP vendor now shares in your success. They only get paid if you are successful and, as a result, they are more likely to help you get to your goal on time. With royalties, both parties need to greatly trust each other under what is considered a partnership, rather than sales.

We now see the broadening of IP products, especially for blocks used for communications. Instead of just supplying the hardware component, the vendor will also supply one or more of the layers of software that support a protocol. This lessens the need to add greater amounts of flexibility in the hardware and transfers the integration challenges into the software that now must be capable of working with multiple operating systems or development environments.

1.4 Who Should Read This Book?

This book should appeal to many different kinds of readers as it attempts to look at certain problems from several different points of view. The two primary views are that of an IP provider and secondly of an IP user, but several other groups of people will also find information that will benefit them.

- IP novices—At some point in time, we were all new to a subject and those around us often assume we know the basics. This can initially make the subject very confusing and can extend the learning period while we try to figure out for ourselves some of the basics. Chapter 2 will investigate the business of IP, chapter 3 will look at some of the legal issues surrounding IP, and chapter 4 will provide a basic guide through the types of IP, such as soft, hard, and analog.
- IP providers—Many of the successful IP providers tend to be large organizations with a portfolio of IP blocks available. The reasons behind this are discussed in detail in chapters 2 and 4. However, there are other ways to add value including techniques discussed in chapters 6, 8, and 9. A company's IP is its livelihood and thus may need

protection as discussed in chapter 5. A new emerging type of IP – verification IP is also discussed in chapter 7.

- IP integrators—Selecting, licensing, integration, and verification of systems containing one or more IP blocks can be a challenge, and it can even make or break a product. Chapter 10 will look at the selection process, and chapter 11 will look at some of the architectural decisions associated with using and creating IP. Chapter 12 discusses many issues associated with the integration and test on systems containing IP and in particular the steps that can be taken to ease issues with post silicon debug.
- Design or IP managers—Any design manager who wants to understand the issues of either creating or using IP will find a lot of information in this book. It also steers away from much of the deep technical issues that can cloud or confuse the issues. We thus believe that most of the chapters in this book will provide managers a good overview of the issues and problems that they are likely to encounter. In addition, the case studies section of the book will give some concrete ideas for ways in which other companies have solved issues that you are likely to face.

1.5 The Book's Structure and How to Read It

The book is divided into four main sections:

- The first section deals with the IP industry in general: what is it, how big and profitable is it, what are the surrounding legal issues, and the various types of IP that exist. For anyone new to the IP industry, these chapters should provide a fairly good basis for understanding the industry and if you are a budding writer for some new exciting IP, this will provide you with some of the issues that you might have to face.
- The second section primarily focuses on the IP provider side of the industry. It deals with some of the issues that IP providers face and the solutions to some of these issues.

- The third section turns the focus onto the user of IP. This section deals with issues such as IP selection and qualification, integration, and verification.
- The final section of the book presents a series of case studies that range in subject matter from physical design issues to the use of electronic system level (ESL) techniques, flow and IP management, and the IP selection process.

Many of the issues presented in this book affect both providers and users, such as the utilization of quality metrics and verification of IP. Just because a chapter is in the provider section does not mean that it does not contain information that is also important to the user. The same is true the other way around as well.

The case studies should be of interest to everyone especially since they are real-life examples of how the industry has solved some of the toughest problems, and where they are still having pain themselves.

1.6 Chapter Listing

In this section, a brief synopsis will be provided for each of the chapters that comprise the book.

Part I: Overview

Chapter 2: The Business of SIP looks at the size and makeup of today's IP market. This chapter will provide a plethora of guidelines and information that will enable well thought-out and educated decisions to be made and may answer any lingering questions on starting one's own IP business.

For a potential IP user, this chapter explores some of the tradeoffs that have to be considered in order to reduce project risk. This chapter explains how to go about finding IP vendors and the standards organization related to various IP issues. The largest IP companies are analyzed in terms of their revenues, customer base, and significant historic milestones in their company histories.

Chapter 3: Building a Legally Sound Intellectual Property Portfolio looks at the role of the legal process for an IP company. It explores the right kinds of protection that may be needed and the stages in development where each of the protection types may apply. It also looks at some of the issues central to licensing, which is one of the most difficult aspects of IP utilization. This should help both IP providers and users ensure that the agreements they create cover all of everything they intended.

Part II: The IP Provider

Chapter 4: Analog, Star and Commodity IP Blocks looks at the various kinds of IP, including analog, star, and commodity, and it also discusses the needs for soft, firm, and hard cores. Analog IP is becoming increasingly important due to the constraints imposed by today's devices, such as power and heat. Analog IP creates very different concerns and consideration compared to digital IP, and this chapter will assist, for example, a digital expert to know the right questions to ask when considering the purchase of some analog IP. Star IP, and particularly processors, often require some additional deliverables beyond those of other IP blocks. It is, in part, the addition of these extra requirements that places them in the star category, as this makes it more difficult for new entrants to commoditize the field. For commodity IP providers, this chapter will help them survive and thrive in a cut-throat environment, and it concludes by looking at some of the emerging issues for IP development.

Chapter 5: IP Protection takes a look at protecting IP through physical, rather than legal, means. Ideas, information, and tactics are all forms of IP, and it is necessary to protect that information so that only the right people can gain access to it. Throughout the ages, various forms of encryption have been used so that information could be protected, none more so than the military needing to protect the information on the battlefield or between dispersed units. However, in the high-tech world, the bigger issue is how to control unauthorized copying, be it at the RTL or silicon levels. This chapter addresses this topic, along with the issues that various people including the IP provider, tool vendor, and users face when dealing with encrypted IP.

Chapter 6: Software turns our attention to this topic. Previous chapters discuss the impact that software can have to make IP easier to use while increasing the total value. This chapter concentrates on the lowest levels of software, namely the hardware abstraction layer and the operating system abstraction layer. Each of these abstraction layers are then broken down into the fundamental capabilities that they provide to make sure that this software is portable across the many environments in which it may be required to operate. It then looks at some basic coding guidelines which will make using and understanding the software a lot easier for customers. Lastly, the chapter briefly addresses the techniques available for running and debugging the software.

Chapter 7: Verification IP talks about a different type of IP, namely verification IP (VIP). In the early days of the IP business, it was not uncommon for a company to buy IP, but only use the verification environment that accompanied it. This provided a test-bench which could be used to verify internally developed circuitry. Today, a number of companies build and sell models, environments, and components for this sole purpose. This chapter looks into the functionality that should be provided in VIP components for various types of IP blocks, along with the benefits that can be obtained from the utilization of such IP. Even if you do not intend to buy VIP, a strong argument is made that similar techniques should be adopted to construct internal verification environments, so its components can be reused and can rapidly reduce the increasing verification costs. This area is still evolving and nowadays even such issues as verification plans can be packaged as VIP components.

Chapter 8: Verification of IP turns the readers' attention to the verification process itself. In many cases, verification now consumes more time than design and for IP, which can be used in many different configurations, verification is an enormous challenge. Many IP users are guided by quality, and only verification can ensure that a consistent level of quality is achieved. This chapter explores the verification of a double data rate (DDR) controller using both internally developed verification components and third party VIP. This chapter addresses the process from creating a verification plan, architecting the verification environment, and creating several layers to the implementation and ways to measure progress. Lastly, the chapter looks at the role that standards play in the verification process.

Chapter 9: IP Quality addresses the dividing line between the IP supplier and user. Quality is a very ephemeral thing that means different things to different people. In the IP industry, it can very easily be defined as the lack of surprises. This means that IP users know exactly what they are going to get, how it is meant to operate, and how it is meant to be integrated. For an IP supplier, this means being able to satisfy these demands from multiple clients, even in the face of requirements that may seem contradictory. The basis for quality is a good methodology that ensures that the IP delivers on its own promises. This chapter addresses a number of ways in which this can be obtained. Perhaps the most important aspect is the communications process between provider, user, and the VSIA QIP standard that provides the communications framework. As a result, this allows information to be exchanged in standard ways such that suitability can be assessed in terms of functionality, fit to internal processes, or based on other factors. Mitigation plans can then be put into place where risks are identified.

Part III: The IP User

Chapter 10: IP Selection addresses questions that arise when a user seeks to obtain some IP for a particular function, such as how an engineer should go about finding what is available and making the selection. This chapter looks at the selection process for both the company producing the IP and the specific blocks of interest. It shows how both Internet searches and more specialized search sites can be used to find suitable blocks based on specific IP attributes. This process is clearly very closely tied to the VSIA QIP framework described in the previous chapter and the more recent FSA selection spreadsheet. The chapter also discusses the evaluation process and some emerging tools to help with the estimation of factors associated with IP block inclusion.

Chapter 11: Block-Level and Platform-Level Reuse looks at the engineering process in great detail and how a successful reuse program can be ensured. Many companies will not only be users of third party IP, but will also be creating blocks to reuse them in future designs. This makes them both users and providers, and in turn they may have an impact of the architectural design of the system. In creating or using platforms, the verification approach also needs to consider the amount of configurability normally built into such platforms since this could cause unnecessary amounts of verification—the very thing that platform reuse is meant to solve.

Chapter 12: Integration, Test, and Debug looks at the issues of integration blocks of IP into a complete chip. The chapter examines the types of required models and the related advantages of the new emerging area of ESL. However, it is important to plan for the post-silicon debug as well, especially with the statistics showing that bug escapes are on the rise despite vastly improved verification techniques. The chapter also looks at how standards can help in this task, not only those at the design level such as bus standards, but debug standards such as the Joint Test Action Group (JTAG) and Nexus. The chapter delves quite deeply into various ways that debug can be performed through on-chip instrumentation including the emerging area of multiprocessor debug.

Part IV: Case Studies

Chapter 13: Back-End Methodology and Techniques for a Multi-Protocol Mixed Signal IP Design looks at many issues associated with the creation of high-speed physical input/output (I/O) interfaces. The authors wanted to create a family of devices that were capable of supporting multiple protocols in different configurations while still maintaining competitive performance characteristics. The design contains both digital and analog blocks that must work over a wide range of processes technologies. The authors likewise explore some of the timing challenges for the design.

Chapter 14: eVC for Functional Simulation of Voltage/Temperature Variation and Noise on an I/O Bus shows how it is possible to create reusable verification components for high- speed I/O interfaces that are able to mimic physical variations which are seen by these kinds of devices. The models enable clock and timing to be adjusted to mimic effects such as skew, drift, jitter, synchronized jitter, duty cycle distortion, Vref distortion, and delay element drift. All these can then be controlled by a pseudo-random verification environment and thus increase the confidence in the I/O block design.

Chapter 15: Toward Automation of IP (Reuse) Quality Measurement discusses the several ways in which this data collection can be automated, along with the impacts that this has on the IP development flow. The establishment of reuse metrics allows the IP quality to be assessed, but this often requires large amounts of data to be provided or collected. Careful consideration must also be given to the rules themselves as the creation of too many can mean that the process will be shunned by engineers. This case study shows how various pieces of the system were built and how a verification environment needs to be constructed so many aspects of the quality metric can be automatically updated.

Chapter 16: Strategies for IP Evaluation, Selection, and Management examines the assessment and selection phase through delivery and support. Designing and verifying IP is one thing, but it also has to work in the complete chip. IP selection must also consider the complete life cycle of the end product and must be able to manage updates from IP suppliers. This chapter proposes ways to treat many types of IP and how to integrate and manage it in a flow.

Chapter 17: Considerations for ASIC Physical Design looks at a flow that was established to help ensure quality through the complete IP development and delivery process including the back end physical design operations. It describes a flow and audit process that makes sure everything stays in step and reduces the risk of bad escapes.

Chapter 18: Case Study, Multiformat Video Decoder IP Design Based on Configurable and Extensible Processor IP examines the creation of a design variant of a reconfigurable processor targeted at the video processing market. The algorithm was described in C and then a suitable architecture for the solution was explored. This resulted in a dual core heterogeneous architecture plus a direct memory access (DMA) engine. The chapter describes the complete design flow used from simulation to emulation and the protection inserted into the design.

References

[1] Brian Bailey, Grant Martin, and Andrew Piziali, *ESL Design and Verification: A Prescription for Electronic System Level Methodology*, Morgan Kaufmann/Elsevier, 2007.

Chapter 2

The Business of SIP

Jack Horgan, Managing Editor, EDACafe

2.1 Introduction

The semiconductor intellectual property (SIP) industry is relatively small. It brings in around $1 billion in annual sales, and ten vendors account for 60 percent of this figure. ARM, the leading vendor, has annual sales two times as much as its closest competitor. Further, there are approximately ten publicly traded SIP vendors, and ARM has 37 percent of their combined revenues.

Anyone seriously seeking to start an SIP firm needs to plan and arrange on financing the operational expenses. Some entrepreneurs may be able to bootstrap a startup from personal assets by borrowing and accepting reduced, or even no, salary through consulting and development contracts. Most founders, however, will require outside funding. The most common source is venture capitalists that provide not only cash but also managerial assistance in exchange for equity often in the form of preferred stock convertible to common. Under this arrangement, the founders and other employees end up with a smaller slice (dilution) of what hopefully is a bigger pie. Venture capitalists (VCs) look for a significant return on investment (ROI) to offset the likelihood that several of their investments in startups will fail. The track record of the founders, the business plan, and a sustainable competitive advantage are among the chief criteria for choosing investments. In fact, VCs cannot achieve their ROI through their share of operating profits but rather by going public though their initial public offering (IPO), by merger, or by acquisition. This goal can ultimately influence operational decisions.

The IPO market was far friendlier to SIP companies several years ago when most of the top 10 public companies had their IPOs. Only two SIP vendors had IP revenues greater than $100 million.

Nevertheless, several firms have consistently lost money. If this reality had been the projection, the IPOs might not have occurred. On the acquisition front, Synopsys entered into an agreement to acquire MoSys for $342 million net of cash in February 2004. Synopsys subsequently changed its mind and paid a $10 million termination fee. In August 2004, ARM acquired Artisan for $913 million which had revenues of $83 million in the previous twelve months. In January 2006, Wipro, the large Indian outsourcer whose 2005 annual revenue reached $1.87 billion, acquired NewLogic Technologies, an Austrian wireless IP developer of Bluetooth and wireless local area network (WLAN), for $56 million. NewLogic had revenues of approximately $17 million in 2005. In June 2006, Motorola acquired TTP for $192 million which had revenues of £37 million in its last full fiscal year.

However, there are exceptions to VC funding. Some research can be carried out at universities possibly to the point of proof of concept and then spun out. An example would be Chipidea Microelectronica, the leader in analog SIP, which started at the Instituto Superior Tecnico in Lisbon, Portugal. In some cases, a research group can spin out of a corporation when it feels the technology is outside their areas of interest or does not represent a sufficient ROI. The parent company may retain an equity stake in exchange for the rights to the technology developed.

The sources of revenue for IP vendors are license fees, technology licenses, development systems sales, royalties, services revenues consisting of design consulting services and revenues from support, and maintenance and training and possibly patent licenses. In the case of the industry leader ARM, licensing fees accounted for over 40 percent of revenue, 38 percent from royalties, and 20 percent from service revenues over the last three fiscal years. Licensing fees grant a customer the right to embed the vendor's IP into their own application specific product. Licenses typically grant rights on a non-exclusive, non-transferable basis. Technology licenses grant the licensee the additional right to create and modify the vendor's IP and to offer it to its own customers. Royalties are incurred when the customer manufactures or sells a product containing the IP. Royalties are set either as a percentage of the licensee's net sale price per chip or as a fixed amount per chip.

Generally, the royalty rate decreases as the volume increases. Royalty revenues are often determined and recognized one quarter in arrears, when a production volume report is received from the customer or foundry. Patent licensing does not require the delivery of any product or service; in fact, some patent holders have no business beyond licensing their patents and vigorously defending their rights. All SIP companies have bulging patents and relevant application portfolios to negotiate leverage with customers and prospects and to erect entry barriers for potential competitors.

Unfortunately, the United States Patent Office (USPO) has insufficient resources to adequately investigate patent applications. The practice has been to largely accept the applicant's position on matters like prior art, grant the application, and rely on those impacted to contest the patents after the fact. In addition, the USPO has been accused of granting overly broad patents which would also have to be contested. Patent holders have an enormous advantage in patent disputes because they may be able to obtain an injunction against those who violate their patent. The same resource limitations make the process of challenging a patent a lengthy, expensive, and unpredictable one. Many firms choose to pay patent licensing fees, rather than fight further, thus strengthening the hand of the patent holders. A well publicized case is Research in Motion (RIM), the makers of the Blackberry, versus NPT, Inc. In 2002, RIM lost a civil case in a jury trial, but they appealed the decision and challenged the validly of the patents. The preliminary analysis by the USPO was in favor of RIM, but the judge refused to wait for a final decision which could have taken years. In March 2006, under threat of an injunction, RIM settled for $612 million despite their position that the patents were invalid.

How does a SIP vendor set a price for its products and services? In the simplest case of a single existing product, there are three basic pricing models. In the first approach, the seller determines the total product cost including the amortization of development costs over the projected number of units and adds on the desired profit margin. This approach is common when the seller is in control due to lack of competition. The second approach is based on the value received by the buyer. The potential buyer always considers value which explains why some would pay thousands of dollars for a watch when a reliable timepiece can be bought for tens of dollars. The third approach is to set the price based upon the competition. If one is to charge more than the competition, there must be a reason such as superior product characteristics (e.g., functionality, speed, and capacity) or better service. Why else would anyone knowingly pay more? When a product becomes a commodity (which involves many competitors), the price plummets regardless of the value received and the providers cost.

In the case of SIP, additional complications come about especially since the vendor must achieve a balance between the certainty of upfront licensing fees and the unpredictability of downstream royalties. Royalties are dependent upon the future success of the customers and not the SIP vendor. It can take anywhere from a year to 18 months for a SIP customer to get a product to market especially when there is too little money on the front end and the SIP vendor struggles to survive while waiting for royalties that may never come. Demanding too much on the front end may either deter prospects or leave royalty money on the table.

2.2 Value Proposition

Prospects' Viewpoints

Potential customers for IP goods and/or services are faced with a "make versus buy" decision. The decision makers must consider several issues to make an informed choice. These issues include:

1. Availability of a qualified supplier—If there is no qualified supplier, then a company has to either redefine its requirements or make it themselves. Locating potential suppliers is not always easy. Further, qualifying and negotiating price, as well as terms and conditions, can be time consuming.
2. Internal expertise—If there is no relevant internal expertise, then the development needs to be outsourced or appropriate technical resources need to be hired for the effort.
3. Development and support costs—One must develop a reasonable cost estimate and support the IP in-house. This estimated cost should be the full burdened cost of the

personnel involved, the cost of any equipment and software, test and evaluation (T&E), and so forth for the entire product lifecycle.

4. Time—The time to develop and test IP in-house can be considerable. If this effort should fall on the overall critical path, it can impact time to market (TTM). A number of studies have shown that a delay in a product can reduce the available market and impact profitability due to the possible, if not probable, product introduction by competing firms. It is not simply a matter of shifting revenue and cash flow to later points in time. In the worse case scenario, the market window may close before the product can be released.
5. Patents—Even if one could develop a technology in-house, it would be useless if it violated another company's patent.
6. Opportunity costs—The resources which might be applied to this effort could possibly be used elsewhere with a greater return on investment. The best people simply can not work on every project.
7. Strategic importance—If a particular IP area is viewed as strategic to the company, it may be in the company's long-term interest to develop it in-house.

These described factors can be boiled down to a single word: risk. Risk is a fact of life, and people differ in respect to their level of risk aversion. Entrepreneurs start businesses despite the high probability for failure, mostly because they find the experience exhilarating and the potential payoff large.

Most people and firms avoid risk, or at least seek to minimize risk, unless there is a proportional reward. Millions of people risk a few dollars on the lottery on a daily basis, but very few, if any, would bet their house on it. While the upside is huge, the probability of wining is infinitesimal. To risk large amounts for a small gain would be considered foolish, and thus many people are more comfortable with the familiar rather than with the unknown.

Any technical project is fraught with risk. The actual time and cost of a project may significantly exceed the estimates. A project may not meet the specified requirements or even fail completely. While there are recognized methods for estimating cost, time, and revenue opportunities, these may be influenced by the people's level of interest in and desire for the project. Some groups suffer from not-invented-here syndrome.

According to a well-cited study by Collett International Research, two-thirds of integrated-circuit (IC)/application-specific IC (ASIC) designs require two or more silicon spins to reach volume. In these cases, logical or functional errors occur over 70 percent of the time. More than 80 percent of these flaws are design errors while less than 10 percent have flaws attributable to external IP.

Another factor is that many high tech companies are organized along project lines. A project leader may be faced with absorbing costs that could, and arguably should, be spread out across several projects. The project leader's risk-reward calculation might be different than what the corporation executives would calculate if they had to make the decision.

Vendor Perspective

Recognizing that prospects are confronting a make versus buy decision, SIP vendors need to articulate a value proposition that address the issues previously raised. This can be summarized as faster TTM and lower lifecycle costs in the context of lower risk.

The challenge becomes convincing prospects that one can in fact deliver on this value proposition. The most effective method is references, that is, testimonials from existing customers, preferably firms similar (same industry, same size, or same application area) to the target prospect. Unfortunately, many firms are reluctant to share with competitors what they may well consider to be a competitive advantage. In addition, the legal or public relations department may object to public statements that appear to endorse a specific product or vendor. There are ways to persuade customers that some form of public acknowledgement of a product may be in their own interest. Alternatively, encouraging technologists within one's user community to publish technical papers which make reference to the tools they use can also be useful. Finally, informal word of mouth can be an effective resource.

2.3 Market Size

There are probably two interpretations of the term "market size." The first is the historical market, and the second is the total available market (TAM). The dollar size of the historical market is simply the sum of the revenues of those who participate in the market. Fortunately, publicly traded companies, at least on US stock exchanges, are required to publish quarterly and yearly financial reports within a short time after the period ends, otherwise, they risk being de-listed from the exchange and might have to pay penalties for inaccurate data. If all the firms in a given market participate only in one market and are all publicly traded, then the task of sizing is straightforward. However, there are two challenges since not all firms in any given market are publicly traded. First, these privately held firms generally refuse to provide any financial data. If they give financial information to an analyst or writer, there is no way to verify the data. While most private companies are relatively small, there are exceptions, such as The MathWorks, a privately held company with over 1,400 employees and estimated annual revenues in excess of $250 million. Second, many companies participate in more than one market or market segment. The segments may be defined by product, geography, revenue source, such as license, royalty, product, services, etc. One should be aware that not every single firm provides segment information, those that do, however, do so in parallel, such as in geographic breakdown across all segments rather than within segments. Again, there is no way to verify information given outside the formal reports.

I have been publishing quarterly and annual summaries of the financial performance of eight (now seven due to acquisition of Artisan by ARM) publicly traded IP firms on EDACafe. These reports do not mention the top three EDA vendors, but rather include ARM, Artisan, Ceva, LogicVision, MIPS, MoSys, Rambus and Virage Logic. The five year combined financial performance of these companies is shown in the *Table 2.1*.

$K	2001	2002	2003	2004	2005
Revenue	537,237	531,711	560,779	686,455	729,657
Delta		-1.0%	5.5%	22.4%	6.3%
Earnings	128,951	82,168	12,624	126,231	146,830
% Rev	24.0%	15.5%	2.3%	18.4%	20.1%

Table 2.1: *Five Year Financial Results of Selected SIP Vendors*

The data in *Table 2.1* is the fiscal year results for each company as reported in the firm's 10K or 20F reports. The fiscal years of some companies are not the calendar year. ARM has been the dominant player much more after its acquisition of Artisan Components at the end of 2004. ARM accounted for 40 percent of the group's revenue in 2001 and 55 percent in 2005. ARM reports its financial results in pound sterling. The conversion from pounds to US dollars is based on the currency exchange rate at the end of the respective calendar year.

The EDA Consortium (EDAC), founded in 1989, is an international association of more than 100 companies that provides tools and services that enable engineers to create the world's electronic products. Since 1994, EDAC has published its market statistics service (MSS) report on a quarterly basis on EDA industry revenue data. The MSS report data is segmented into revenue type, application, operating system, and geographic region, with many subcategories of detail provided. The data is voluntarily and confidentially submitted, and no individual company statistics are disclosed.

In April 2004, EDAC announced that it would expand its SIP revenue reporting by adding over 40 new categories from the original six based upon the Virtual Socket Interface Alliance (VSIA)'s virtual component transfer (VCT) product classification. VSIA, Inc. was an open, international organization that included representatives from all segments of the systems-on-chip (SoC) industry. This change makes the comparison with earlier revenue figures quite misleading.

$M	2004	2005
Semi IP	731	816
Y-Y Delta		11.6%
% Total	16.4%	17.8%
Total EDA	4,451	4,574

Table 2.2: *EDAC Data on SIP*

Various industry analysts track the performance of individual firms in the SIP category. For example, Gartner Dataquest used to track the entire industry, and in August 2006, they announced the projected worldwide IP revenue to total $1.8 billion, a 24.9 percent increase from the 2005 revenue of $1.4 billion. By 2010, the firm projects worldwide semiconductor IP revenue to surpass $2.7 billion.

Table 2.3 shows Gartner's estimate on the size of SIP market based on press announcements over the years. According to Gartner, the top ten vendors accounted for over two-thirds of the revenue.

$M	2001	2002	2003	2004	2005	2006P
Revenue	891	934	1,056	1,274	1,400	1,800
Delta		4.8%	13.1%	20.6%	9.9%	28.6%

Table 2.3: *Gartner Dataquest Estimate of SIP Market Size*

The difference between the three estimates presented on the SIP market size is due to the number of firms included in the estimates. Gartner includes SIP revenue contributions from Mentor Graphics and Synopsys.

The second interpretation of market size is the TAM. This is the potential number of units (or dollars) that could be sold (or spent). The dollar value of the TAM is the number of units multiplied by the average selling price. Since the selling price will vary over time, it is easier to estimate the number of units. The total available market for a consumer product or service is the number of people or possibly the number of households that fit the target demographics (e.g., age, gender, income level, geography, occupation, education, race, religion, psychographics, etc.). For example, the TAM for cell phones is the roughly 300 million people in the United States minus the very young (unable to use), the elderly (unlikely to adopt) and the lower income (cannot afford). In the case of EDA tools, the number of semiconductor firms in a given area (analog, digital signal processor [DSP], and microprocessor) is a starting point. Let us now combine this figure with the usage model (shared versus dedicated, and architect versus designer versus verification engineer) to determine the number of potential units.

In the case of SIP, there are two measures of the TAM. The first is the number of firms and persons within those firms that would use the SIP in developing their own products. The second measure is the number of units that the end user firm will eventually ship with the embedded SIP and pay royalties.

There is another interpretation of the TAM acronym where the A stands for accessible. This is a lower figure which considers the practicality of going after the total available market. For example, cell phone carriers may choose to ignore rural areas with low population as being economically unviable at least without government subsidies. SIP vendors may choose to concentrate on the top five or ten semiconductor firms rather than on small firms where the likely of large royalties is less.

The market for a given product or technology does not develop overnight especially as it can take months, years, and even decades to fully develop.

The well-known book, *Crossing the Chasm*, first published in 1991 by Geoffrey Moore, describes the rate of technology adoption by a bell curve where the number of adopters versus time is plotted. The total number of adopters and the time line varies from technology to technology. The adopters can be grouped as pioneers, visionaries, early majority, late majority, and laggards. The early and late majorities combine to form the mainstream market and account for two-thirds of the total. The pioneers and visionaries combine to form the early adopter segment which is the same size as the laggard or skeptic segment or approximately 16 percent. The early adopters are willing to deal with the inevitable problems of a new technology and require less hand-holding than later adopters. The pioneers, or enthusiasts, will most likely buy low serial numbers of any new technology relevant to their interests. The visionaries see the potential or strategic importance in a new technology and seek to exploit it. The early majority are pragmatists, as they want a solution that works. This group has a herd mentality, but they distrust the references from the early adopters. As a result, there is not a nice smooth transition between the two groups. There is in reality a "chasm" that must be crossed, and it requires different marketing and sales approaches.

The same point can be made from a different perspective as was done in the 2000 book *The Tipping Point* or *How Little Things Make a Difference* by Malcolm Gladwell. According to the thesis of the book, the adoption of a new technology is like the spread of a virus which when it reaches a certain critical point (or, the tipping point) becomes an epidemic. Some technologies require a certain supporting infrastructure for rapid adoption. There are an increasing number of ways to view on demand audio and video entertainment including cable, Internet, and cell phones. The tipping point for consumer adoption of this technology is the amount of available content. When Apple first introduced its video iPod, content was limited to some Pixar animation films and a few episodes of *Lost* and *Desperate Housewives.* The spread of the Internet was a tipping point for the personal computer that had been limited to hobbyists and professionals. The Internet has also revolutionized the way products are sold (e.g., Amazon.com and eBay), marketed, and distributed.

An obvious question is whether SIP has crossed the chasm or has reached a tipping point. According to EDAC, SIP accounted for $816 million in revenue for 2005 or nearly 16 percent of total EDA revenue. SIP represents an even smaller percentage of worldwide semiconductor sales which were $228 billion in 2005 according to the Semiconductor Industry Alliance. Seventy-seven percent of this figure is outside the United States.

If the entire SIP industry were a single firm, it would not make the Fortune 1000 list. However, the same can be said for the EDA industry. The unit numbers are more impressive than the revenue figures. In the last fiscal year, product shipments with embedded SIP were 11.9 million for Imagination Technologies, 135 million for CEVA, and 1,662 million for ARM.

The answer to the question is not determined by the size of the industry or the number of units shipped but rather the degree of acceptance by the target industry. The list of SIP customers reads like a veritable *who's who* of semiconductor and related companies. The only reservation is that most of the SIP vendors have a small number of key customers albeit likely multiple groups within the same companies. For example, for the last fiscal year, Rambus reported that 75 percent of its revenue comes from five customers, Saifun reported that three licenses accounted for 87 percent of licensing and service revenue and MoSys reported that 52 percent of its revenue comes from its two largest customers NEC and Fujitsu. This concentration is due partly to the nature of the market and to the vendors' limited resources coupled with their strategy to target mega deals. The fabless semiconductor industry business model has demonstrated the willingness to rely on outsourcing of critical function.

2.4 How to Find IP Vendors

How does one discover who the IP vendors are? Search engines are a good source of information. One can scan lists of vendors that exhibit and present at such tradeshows and conferences as the Design Automation Conference (DAC) and Design, Automation and Test in Europe (DATE), the partners list of the major EDA companies and membership in IP relevant consortiums and alliances. There are also several IP portals, and SIP vendors would be well advised to participate in these areas to gain visibility.

Design & Reuse (D&R) was founded in October 1997 as a Web portal for added-value information in the field of electronic virtual component. In 2001, D&R put in place a data intranet import gateway feature for feeding automatically intranet IP/SoC resource catalogs for large companies with supplier confidential data from a gateway external to the firewall.

D&R licenses technology and services to allow IP sharing and reuse worldwide and in each electronic design enterprise through three major products. To IP providers, D&R offers IP provider stations for publishing their offers through electronic catalogs, packaging their IPs, and optimizing their delivery to customers while providing outstanding customer support. For large intranet corporations, D&R offers IP consumer Intranet Station to organize intranet catalogs for pre-qualified external supplier and installing IP delivery stations for external suppliers under corporate contract. Finally, D&R delivers to large corporations the IP reuse Intranet Station so designers can exchange their reusable blocks with close to no overhead.

Chip Estimate Corporation, formerly Giga-Scale Integration Corporation, is an EDA company that delivers streamlined access to integrated chip project planning solutions. The company develops and markets the InCyte chip estimation systems, into which it integrates IP and manufacturing process models for the most accurate estimation possible. InCyte performs estimations of chip die size, power, leakage, yield, and cost and allows for rapid "what if" analysis at the architectural stage in the design flow. InCyte systems range from the free InCyte Lite to InCyte Enterprise, used by leading semiconductor companies to refine chip size, power and cost estimates throughout chip implementation flows. Chip Estimate partners with leading IP vendors and semiconductor manufacturers who share its mission of early and accurate chip planning for mutual customers' ultimate design success. Since the 2005 launch of ChipEstimate.com, over 9,000 users have performed over 40,000 chip estimations with InCyte.

Alliances

The VSIA was an open, international organization that included representatives from all segments of the SoC industry such as system houses, semiconductor vendors, EDA companies, and IP providers. VSIA's mission was to dramatically enhance the productivity of the SoC design community by providing leading-edge commercial and technical solutions and insight into the development, integration, and reuse of IP. VSIA had wide industry participation with more than 70 member companies from around the world.

VSIA created working groups based on three pillars—quality, protection, and transfer—to work on standards issues to improve productivity for SoC and IP developers and users.

The IP quality pillar created the VSIA QIP Metric tool which can aggressively reduce the time typically required to make an IP purchase decision and to integrate the core. The Metric helps the IP vendor and consumer communicate based on an objective foundation. Besides setting up the basis for measuring a core's characteristics against an industry-approved list of attributes, the Metric provides a view of the IP vendor's general approach to IP development. This enables an integrator to evaluate similar cores from competing vendors. With the closure of VSIA, this work has been transferred to the Institute of Electrical and Electronics Engineers (IEEE).

The SPIRIT Consortium is a global organization focused on establishing multi-faceted IP and tool integration standards that drive sustainable growth in electronic design. It is comprised of leading EDA, IP, system integration, and semiconductor companies dedicated to the adoption of a unified set of specifications for configuring, integrating, and verifying IP in advanced SoC design tool sets.

The Open Core Protocol International Partnership (OCP-IP), launched in 2001, seeks to proliferate a common standard for IP core interfaces, or sockets, that facilitate "plug and play" SoC design. OCP-IP provides the tools and services to its members that are necessary for convenient implementation, maintenance, and support of the standard socket interface. OCP-IP has over 120 member firms and universities. It also offers its members the use of an EDA tool, CoreCreator, to automate the tasks of building, simulating, verifying, and packaging OCP-compatible cores.

Partners

SIP vendors simply cannot meet all the needs of their customers. Just like the leading EDA firms, SIP vendors should and, in fact, do create their own ecosystems by forming partnerships with foundries, independent device manufacturers (IDMs), EDA vendors, design service firms, and so forth. The ARM Connected Community is the best example. Other examples of partner programs include those of CEVAnet, Virage Logic VIP, and LogicVision LVReady.

2.5 Publicly-Traded Companies

Partners

ARM is a public limited company incorporated under the laws of England and Wales. The company was formed in October 1990 as a joint venture between Apple Computer (UK) Limited, Acorn Limited Computers Limited, and VLSI Technology, and they operated under the name of Advanced RISC Machines Holding Limited until March 1998, when its name was changed to ARM Holdings, plc. Its initial public offering was in April 1998. At the end of 2005, the company has around 1,250 employees, 783 of which are in research and development (R&D). The firm has facilities in 12 countries on three continents.

£K	2001	2002	2003	2004	2005
Rev	146,274	150,922	128,070	152,897	232,439
Delta	44.6%	3.2%	-15.1%	19.4%	52.0%
Net Inc	34,002	31,637	13,016	27,977	41,880
NI/Rev	23.2%	21.0%	10.2%	18.3%	18.0%

Table 2.4: *ARM Financial Performance (£K)*

The annual revenue for 2005 was about $400 million, and for 2004 around $290 million.

On December 23, 2004, ARM acquired Artisan Components, Inc. a provider of physical IP components for the design and manufacture of complex SoCs. The total consideration paid was $926.9 million. The acquired company was renamed ARM Physical IP, Inc ("PIPD"), and its products include embedded memory, standard cell, input/output (I/O) components, and analog and mixed signal products. In 2005, PIPD generated over $50 million in revenue.

$K	2001	2002	2003	2004
License	21,136	27,728	58,389	52,066
Royalty	4,621	9,518	18,357	31,809
Total	25,757	37,246	76,746	83,875
Delta	26.9%	44.6%	106.1%	9.3%
Net Inc	(13,972)	2,115	8,887	15,536

Table 2.5: *Artisan Financial Data*

PIPD has licensed its intellectual components to over 2,000 companies involved in IC design. These customers include foundries, IDMs, system manufacturers, and fabless integrated circuit companies. The company counts 19 out of top 20 semiconductor vendors as customers. PIPD makes the core set of its products available to licensed IC designers at no charge upon receipt of the IP. However, the foundries, where their designs are manufactured, have paid a license fee, and will pay royalties to PIPD when the silicon containing PIPD IP is shipped. PIPD also provides customized IP components and services to its licensed customers on a separate fee basis.

ARM designs and licenses reduced instruction set computer (RISC) microprocessors, physical IP, system components, software and development tools, which enable the design of embedded microprocessor solutions for use in a wide variety of applications. ARM offers a family of 32-bit RISC microprocessor "cores" based upon a common architecture and spanning a wide performance range. ARM also offers physical library solutions on a variety of processes at leading foundry providers. It has designed standalone microprocessors based upon its microprocessor cores and licensed semiconductor partners to manufacture its standalone microprocessors. In addition, ARM provides the necessary development boards, software development toolkits, and software debug tools that facilitate both the system design and development of system solutions.

The company has five target market segments, namely home solutions, mobile solutions, enterprise solutions, embedded solutions, and emerging applications. In 2005, these segments accounted for the following percentages of total revenue: home 9 percent, mobile 65 percent, enterprise 20 percent, and embedded 6 percent.

On a geographic basis, the United States accounted for 42 percent of total revenue, Japan 18 percent, Asia Pacific 25 percent, and Europe 14 percent in 2005. The US percentage was around 51 percent in 2004 and 2003, while Asia Pacific percentage was around 13 percent.

ParthusCeva

ParthusCeva was formed through the combination of Parthus Technologies, plc and ParthusCeva (formerly known as Ceva, Inc.) on November 1, 2002. Prior to that date, the DSP cores licensing business was part of DSP Group, Inc. of Tel Aviv, Israel. On November 1, 2002, the DSP Group contributed the DSP cores licensing business to ParthusCeva, which was then its wholly owned subsidiary. The DSP Group distributed all of the existing common stock of ParthusCeva to the stockholders of DSP Group. Immediately thereafter, ParthusCeva combined with Parthus Technologies plc of Dublin, Ireland. In November 2003, the company changed its name to CEVA, Inc. which coincided with the launch of the company's pioneering new DSP architecture, called CEVA-X.

CEVA addresses the requirements of the embedded communications and multimedia markets by designing and licensing programmable DSP cores, DSP–based subsystems, application-specific platforms, and range of software components which enable the rapid design of DSP–based chips or application-specific solutions for developing a wide variety of applications. The services division assists customers in the deployment of their CEVA–based solutions. In addition, the company offers design services to its customers.

CEVA is the leading licensor of DSP cores. In 2005, Gartner Dataquest reported CEVA's share of licensable DSP market at 53 percent. The portfolio of cores includes CEVA-X, a scalable very-large instruction word/single instruction multiple data (VLIW/SIMD) DSP architecture delivering very high levels of performance at low power, and CEVA-Teak and CEVA-TeakLite-II, 16-bit fixed-point general-purpose DSP cores.

$M	**2003**	**2004**	**2005**
License	25.7	26.2	24
Royalty	4.1	6	6.8
Other	7	5.5	4.8
Total	36.8	37.7	35.6
Shipped Units	**55M**	**106M**	**115M**
Net Inc	(11,991)	1,650	(2,266)

Table 2.6: *CEVA Financial Performance*

CEVA had revenues of $21 million in 2001 and $14.8 million in 2002. The following three years were essentially flat although the number of units shipped by the firm's licenses doubled. On a geographic basis, the United States accounted for 47 percent of total revenue in 2003 but only 35 percent in 2005. During the same three-year period, EMEA revenue decreased from 31 percent to 22 percent while revenue from AP grew from 22 percent to 43 percent. Twenty-three customers were producing royalty revenue at the end of 2005. The five largest customers accounted for 36 percent of total revenues in 2005, 46 percent in 2004, and 39 percent in 2003.

At the end of 2005, CEVA had approximately $35 million in cash and cash equivalents and $26.5 million in deposits and marketable securities.

Imagination Technologies Group

In September 1999, VideoLogic Group, plc was renamed Imagination Technologies Group plc. They are publicly traded on the London Stock Exchange, and they both develop and license innovative silicon and software IP for SoC devices which target multimedia and communication applications.

Imagination employs over 350 people which includes over 270 engineers. The firm has its headquarters in Kings Langley, Hertfordshire, England and has offices in Chepstow, Leeds, Tokyo, California, Taipei, and Seoul.

As of April 2005, it operated as two business segments: the technology business, which comprises licensing and royalty revenues, and the systems business. The technology business unit accounts for over 40 percent of total revenue and all of the operating losses.

Imagination Technologies' three IP development groups, PowerVR, Metagence and Ensigma, each focus on silicon and software IP generation in a specific area. Each develops a complementary licensable technology family for multimedia processing: PowerVR for graphics, video and display; META for RISC and DSP; and the Ensigma universal communications co-processor (UCC) for multi-standard communications and broadcast.

In March 2000, the company acquired Ensigma Limited, a private company specializing in DSP algorithms and technologies for entertainment, communication and speech applications, for a maximum consideration of £5.0 million.

£K	2001	2002	2003	2004	2005	2006
Rev	17,821	13,504	19,568	31,215	30,583	35,273
Delta		-24.2%	44.9%	59.5%	-2.0%	15.3%
Net Inc	1,902	(5,547)	(5,327)	(3,266)	(5,633)	(7,360)

Table 2.7: *Imagination Technologies Financial Performance*

In the last fiscal year, Europe and the United Kingdom accounted for 67 percent of revenue, Asia 16 percent, and North America 14 percent.

LogicVision

LV Software started in California in July 1992 and was principally engaged in research and development activities through 1994. The company first generated meaningful commercial revenues from the license of initial embedded test products in 1995. In June 1996, the firm changed its corporate name to LogicVision, Inc. and was reincorporated in Delaware in September 2000. The company's

IPO in October of 2001 raised about $40.5 million. In November 2004, the company completed the acquisition of SiVerion, Inc. for a total purchase price of $7.4 million.

LogicVision's approach to test and measurement is to insert test structures, called "eyes in the die," at strategic locations throughout the device. These "eyes" are design IP components specialized for the test and measurement of different portions of the design. "Eyes" exist for high-speed logic, embedded memories, mixed-signal components, and very high-speed serializer/deserializer (SerDes) I/O. All "eyes" run at full functional speed and are accessed from a common low-speed standard five-pin test interface (JTAG). The "eyes" monitor for defects in the chip life, from initial design debug through board and system test and into the field.

The products are used in design, post-silicon manufacturing, test and parametric yield analysis of large digital and mixed-signal ICs, especially those using leading-edge processing nodes. This includes logic, memory, phase-locked loop (PLL) and high-speed I/O. The products cater to high-volume or yield-sensitive markets where minimal defects per million (DPM) counts, low field returns, quick silicon bring up and low cost of test are key requirements.

$K	**2001**	**2002**	**2003**	**2004**	**2005**
License	12,079	10,247	4,740	5,686	6,735
Service	5,153	5,308	3,987	4,095	3,951
Product			275	313	196
Total	17,232	15,555	9,002	10,094	10,882
Delta		-9.7%	-42.1%	12.1%	7.8%
Net Income	(6,143)	(8,504)	(12,015)	(8,388)	(9,973)

Table 2.8: *LogicVision Financial Performance*

LogicVision currently offers embedded test circuits, design software, and manufacturing software in a variety of product bundles under the product family name LV2005™ with the product sub-families called ETCreate™, ETAccess™, and SiVision™.

ETCreate is the product sub-family formerly known as the integrated-circuit built-in self-test (icBIST), which consists of embedded test IP and corresponding design automation software that provides embedded test solutions for different ASIC or SoC design components.

ETAccess is the product sub-family which consists of hardware and software products for use with third party test platforms.

The SiVision product sub-family consists of parametric analysis and visualization software that uses semiconductor manufacturing process and test data to help assess parametric yields and identify parametric yield limiters.

In 2005, ETCreate accounted for 82 percent of total revenue, ETAccess for 17 percent, and SiVision for 1 percent. On a geographic basis, the United States accounted for 82 percent, Japan accounted for 12 percent, and the rest of the world for 6 percent. Four customers (Agere, Broadcom, LSI Logic, and Matsushita) account for nearly 60 percent of total sales.

At the end of 2005, LogicVision had $3.6 million in cash and cash equivalents and $7.1 million in short-term investments.

MIPS Computer Systems, Inc.

MIPS Computer Systems, Inc was founded by Stanford University engineering professor John Hennessy, who is now its acting president. In 1992, Silicon Graphics, Inc. (SGI) acquired the firm and incorporated MIPS Technologies, Inc. in Delaware. It completed its IPO in July 1998 and continued to operate as a SGI majority-owned subsidiary. In 2000, SGI spun off its shares of the firm to its shareholders as a dividend.

For more than twenty years, MIPS has developed and licensed a wide range of RISC processor architectures and cores. The flexibility, performance, and cost of these architectures and cores have resulted in broad usage, and a large number of suppliers have designed their products, usually software or tools, to operate with the MIPS architecture. MIPS is the hub of an environment—namely, an "ecosystem"—that includes these suppliers, together with semiconductor licensees and their system level customers. This ecosystem has been responsible for mutually reinforcing innovations across a broad spectrum of markets and applications.

MIPS has developed standards for both 32-bit and 64-bit computing. They license industry-standard MIPS32 and MIPS64 instruction-set architectures, application-specific extensions (ASEs), core designs, and other related intellectual property to semiconductor companies and system original equipment manufacturers (OEMs). Together with their architecture and core licensees, they offer a broad variety of performance-oriented embedded processors that scale across multiple markets in standard, custom, semi-custom and application-specific products. MIPS currently has more than 155 license agreements with more than 105 global companies that offer more than 280 processor–based chips for the embedded market.

$K	2001	2002	2003	2004	2005
Royalty	41,931	16,791	15,693	23,439	29,988
Contract	42,978	30,970	23,397	24,446	31,231
Total	84,909	47,761	39,090	47,885	61,219
Delta		-43.8%	-18.2%	22.5%	27.8%
Net Income	19,062	(9,390)	(28,907)	(1,531)	14,909

Table 2.9: *MIPS Financial Performance*

$K	2003	2004	2005
US	18,440	25,546	37,606
Japan	9,520	13,269	9,671
Pac Rim	6,850	4,544	3,614
Europe	2,937	3,114	6,622
ROW	1,343	1,412	3,706
Total	39,090	47,885	61,219

Table 2.10: *MIPS Revenue by Geography*

On a percentage basis, the United States accounted for 61 percent of total revenue in 2005 compared to 47 percent in 2003. During the same period, revenue from Japan declined from 24 percent to 16 percent and revenue from the Pacific Rim from 17 percent to 6 percent.

MOSAID Technologies, Inc.

MOSAID Technologies Inc., based in Ontario, Canada, was founded by Richard C. Foss and Robert Harland in 1975. MOSAID develops and licenses IP, along with the supplies memory test and analysis systems. MOSAID became a publicly traded company in October 1993, and it is currently listed on the Toronto Stock Exchange. In October 2005, MOSAID acquired Virtual Silicon Technology Inc. for $5.35 million. Virtual Silicon, founded in 1997, offers complete SIP solutions, and since its inception, it has received over $40 million in VC financing. In its fiscal year ending March 31, 2005, Virtual Silicon generated revenues of about $7 million. Virtual Silicon's main product line, Mobilize, addresses a growing market requirement for low power consumption in portable electronics to extend battery life.

MOSAID patent portfolio contains over 500 issued or pending patents in the areas of dynamic random access memory (DRAM), synchronous DRAM (SDRAM), multi-level DRAM, embedded DRAM, networking, computational memory, and graphics.

MOSAID SIP products include Mobilize, a suite of standard cells, compiled static RAM (SRAM) and I/O SIP specifically targeting low power applications, Maestro a sophisticated and programmable Delta-Sigma Fractional-N PLL digital frequency synthesizer product line and Memorize a suite of memory interface related I/O libraries and memory controllers.

$K	2001	2002	2003	2004	2005	2006
Rev	82,926	51,861	37,981	29,396	49,743	63,899
Delta		-37.5%	-26.8%	- 22.6%	69.2%	28.5%
NI	7,002	(24,686)	(19,897)	(8,581)	37,585	15,483

Table 2.11: *MOSAID Financial Performance*

The year 2002 saw a considerable drop-off in sales and a major loss. Yet, 2001 was the worst-ever year for the DRAM industry: revenues from 2000 declined by 62 percent to $11.9 billion. Through the four quarters of 2001, DRAM revenues continued to drop, followed by a brief rebound in the first quarter of 2002.

MOSAID has two business segments, namely IP and systems. The IP segment accounted for 66 percent and 54 percent of the total revenue in fiscal 2006 and 2005, respectively. In 2006, IP revenue grew 56 percent over 2005. The systems segment accounted 34 percent and 46 percent total revenue in fiscal 2006 and 2005, respectively. Systems revenue dropped nearly 6 percent in 2006 relative to 2005. In the last fiscal year, Korea accounted for 64 percent of total revenue, Taiwan 16 percent, United States 7 percent, Asia 8 percent, and Europe 5 percent.

Monolithic System Technology, Inc.

Monolithic System Technology, Inc. was incorporated in California on September 16, 1991 to design, develop, and market high performance semiconductor memory products and technologies used by the semiconductor industry and electronic product manufacturers. On September 12, 2000, the stockholders approved the company's reincorporation in Delaware, and it received about $51 million in net proceeds from its IPO in June 2001. In May 2006, the company changed its name to MoSys, Inc.

MoSys has developed a patented semiconductor memory technology, called 1T-SRAM, which offers a combination of high density, low power consumption, and high speed at performance and cost levels that other available memory technologies do not match. The firm licenses this technology to companies that incorporate, or embed, memory on complex integrated circuits, such as SoC. The company had also sold memory chips based on 1T-SRAM technologies, but in 2004, ceased shipments of 1T-SRAM memory chips.

On February 23, 2004, Synopsys entered into a definitive agreement to acquire MoSys in a cash and stock transaction valued at approximately $453 million. In April 2004, Synopsys exercised the right to terminate the merger agreement and paid MoSys a $10 million termination fee.

In 2005, MoSys began delivering a new family of 1T-SRAM CLASSIC memory macro products to licensees. These macros are silicon-proven, high-density solutions offering customers rapid memory block integration into their SoC designs. They are pre-configured and require minimal additional customization.

$K	2001	2002	2003	2004	2005
Product	12,991	2,924	1,904	952	10
Licensing	6,053	10,523	10,418	4,544	7,725
Royalty	3,446	14,344	6,911	5,325	4,547
Total	22,490	27,791	19,233	10,821	12,282
Delta		23.6%	-30.8%	-43.7%	13.5%
Net Inc	6,991	12,367	2,508	(1,907)	(2,982)

Table 2.12: *MoSys Five Year Financial Performance*

Based on total revenue, North America accounted for 30 percent in 2005 compared to 40 percent in 2003. During the same period, total revenue from Japan increased from 37 percent to 62 percent while revenue from the rest of Asia decreased fro 21 percent to 8 percent of total revenue. In 2005, NEC and Fujitsu accounted for 35 percent and 17 percent of total revenue, respectively. In 2004, NEC, Fujitsu, and Marvel accounted for 19 percent, 17 percent, and 11 percent, respectively. In 2003, NEC, UMC, and Sony accounted for a combined 42 percent of all revenue.

As of December 31, 2005, MoSys had 76 full-time employees of which 51 were in R&D. At the end of 2005, MoSys had nearly $69 million in cash, cash equivalents, and short-term securities.

Part of the MoSys strategy is to defend their technology with patents and they have large numbers of both US and foreign patents, with more pending.

Rambus, Inc.

Rambus, Inc. was founded in 1990 and reincorporated in Delaware in March 1997. The firm's IPO in May 1997 raised $33 million in company capital.

$K	2001	2002	2003	2004	2005
Rev	117,160	95,565	118,203	144,874	157,198
Delta		-18.4%	23.7%	22.6%	8.5%
Net Inc	31,271	24,704	23,221	33,559	33,677

Table 2.13: *Rambus Five Year Financial Performance*

The year-over-year revenue increases for 2003, 2004, and 2005 were 24 percent, 23 percent, and 8.5 percent, respectively. Royalties accounted for about 84 percent of revenue in this time period. Net income was around 20 percent of total income. As of December 31, 2005, the company had 333 full-time employees and $355 million in cash, cash equivalents, and marketable securities. Litigation expenses were $38 million, $23 million, and $20 million in 2005, 2004, and 2003, respectively.

Japan accounted for 61 percent of total revenue in 2005 up from 46 percent in 2003. During the same period the US share of revenue dropped from 36 percent to 29 percent, and Korea dropped from 17 percent to 5.4 percent.

For the twelve months ended December 31, 2005, 2004, and 2003, revenues from the top five licensees accounted for approximately 73 percent, 74 percent, and 75 percent of revenues, respectively. For the twelve months ended December 31, 2005, revenues from Intel, Elpida, Toshiba, and Matsushita each accounted for greater than 10 percent of the total revenues.

Rambus' chip interface technologies are covered by more than 440 US and international patents. Additionally, the firm has over 480 patent applications currently pending.

In November 1996, Rambus entered into a development and license contract with Intel. The contract allowed the parties to cooperate in developing a specification for Direct Rambus next-generation 64 Mbit RDRAMs, which was targeted at the personal computer (PC) main memory market segment. The contract also called for Intel to use reasonable best efforts to develop a PC main memory controller designed for use with these RDRAMs. The two companies set a target date of 1999 to produce a Rambus chip that could operate at a rate of 800MHz. Intel had warrants to purchase up to four million shares of Rambus stock if at least 20 percent of chipsets sold by Intel in two consecutive quarters within a given year would be geared to work with Rambus memory.

In September 2001, Intel and Rambus signed a five-year patent cross-licensing deal, under which Intel will pay Rambus $10 million per quarter over the length of the agreement.

Rambus has certain patents related to SDRAM. Rambus entered into license agreements with several major DRAM manufacturers under which Rambus was entitled to collect royalties in the range of $50 to $100 million per year. Some memory manufacturers (Infineon Technologies, Hynix Semiconductor, Inc., and Micron Technology, Inc.) opposed Rambus's royalty demands, choosing instead to dispute Rambus's patent rights in litigation.

In 2000, Rambus filed suit for patent infringement against Samsung, Infineon Technologies, Micron Technology, and Hynix Semiconductor. The suit against Infineon was dismissed on the grounds that Rambus destroyed key documents, but Infineon regardless agreed to pay Rambus license fees.

On June 19, 2002, the Federal Trade Commission (FTC) charged Rambus with violating federal antitrust laws in particular by deceiving industry-wide standard-setting organization, namely the JEDEC Solid State Technology Association. JEDEC has maintained a commitment and has implemented procedures to avoid, wherever possible, the incorporation of patented technologies into its published standards, or at a minimum to ensure that such technologies, if incorporated, will be available to be licensed on royalty-free or otherwise reasonable and non-discriminatory terms. According to the FTC's complaint, Rambus participated in JEDEC's SDRAM-related work for more than four years without ever making it known to JEDEC or its members that Rambus was actively working to develop, and did possess, a patent and several patent applications that involved specific technologies proposed for, and ultimately adopted in, the relevant standards.

The judge ruled that the complaint counsel failed to prove the facts they alleged in the complaint and that the legal theories advanced by compliant counsel failed to demonstrate that a violation of the FTC Act had occurred.

In April 2006, Rambus was awarded $306 million in its patent infringement suit against Hynix Semiconductor. Three months later, Rambus announced that it elected remittitur of the jury's $306.9 million award, announced April 24, 2006, to $133.6 million. On July 14, 2006, the judge in the Hynix patent infringement trial granted Hynix's motion for a new trial on the issue of damages unless Rambus elected to accept this lower amount for damages. The damages award results from the jury's finding that Hynix's SDRAM, double data rate (DDR), and DDR2 memory products sold in the United States between June 2000 and the end of 2005, infringed Rambus patents. During the trial, ten of Rambus' patent claims were found valid and infringed.

On December 31, 2005, Rambus entered into a five-year patent license agreement with AMD. The firm expects to recognize royalty revenues of $18.8 million in 2006, $15 million from 2007 to 2009, and $11.3 million in 2010 under the AMD agreement. During 2006, Rambus signed licensing agreements with Fujitsu, Matsushita, and Toshiba.

The Intel patent cross-license agreement represented the second largest source of royalties in the twelve months ending in December 31, 2005, 2004, and 2003. Rambus expects to continue recognizing revenue from this agreement through June 30, 2006, at which time, Intel will have a paid-up license for the use of all of the patents that Rambus owns claiming priority prior to September 30, 2006 and for which they own applications as of that date.

In October 2006, Rambus announced that its audit committee of the board of directors (BoD) had determined that a significant number of the stock option grants were not correctly dated or accounted for. The vast majority of incorrectly dated grants, both in terms of number of shares of common stock and financial accounting impact, occurred between 1998 and 2001. Rambus preliminarily estimates that the aggregate pre-tax, non-cash stock–

based compensation charges in connection with these stock option grants will be in excess of $200 million.

On August 15, 2006, Rambus announced the resignation of Geoff Tate, former chief executive officer of Rambus from 1990 through 2005, from its BoD. Mr. Tate was chief executive officer (CEO) and the sole member of the Stock Option Committee during the period that the audit committee preliminarily concluded that the majority of stock option irregularities occurred.

Saifun Semiconductor, Ltd.

Saifun Semiconductor Ltd. began in Israel in November 1996 and began operations in July 1997. This same year, they entered into a license agreement with Tower Semiconductor to implement their technology. In 1999, they generated initial fees from Advanced Micro Devices and Fujitsu, and in the proceeding year, they entered into a license agreement with Macronix. In 2001, further, they established a joint venture with Infineon Technologies. The firm's 49 percent ownership interest in the joint venture was reduced to 30 percent in January 2003. Saifun entered into license agreements with Spansion in 2002, with Matsushita in 2003, and with Sony Corporation in 2004. In December 2004, Saifun sold their interest in Infineon Technologies Flash to Infineon Technologies and, in January 2005, entered into an amended license agreement and a basic agreement on development orders with them. In 2005, they entered into a license and service agreement with Semiconductor Manufacturing International Corporation (SMIC). The firm's initial public offering and listing on the Nasdaq National Market happened in November 2005. Since the IPO, the firm has amended or extended existing agreements with key licensees, including Macronix, SMIC, and Matsushita.

Saifun has invented and patented a technology referred to as nitride read-only memory (NROM). It doubles the storage capacity of each memory cell on a silicon wafer by enabling the storage of two physically-separated bits of information within a single cell. This results in a significantly lower cost per bit. In addition, the company's next-generation NROM technology, referred to as QUAD NROM, enables the storage of four bits of information in a single cell.

US$K	2001	2002	2003	2004	2005
Licenses	2,468	3,170	7,817	22,640	65,790
Services	1,914	3,438	6,639	7,926	12,811
Total Rev	4,382	6,608	14,456	30,566	78,601
Delta		50.8%	118.8%	111.4%	157.2%
Net Inc	(6,795)	(12,228)	(15,190)	(37,904)	44,507

Table 2.14: *Saifun Financial Performance*

Three licensees, namely Macronix, Infineon, and Spansion accounted for 90 percent of licensing and service revenues in 2004 and 87 percent of licensing and service revenues in 2005.

In 2005, Germany accounted for 57 percent of licensing and service revenue compared to 4 percent in 2004 and 0 percent in 2003. Israel accounted for 45 percent of revenue in 2003 due to the joint venture with Infineon Technologies and contributed no revenue in 2005. Taiwan accounted for 18 percent of total revenue in 2005, down from 36 percent in 2003.

In November 15, 2005, Saifun announced the closing of its initial public offering from which it expects to receive total net proceeds of $121 million. In April 5, 2006, Saifun announced the closing of another public offering through which it expected to receive net proceeds of approximately $9.4 million and $2.2 million from the exercise of options by certain selling shareholders.

Silicon Image, Inc.

Silicon Image, Inc. is a leader in the secure distribution, presentation, and storage of high-definition content. Silicon Image offers robust, high-bandwidth semiconductors in the global PC/display, consumer electronics, and storage arenas based on its innovative digital interconnect technology.

Silicon Image creates and drives industry standards for digital content delivery such as digital visual interface (DVI), high-definition multimedia interface (HDMI), unified display interface (UDI) and serial advanced technology attachment (SATA), leveraging strategic partnerships with global leaders in consumer electronics and personal computing.

The company broadens the market adoption of DVI, HDMI, UDI, and SATA interfaces by licensing its IP cores to companies providing advanced SoC solutions that incorporate these interfaces.

$K	2001	2002	2003	2004	2005
Product	51,966	74,884	89,335	152,350	193,923
IP	0	6,655	14,190	20,809	18,476
Total	51,966	81,539	103,525	173,159	212,399
%IP		8.2%	13.7%	12.0%	8.7%
NI	(76,108)	(40,092)	(12,810)	(324)	49,549

Table 2.15: *Silicon Image Five Year Financial Performance*

Virage Logic Corporation

Virage Logic Corporation was incorporated in California in November 1995, and it was reincorporated in Delaware in July 2000. The initial investment came from company employees followed by a $3.5 million round of private financing in 1998 and another $10 million in January 2000. In 1999, the company acquired Mentor Graphics' Physical Library Business, and it received net proceeds of about $51 million from its IPO in August 2000. In 2002, the firm acquired InChip.

$K	2001	2002	2003	2004	2005
Revenue	31,763	45,613	40,657	53,003	53,389
Delta		43.6%	-10.9%	30.4%	0.7%
Net Inc	(2,054)	225	(3,850)	1,910	(315)

Table 2.16: *Virage Logic Five Year Financial Performance*

In 2005, revenue from North America accounted for 37 percent of total revenue down from 49 percent in 2003. In the same time period, revenue from Taiwan increased from 17 percent of total revenue to 22 percent. In 2003 revenue from 180nm and 130nm accounted for 84 percent of the total revenue. This dropped to 62 percent as revenue from 90 nm rose from 7 percent of total revenue to 36 percent.

The combined revenue for the publicly traded firms covered in the preceding paragraphs is presented in the table below. Only Silicon Image's IP revenue is included.

Total $K	567,531	566,716	624,452	797,763	879,239
Delta		-0.1%	10.2%	27.8%	10.2%

Table 2.17: *Combined Revenues of Covered Publicly Traded Companies*

2.6 The Big Three EDA Companies

The three major EDA companies Cadence Design Systems, Synopsys, and Mentor Graphics dominate the industry accounting for over $3 billion in revenue in 2005 or over three-quarters of the total EDA industry revenue. Their involvement with SIP is described in the following section.

Synopsys, Inc.

Synopsys' IP solutions consist of the DesignWare foundation library, DesignWare verification library, and DesignWare cores. DesignWare foundation library is an extensive library of basic chip elements (e.g., adders and multipliers) in which the design compiler incorporates into the design during the logic synthesis stage. The DesignWare verification library is comprised of popular chip function models used to create testbenches for the verification process. DesignWare cores are pre-designed and pre-verified digital and mixed-signal logic blocks that implement many of the most important industry connectivity protocol standards, including universal serial bus (USB) (1.1, 2.0 and on-the-go), peripheral component interconnect (PCI) (PCI, PCI-X, and PCI Express), Serial ATA, Ethernet, and Joint Photographic Experts Group (JPEG). According to their marketing literature, they believe they offer the industry's most complete PCI Express portfolio.

As shown in *Table 2.18*, IP represents between 6 percent and 7 percent of total revenue for the last four years. This amount is still sufficient to rank in the top tier of IP providers.

$K	2001	2002	2003	2004	2005
IP Rev	64,163	59,956	74,096	72,100	70,700
Delta		-6.6%	23.6%	-2.7%	-1.9%
% Total Revevenue	9.4%	6.6%	6.3%	6.6%	7.1%

Table 2.18: *Synopsys Five Year Financial Performance*

In February 2004, the company completed the acquisition of Accelerant Networks, Inc. for a total of $23.8 million in order to enhance Synopsys' standards–based IP solutions. In October 2004, the company completed the acquisition of Cascade Semiconductor Solutions, Inc. for total upfront consideration of $15.8 million and contingent consideration of up to $10 million. The company acquired Cascade, an IP provider, in order to augment Synopsys' offerings of PCI Express products.

On February 23, 2004, Synopsys entered into a definitive agreement to acquire MoSys in a cash and stock transaction valued at almost $453 million. In April 2004, Synopsys exercised the right to terminate the merger agreement and paid MoSys a $10.0 million termination fee.

In November 2004, Synopsys announced that it had acquired certain assets and hired the engineering team of LEDA Design, a developer of mixed-signal IP. LEDA Design had a team of more than 80 experienced digital and mixed-signal IP design engineers and support personnel located in Yerevan, Armenia who joined the Synopsys DesignWare IP engineering team.

Mentor Graphics Corp.

In it various financial releases and Securities and Exchange Commission (SEC) filings, Mentor Graphics Corporation does not state the exact size of its IP business. Industry analysts put it between $20 million and $30 million for the last several years. This translates into a little more than 3 percent of the total revenue.

Mentor's digital IP offering includes a large portfolio of Ethernet, USB, and storage interfaces available in Verilog RTL source code. In the area of mixed-signal IP the company offers hard macro cores for embedded PHY applications available. In the storage arena Mentor has offerings supporting Serial ATA, Parallel ATA and Personal Computer Memory Card International Association (PCMCIA).

In August 23, 2004, Mentor acquired Palmchip Corporation's complete parallel and serial ATA IP business. In July 2003, Mentor Graphics acquired Alcatel's complete line of Ethernet and MAC IP. At the time, Alcatel's Ethernet IP was being used in approximately 80 percent of the 10/100 Mbit/s ports and over 50 percent of the Gigabit Ethernet ports shipping worldwide.

Cadence Design Systems

The Cadence OpenChoice program enables interoperability and facilitates open collaboration with leading IP providers to build, validate, and deliver accurate models for Cadence design and verification solutions. The program aims to ensure IP quality, integration, and provides engineers access to a broad IP offering through a complete IP catalog. They claim that this optimizes the electronics design chain and accelerates customer time to market. Members of OpenChoice include ARM, CEVA, MIPS, Rambus, the Taiwan Semiconductor Manufacturing Company (TSMC), and Virage Logic.

2.7 Privately-Held SIP Companies

There are several dozen privately help SIP companies which include: CAST, Inc., Chipidea Microelectronica S.A., Denali Software, Inc., Impinj, Inc., Kilopass Technology, Inc., Qualcore Logic, Inc., Tensilica, Inc., and True Circuits, Inc.

No matter how much effort to compile a list and regardless of how long the list might become, there can inevitably be omissions. Space does not permit providing company and product information on the firms mentioned.

2.8 Non-SIP Companies

There are public and private companies that participate in the SIP market but whose primary source of revenue lies elsewhere. Large companies like IBM and Texas Instruments, at a minimum, offer patent licenses for technologies that may well fall into the SIP category. However, they generally do not report the total revenue from patent licenses much less break that down into segments like SIP. IBM, who every year is awarded more US patents than any other firm, did report $948 million in income from IP and custom development. Of this amount, $296 million was from sale or transfer of IP and $367 million was from licensing and royalty–based fees. This does not include cross-licensing agreements.

2.9 Entering the IP Market

Budding entrepreneurs might well ask if the SIP market is a good one to enter, but there are several challenges involved. ARM, the industry leader started as a joint venture between Apple Computer (UK) Limited, Acorn Limited Computers Ltd, and VLSI Technology. The number two vendor, Rambus, had early success after being "anointed" by Intel as having the preferred memory solution. A rawboned startup would not have these advantages. With the exception of ARM and Rambus most of the other vendors profiled in this chapter have had cumulative losses over the last five years.

The SIP market is a global market. Not only are several of the companies headquartered outside of the United States (ARM is in the United Kingdom, MOSAID in Canada, Saifun in Israel, and Chipidea in Portugal), but more importantly the leading foundries, IDMs, and many semiconductor companies are located in the Asia Pacific region. This means a SIP company from the beginning needs to have a global presence in both sales and field support. They cannot start in the United States and grow over time to become global.

The manufacturers of such end-use products as cell phones and computers have long sought components including microprocessors, DSPs, and ICs from third-party suppliers. The difference with SIP is that the product does not arrive in a box at the loading dock ready to plug into the final product. The decision to use SIP lies more in the hands of the design team than the purchasing department. The decision makers are very knowledgeable, and good presentations and slick salesmanship will not win the day. A vendor must be able to convince prospects that they should buy rather than make by the criteria set out earlier in this chapter. In a chicken and egg situation, one needs credible success stories to convince prospects to risk their projects. The sales cycle to convince prospects is long. Furthermore, it can take 12 to 18 months after opting for SIP components for a customer to get a product with the embedded IP to market and begin to generate a royalty stream for the vendor. During this period, the SIP vendor has to provide first class support.

The list of companies that have opted for SIP reads like a *who's who* of semiconductor companies. The idea of using SIP is gaining acceptance, and this is clearly a plus. However, most of the SIP vendors either have a small number of customers or only a few of their customers account for the bulk of their revenue. This gives the customers and prospects leverage in what is a complex negotiation for licensing fees and royalties.

The development of SIP for standard interfaces like USB is possibly the easiest from a technology point of view but would be most susceptible to become a commodity item. Memory has been a fruitful area for several of the SIP vendors including Rambus, Saifun, MoSys, and MOSAID. In any business endeavor, one must identify and develop a competitive advantage such as price, performance, capacity, quality or technology. Patentable superior technology is an obvious basis for sustainable competitive advantage. All SIP vendors have bulging portfolios of patents and patent applications that they vigorously defend.

SIP firms need to work closely with the foundries, first, to validate the IP and second, to use the foundries to protect the IP when it is used by companies in countries where there has been little respect and enforcement for IP rights. The foundries are in an excellent position due to financial assets, knowledge of their own fabrication processes, engineering talent and customer base to enter the SIP arena either as a source of revenue or as a competitive advantage. If one starts, the others will follow.

Chapter 3

Building a Legally Sound Intellectual Property Portfolio

Dennis Fernández, Managing Partner, Fernandez & Associates, LLP

Behfar Bastani, United States Patent Agent

Susan Fintz, Attorney, Wilson, Sossini, Goodrich, and Rosati

3.1 Introduction

In the system-on-chip (SoC) design business, the strategic value of building legally sound intellectual property portfolios as barriers to competition or as trading cards for cross-licensing deals has the greatest importance. Although many design companies have been labelling their wares as intellectual property, it is not always clear to what extent legally sound or enforceable intellectual properly rights apply to them. It is especially unclear when the primary product consists of electronic files containing circuit designs.

The term "intellectual property rights" generally refers to federal and state legal protections excluding unauthorized users from exploiting a technology. Such protections may also be extended to other countries. Unless a technology is properly protected by some form of intellectual property right, it generally falls into the public domain and may be used by anyone, without license.

Companies that want to build a legally sound intellectual property portfolio must systematically recognize and protect their claims. During system or chip design, valuable intellectual property rights must be carefully managed in order to avoid forfeiture. Companies should refine their interests in these rights with internal policies and practices that maintain the confidentiality of proprietary information and keep records to file applications for protection in the US Patent and Trademark Office (USPTO).

3.2 Forms of Protection

While patents may be the most well-known and appropriate form of intellectual property protection for electronic design automation (EDA) technologies, there are several other forms of protection with associated trade-offs in speed, cost, and protection strength. Each item in the following list represents a separate re-licensable intellectual property right.

Property right	Scope of Protection	Advantages and Disadvantages
Patent	Non-obvious apparatuses and methods	• Expensive and time-consuming • Best for protecting ideas One year to file after first offer for sale
Copyright	Original literal works, software	• Easy • Fast to obtain Appropriate against piracy but not against reverse engineering
Trade secret	Secret business and technical data	• Requires effort to keep secret, such as generating non-disclosure agreements A fragile form of protection
Maskwork	Chip layout information	• Requires federal registration • Does not protect netlists Companies have two years to file after commercialization
Trademark	Distinctive product and company names	• Relatively easy to change Does not protect technology or descriptive terms

Figure 3.1: *Intellectual Property Rights for Protecting Chip Designs*

Patents

Patents may be the most appropriate form of intellectual property protection for EDA technologies, such as SoC designs, computer-aided design, or engineering tools, and testing or manufacturing process technologies. In general, patents protect concepts, such as novel methods, processes, apparatuses, or systems. Obtaining patents can be very time-consuming and costly, especially considering the USPTO and patent attorney fees. A patent covering a relatively complex circuit or software technology can take a few years to issue and usually costs more than $15,000 to prepare and submit. In the United States, patent applications must be filed within a year of the first attempt to sell the inventive product.

Let us note that a US patent only provides protection from infringement within the country. While international treaties exist for using international patent applications to allow for grace periods between filing for a patent in the United States and filing in other countries, eventually such an international patent application must be filed, examined, issued, and become enforceable in the individual countries where enforcement is sought. This means that a US patent grants protection to the patent owner in the country (including against importation of the technology from a foreign country into the United States) but no enforcement in foreign countries such as Japan or Germany, unless foreign patents are filed for and obtained in those countries.

Copyrights

In comparison with patents, copyrights are both cheaper and faster to obtain, and copyright protection is most appropriate for protecting literal (that is, expressed in letters), software code (such as netlist files), and graphical computer display images such as schematics and other engineering drawings. Copyright protection instantly arises when original copyrightable subject matter is fixed in a tangible medium, such as digital storage; registration and notice are recommended rather than required. Unfortunately, copyright protection may not prevent competitors from reverse-engineering software to develop code that is independently capable of providing the underlying functionality.

Trade Secrets

Trade secrets, like copyrights, are protected automatically, here, if a company can show that information has been developed in confidence, sought to be kept confidential, and that it provides a competitive advantage.

Accordingly, the diligent use of nondisclosure agreements and other proprietary warnings on sensitive documents is highly recommended. This kind of protection ends when information is no longer secret. Pending patent applications are normally regarded as trade secrets until the patent issues publicly. There is no formal filing procedure for registering trade secrets.

Trademarks

Trademarks are another form of protection, and they cover distinctive symbols and names of products, services, and companies. Trademarks chosen and filed for registration must have a distinctive meaning such that they are neither descriptive or have generic terms. Trademarks usually extend legal protection, not to technologies, but to company good will and to the quality associated with recognized names or symbols.

Trademarks

Semiconductor chip products can be protected when the chip layout includes original circuit designs that may be pieces of a final chip. Maskworks specifically protect companies against the unauthorized copying of chip layout information, although this protection does not extend to pre-layout design information, such as netlists or hardware description files. An application for federal registration must be filed within two years of the first commercialization of the relevant chip product.

3.3 Design States

During the conventional application-specific integrated circuit (ASIC) or SoC design process, intellectual property rights typically arise at every point in the chip prototyping and production flow process, particularly during specification, system architecture, circuit design, simulation, layout, manufacturing, and distribution.

Each design stage can then generate licensable rights and produce revenue streams when properly formulated and protected.

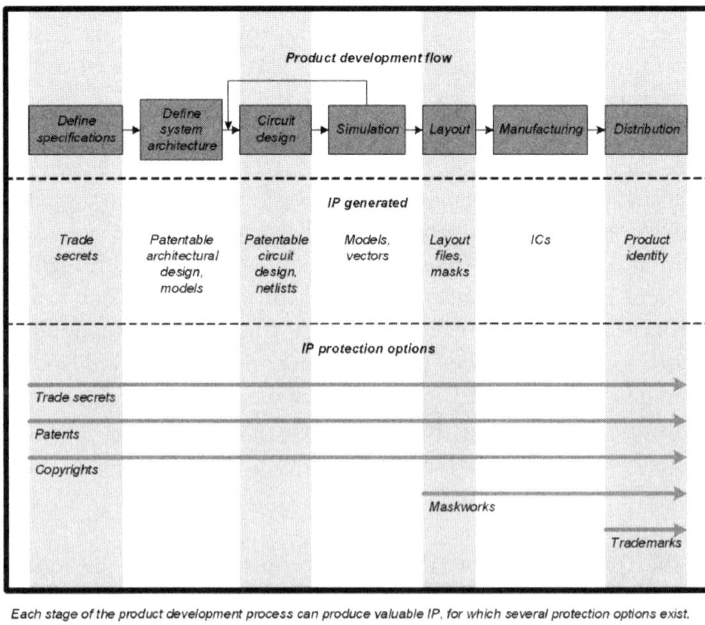

Figure 3.2: *Stages of Protection in the Development Process*

Chip Specification

Initial chip or system specifications, although almost always preliminary, may contain copyrightable matter, protectable trade secrets, and patentable ideas. The specification document, including any original text and graphics, can of course be copyrighted, and if it could provide a competitive business advantage—such as details of a company's business, technical or marketing strategy, and implementation—it may also have protectable trade secrets. Perhaps more important from the standpoint of protecting proprietary technology, patent coverage may be available for any apparatus or method named in the specification or contemplated before or during its preparation.

To protect patent rights, it is necessary to file and prosecute them in the patent offices of all countries where the applicant seeks protection. In the United States, inventors generally do not have to reduce innovative products or methodologies to practice to obtain patent protection, although the patent application must describe them in sufficient engineering detail to permit someone of ordinary skill in that "art" to design and operate the invention.

It is therefore possible, and often strategically desirable, for applicants to seek patent protection for future competitive technical solutions—improvements that applicants are the first to invent and define in sufficient implementation detail, although they may later be designed into competing products. Another strategic approach to building valuable patent portfolios is to apply for patent coverage in the path of emerging industry standards. Companies that succeed in doing so have exclusive rights against unwitting or knowing infringers who subsequently use such standards when they are adopted.

System Architecture

At the next stage of the chip prototyping and production process—architectural, behavioral, or some other form of top-level system definition—the same intellectual property rights are available: patents for non-obvious apparatuses or methods, copyright for literal works, and trade secrets for confidential business or technical information.

System and design engineers often decide on important functions, typically defined in high-level software or hardware description files, at this point in the process. Novel systems or chip architectures are usually best protected with patents, since copyrights of architectural description files are relatively easy to undermine simply by legal reverse engineering and trade secret status; although applicable in the early stages of the design process, this may be unobtainable when products and architectures are publicly announced. Companies often cannot use patents to protect system architectures since busy design teams inadvertently fail to file United States patent applications within a year of the first effort to commercialize or offer inventive subject matter for sale.

Circuit Design

Further detailed system and chip functionality is defined at the stage of circuit and logic design. Here as well, the usual forms of intellectual property rights such as patents, copyright, and trade secrets may be created. Advanced computer-aided design tools are usually used to generate circuit definition files with valuable IP for producing core circuits. Specific circuit and logic designs and associated netlist files may be protectable through patents, copyright, or trade secrets.

Since consultants and other third parties are often employed to design and verify circuit functions at this detailed and fast-moving stage, it is very important to assess and record innovative features and thus ensure that appropriate legal measures are taken to perfect a company's rights. Otherwise, valuable IP will probably be left unprotected, or worse, owned exclusively by competitors. It is particularly important to prevent leaving rights unprotected when companies develop or exploit reusable complex core functionality, standard cells, logic blocks, macros, and other valuable commercial design libraries and building-block virtual circuit components.

Simulation

The system or chip definition may be verified and tested throughout the design process at various levels of functionality such as at behavioral, logic, and circuit levels of abstraction. As in the previous stages, copyrights, trade secrets, and patents may be used to protect relevant technology, now typically taking the form of simulation or test vectors, models, and other related software and firmware. Such examples of such patentable inventions include novel pieces of automated test equipment (ATE), novel software simulation models, and novel testing processes relating to the systems or chip definitions at hand.

Layout

Physical chip definitions can generally be protected under patent, copyright, trade secret, and maskwork rights, but mostly the latter. However, to obtain maskwork protection for any part of a chip, it is necessary to follow proper federal registration procedures for original semiconductor chip products within two years of its first commercialization. Applicants must complete and file maskwork application forms typically including actual chip samples and layout drawings.

Manufacturing

Patent, copyright, trade secret, and maskwork protection are also available when chips are manufactured. At that point, new fabrication-related issues may conceivably arise, typically in the form of novel device structures, process improvements, or other innovative circuit dependencies on particular fabrication conditions or steps. Yet, most systems and chips are developed with the expectation that they will ultimately be produced conventionally—for example, with the standard complementary metal-oxide-semiconductor (CMOS) process. In other words, the manufacturing stage, for either prototype or production runs, typically does not give rise to new intellectual property rights.

Distribution

When products are finally distributed, a new form of protection, namely trademarks, may enter the picture, taking its place beside patent, copyright, trade secret, and maskwork protection. Intellectual property rights pertaining to the distribution and sale of chip products are typically governed by license agreements defining such standard contract terms as "exclusivity," "duration and scope of business relationships," "indemnification against possible infringement risks," "warranties," and "technical support covering product defects." Since most, if not all, of the standard terms may apply to vendors of whole chips and specialized blocks alike, it is important for the business models of both to define such intellectual property and other business-related contractual terms carefully. A common licensing problem arising at this stage is accountability for, and ownership of, functional defects discovered after products have been distributed.

3.4 License Terms, Risks, and How to Mitigate

For the purpose of this licensing agreement study, we assume that the owner of a patent wants to enter into a licensing agreement with a licensee. On one side of the transaction is the owner of a patented technology who is eager to obtain a monetary return on the investment that resulted in the patent. On the other side is the licensee who is ready to bring a product or service to market but is in need of a certain technology to accomplish a business goal.

When entering into a licensing agreement, the parties must think beyond the closing of the deal, since the particular language of the licensing agreement will produce future consequences and shift business risks. In the following, we highlight some examples of common issues to be resolved by the parties.

Rights Granted and Rights Reserved

In a licensing agreement, all rights granted by the owner to the licensee must be specifically mentioned in the agreement. Any rights not specifically granted to the licensee are reserved for the owner, and the licensee should not assume that any other rights are included.

For instance, if the license language only covers the rights to make and sell a patented technology in the United States, a licensee wishing to have rights to advertise and lease the technology should ensure the agreement clearly spells out these additional rights. By having more in writing at the onset, the licensee avoids having to re-negotiate with an owner for additional rights at a possibly higher cost. Therefore, it is more advantageous for the licensee to negotiate rights upfront, rather than later from a possibly weaker bargaining position.

Exclusivity and Enforcement Obligations

A license can be granted either on a non-exclusive or exclusive basis. The type of grant chosen influences the value of the license, the balance of power between the parties, and the cost that a party may later be contractually obligated to pay to enforce patent rights. In terms of value, an exclusive license tends to be more valuable because the licensee is the only one who can make use of the technology. As a result, exclusive licenses tend to be more expensive since the technology owner can only obtain royalties from one licensee.

In regards to rights and enforcement decisions, a non-exclusive license gives the licensee the right to "march in" or, in other words, to operate in a certain market. As a result, a licensee does not usually have much say when it comes, for example, to deciding whether or not to sue another licensee. Enforcement decisions against others, in a non-exclusive license, are usually reserved for the owner. Thus, the owner decides whether or not a second licensee in breach should be sued.

In contrast, an exclusive licensee has the right to exclude all others from making use of the rights, according to the terms of the agreement with the owner. Since exclusive licensees usually have the ability to sue any infringer, this option is more attractive to a licensee who wants to avoid any interference with their rights. Regardless, exclusive licenses are not always available, are typically more costly to obtain, and it is often followed by an obligation to defend the granted rights. Both types of licenses can overall create potentially good business opportunities for the parties.

Proprietary Rights: Joint versus Sole Ownership

According to US federal law, "in the absence of agreement to the contrary, each of the joint owners of a patent may make, use, offer to sell, or sell the patented invention within the US, or import the patented invention into the US, without the consent and without accounting to the other owner" (35 U.S.C. 262). This becomes an issue when two or more joint owners disagree on how to explore patent rights.

One common problematic scenario is when two co-inventors hold a patent and one wants to sell or license his rights and the other does not. In this scenario, each inventor would have complete freedom to operate despite the disapproval of the co-inventor. A preventive measure to avoid this problem would be to write, as early as possible, an agreement between co-inventors to establish the decisions that must be made jointly.

A second problematic scenario is when co-inventors agree to sell or license, but do not agree on terms such as to whom, for what price, and in what circumstances. As a consequence, the co-inventors may proceed to negotiate separate agreements with a licensee and the licensee benefits by bargaining for price and terms with each of the co-inventors. The co-inventor who makes the worst deal obtains an agreement, and the co-inventor who refuses the deal ends up with nothing. Both inventors most likely give away value they would have gotten, if they had negotiated together. It is always recommended that co-owners negotiate together, and if they can not agree, perhaps the joint ownership should be dissolved with one owner buying out the other.

Technology Improvements

Let us know consider the issue of assigned and reserved rights on an improved technology. A licensee only has the rights that are specifically granted in the agreement, and the owner cannot depend on any right not specifically granted to generate a stream of royalty revenue.

Furthermore, the parties must attend to substantial improvements to the technology made by the licensee. The owner seeks to acquire improved technology as a right granted, so that once improved by the licensee, the owner receives royalty revenue on this. The licensee, on the other hand, may argue that since the owner did not actively participate in improving the technology, revenue on the improvement thus can not be warranted.

The parties must also decide the consequences to the agreement when the licensee creates major improvements that render the initial technology obsolete. If this occurs, the licensee may want to stop using the originally licensed technology and stop paying royalties for it. The arguments here are similar as before in that the licensee does not want to pay the owner for something independently created. The owner, on the other hand, may argue that without the grant of the initial technology, the licensee would not have had an opportunity to improve it, and as a result, the original technology has some value in, if nothing else, paving the way for the licensee's improvements.

The solution, in both cases, is to write a balanced agreement early on which satisfies both parties with respect to payments for improvements and use of the improved technology by the owner, the licensee, or both.

Sublicenses

The agreement between an owner and a licensee should explicitly establish if the licensee can grant a sublicense and under what conditions. The issue can be framed as one of control. The owner may not want to grant a licensee this right because the owner is afraid of losing control over the invention. The fear is that a stranger, not chosen by the owner, can obtain a license granted by licensee with better terms than those the owner would have offered. Another concern of the owner is that a licensee may grant rights to a sub-licensee who is neither creditworthy nor commercially capable.

However, there are many scenarios which may compel the licensee to introduce a new party, other than simply trying to side-step the owner and grant rights to a new party to exploit the patent. As an example, the licensee may change its corporate structure, as often occurs when companies are bought and sold. In addition, from the licensee's perspective, it is common that the owner be happy that a new party is introduced since a sub-licensee increases the stream of royalty payments for the owner.

The solution is to write the agreement with proper and balanced language. One way to achieve this result is for the owner to allow the licensee to create sublicenses, but hold the licensee responsible for the maintenance of certain conditions imposed by the agreement between the owner and the licensee.

Royalties

In a technology licensing agreement, there are many ways for the parties to structure how payments are made in exchange for the use of the technology. One important issue is the decision between a lump sum payment versus periodic payments over time. Let us examine the components of this decision in more detail. A lump sum payment can be a good choice since the amount to be paid is clear and free of manipulation. However, it is often hard to quantify upfront what amount the owner should receive and, sometimes, this amount can be grossly under or overestimated by the parties. Further, when drafting a lump sum agreement, it is important to avoid drafting a licensing agreement that, by its terms, can be best qualified as a sale.

Periodic payments are frequently based on factors such as the percentage of licensee's revenue, fixed price per unit, percentage of selling price per unit, a combination of percentages, fixed floor, etc. One problem with periodic payments is that, depending on its overall structure, the licensee can easily manipulate the amount. Manipulations are carried out, for example, by creating promotions that vary the price of the product. An owner who is interested in receiving periodic payments should specify in the agreement what records must be kept by the licensee which can later be inspected for accuracy.

Reports and Records

In regards to record keeping and reporting, the parties must address the right of an owner to audit and the licensee's need to the confidentiality of corporate records.

The justification for the owner's right to audit is that even the most trusting partner needs to have the means to check for errors, underpayments, and assure the honesty of the other party. The problems for the licensee include the fact that constant monitoring will cause disruption of normal activities, will increase the cost of compliance, and will give access to highly confidential information. Also, the licensee may fear that the owner may unjustly benefit from the information by either using it, sharing it with a third party, or allowing the information to become public, thereby hurting the interests of the licensee.

Licensing agreements should cover this subject in detail in order to deal with these specific risks. The parties should be prepared to negotiate the frequency of reports, conditions for inspection, information that will be accessible to the other party, the party responsible for paying for the audit, and disposition of confidential documents.

Warranties

Warranties and representations are all about risk assumption. Other than the value that a technology can bring to a business, one must consider the risks that are taken in connection with the license.

It is fair to state that both parties want the patent to be valid, useful, and enforceable. The licensee wants to have a commercially profitable use for the technology covered by the patent. The owner wants to generate a royalty revenue stream. However, there are some threshold questions that must be addressed by the parties. What happens if another product infringes on the patent, if the licensee's product infringes the rights of others, or if the patent is declared invalid? In such cases, who will be responsible for filing for or defending against infringement, misuse, and misappropriation? Who will bear costs, legal fees, and losses?

There is no correct and absolute answer to these questions. Risk assumption is negotiable, and it highly influences the value of the agreement. Therefore, before stating a price for a technology, it is important to understand and negotiate the risks each party will take.

Termination

There are many possible grounds to terminate a licensing agreement such as a change in the licensee identity (corporate structure), an acquisition of the license by a competitor of the licensee, non-payment of royalties by licensee, failure to achieve royalty goals by licensee, among others.

A licensing agreement should strive to forecast the problems and provide for solutions. For example, if a license has a royalty provision where royalties are based on sales, termination caused by the failure to pay minimum royalties when the business is slow could perhaps be cured if the agreement provided for a minimum royalty is not based on sales. Conversely, if a harsh agreement provides for automatic termination in case of failure to pay minimum royalties (a "use it or lose it" license), such a license may be giving a licensee who wants termination an easy way out. A harsh agreement that at first glance seems to protect the owner may in fact hurt the owner. Termination is a complex issue that must be analyzed in connection to the other terms of the agreement.

In conclusion, the negotiation of a licensing agreement is highly complex and should be carried out on a case-by-case basis. Some of the factors affecting the negotiation of such agreements include industry standards and procedures, the bargaining power of the parties, the true level of interest of each party to a particular deal, and the existence and desirability of alternatives to both parties.

While this chapter brings some valuable information and lessons to the reader in regards to building a valuable intellectual property portfolio and licensing negotiations, it is important to consult with an attorney for specialized advice that is tailored to the specifics of a business entity.

Part II:
The IP Provider

Chapter 4

Analog, Star, and IP Blocks

Henry Chang, Vice President and Co-Founder, Designer's Guide Consulting, Inc.

Larry Cooke, Engineering Consultant

Grant Martin, Chief Scientist, Tensilica, Inc.; Senior Member, IEEE

4.1 Introduction

Intellectual property (IP) has traditionally been classified as hard, soft, or firm. Hard IP blocks have physical mask data, and thus, they have a physically estimated size, dimensions, power, and performance. They are usually targeted to either a generic process that is available in a number of fabrication facilities, or they are targeted to a specific foundry with a specific integrated-circuit (IC) process. As a clear advantage, they are pre-tested and verified, in many cases in real silicon, and therefore can very well function as specified when embedded in larger IC designs. However, they have the disadvantage of being inflexible in all their physically-defined attributes.

By contrast, in soft IP blocks, the logical data is transferred to the customer. Many have scripts and other details to direct IP implementation, and they may have been physically implemented, even to actual silicon, as part of their verification process. Yet, that implementation is generally not part of the data transferred to the IP user. Soft IP has the advantage of being flexible in size, power, performance, and function, but it is inaccurate in these same characteristics, relative to hard IP. Furthermore, since it is unlikely that the specific user implementation has been executed, the range verification of user implementations is typically less than hard IP, and as a result, under rare circumstances, user implementations may be difficult, if not impossible.

Firm IP is most conveniently defined as everything else. Various definitions of firm IP include IP that is soft, but contains hard blocks; IP that has a soft component, but includes specific scripts for implementation of parts; all of the IP; or at the other extreme, hard IP with some generated soft or semi-soft features. In short, there is no specific definition for firm IP, other than everything that is not soft or hard IP. Firm has not proven successful in the past decade of IP industry development, but it may soon as design for manufacturing (DFM) considerations at 65 nm and below may require more information about, and control of, block layout to accompany the soft register transfer level (RTL) description.

While this first-level taxonomy has been the traditional way to classify IP, much of the IP being developed today has aspects of all three types. As a result, we feel that more insight may be gained by looking at IP blocks in light of their functional characteristics, design methodologies, and market conditions. The three most significant areas that stand out are analog, star and commodity IP.

Analog IP is typically hard, increasingly with firm or soft digital interfaces. The tools and resulting methodologies are very different from digital IP, with far more design activity focused on the circuit, layout design, and verification functions.

If commodity IP is, as the name suggests, a commodity, then star IP, is the market antithesis. Star IP, thus the name, commands a stellar price over a much longer period of time than commodity IP. Star IP generally contains either protected proprietary IP or is created through protected proprietary techniques, and it has a growing coupled user-defined base of user-designed IP. Yet, commodity IP may have one of these characteristics, but not both.

These two conditions make all the difference. Without proprietary IP, many IP providers can, and do, build functionally equivalent IP; without a corresponding base of user-defined IP, there is nothing that ties the user to any particular IP supplier's solution, particularly as marketplace competition decreases the prices.

Yet, this chapter describes these types of IP blocks from the IP provider's perspective. In the following sections, we are attempting to answer such questions as: What are the tools and methodologies the IP provider employs? What are the customer demands? And what are the problems the IP provider must solve?

Clearly, the answers to these and other questions, from the IP provider's point of view, vary widely between analog, star, and commodity IP providers, but they also vary, though to a lesser extent, on the target technology for the IP. Cutting edge processes create more issues for analog IP developers than older more stable processes. Similarly, digital IP targeted for field-programmable gate arrays (FPGAs) has different market and technical constraints than IP targeted toward application-specific standard products (ASSPs). For example, in the latter case, hard digital IP may be appropriate for ASSPs, but soft or firm are the only applicable forms of IP targeted to FPGAs. Further, there are far more designs and generally less volume per design with FPGAs than in ASSPs, which creates very different user expectations about the price of IP. The following sections also discuss various aspects of implementation in regards to the previous questions.

Lastly, since quality is one of the most important problems the IP provider must solve, the rest of this book discusses quality in both a limited and peripheral manner. For example, while we cover verification as part of the IP provider's design methodology, we do not specifically cover IP quality. We depend on the readers to integrate the methodologies and marketplaces presented in this chapter with the quality metrics, measures, and methods presented later in the book.

4.2 Analog IP

Introduction

Analog IP is a fact of life for today's systems-on-chip (SoC). An architecture having a generic high bandwidth, high-resolution analog-to-digital (A/D) converter as a front-end and digital processing as a back-end to implement analog functions is, in general, prohibitively expensive in terms of power consumption. This is certainly true in handheld devices, but often in set-top boxes as well, not so much due to power usage, but rather to heat dissipation. During the past decade, analog IP has actually become much more complex, as these IP blocks need to continually hit higher performance targets while operating at lower power.

There are several key types of analog IP. There are, of course, the real world interfaces such as radio frequency (RF) transceivers, audio coder/decoders (CODECs), and video drivers. Digital interfaces are becoming increasingly analog to achieve higher bit rates, and a classic example is a serializer/deserializer (SERDES). Such examples of SERDES include serial advanced technology attachment (SATA), universal serial bus (USB) 2.0, and Firewire. Power management especially on handheld devices has become a complex task, and its numerous functions include managing multiple power domains, the ability to control battery charging, taking power from an attached USB line, driving light-emitting diodes (LEDs) and speakers, among others. The control aspects of these chips are digital, but the regulators, actuators, and sensors themselves are analog. Finally, all SoCs have ever more complex clocking circuitry. Phase-locked loops (PLLs) are considered analog and are often purchased from analog and library IP vendors.

There are many factors that increase the difficulty of analog IP design which include designing in digital processes at small geometries, working at low supply voltages, and in the context of noisy digital circuits. Analog designers have met these challenges by innovating new IP-level algorithmic architectures such as using delta-sigma techniques to create a fractional-N PLL, clocked circuitry on traditional analog functions, calibration techniques, and also by developing new circuit-level architectures such as ones that exploit complimentary metal oxide semiconductor (CMOS) switch usage. Fortunately, for the IP buyer, the analog designers have met these performance, low power, and integration challenges through clever innovation. The result, however, is a state-of-the-art, highly complex, and process-specific hard analog IP block.

The key questions that today's analog IP buyer face are:

- Did my IP vendor design the block correctly?
- Do I fully understand how to use the IP block?

- In my final implementation, did I hook everything up correctly to that analog IP block?

The underlying reality is that the typical analog IP buyer does not have the expertise nor the time to verify the IP by themselves. Even within a company where internal IP is being consumed, there is usually a large divide in terms of technical knowledge and organization groups between the consumers of analog IP and the developers. One might be surprised to learn that a simple analog IP block such as an audio CODEC found in any cellular phone or MP3 player could have hundreds of control pins implementing more than 500 modes of operation which include all the settings, knobs, calibration, power-up/down, tuning, and test modes. The criticality of guaranteeing analog IP functionality in the first spin is often very high. For example, if the PLL does not work, the chip is dead on arrival. If the radio on the chip is dead, prototype chips cannot be put into systems for software development and initial testing. Usually, small errors in performance can be fixed in future spins, but a system company buying a SoC that finds that they cannot begin testing or writing the software for their system quickly loses patience with the IC vendor and moves onto a different supplier. Thus, these key questions faced by the analog IP buyer can be critical for the product and perhaps even for their company, to such a point that we have encountered several instances where the analog IP buyer will hire a different outside vendor, usually a consultant, to independently verify the quality of the purchased analog IP, along with the proper connection of the SoC to the IP block. At times, one outside consultant will not have enough expertise to fulfill these tasks, and they will work with multiple vendors for this verification effort.

We will describe a methodology to address these issues in the following section.

Methodology

Fundamentally, the answer to the IP buyer's questions revolves around the creation of a hardware description language (HDL) model of the analog IP block. It is typically written in Verilog, VHDL, or perhaps in SystemVerilog in the near future. This model serves as the primary communication vehicle between the analog and digital designers. The HDL model serves as an executable description of the analog block where the specifications are often stale or are left unread. Since it is meant for the digital designers, it is written with the digital version of these languages, and not the analog version, for example, Verilog-A. For some design groups where analog understanding is a bit higher, languages which are supersets of the digital language such as Verilog-analog mixed signal (Verilog-AMS) may be appropriate for the analog IP model representation. We present the very basics of how to create this HDL model in this section.

The creation of this HDL model is simpler said than done, especially since there are no automated tools and nothing close in research to automatically create this HDL model. It is a manual and error prone process, and there are also no automated tools to verify the models. No such thing as "equivalency checking" exists for this task. An analogy in the digital world would be verifying that a high level SystemC model is equivalent in certain respects to a complete RTL description. In the digital world, using extensive verification environment is often the only methodology available[1]. We approach the analog model verification problem the same way.

For both the analog IP buyer and developer, those three key questions boil down to a single question: is the specification equivalent to the transistor level implementation and to the HDL model being delivered to the IP buyer? Our advice to any buyer is to ask two simple questions of any analog IP supplier. The first one is: do you supply an HDL model of your analog IP? In many cases, the answer will be "yes." When it is "no," the frequently cited reason is that the analog IP supplier does not know anyone who has modeling or HDL language expertise. The next question, to which usually gives away to an unsatisfactory answer, is: how do you know that a HDL model is functionally correct in all its interfaces, modes of operations, etc.? It is very risky to use that IP

[1] Some "sequential equivalence checking" tools are beginning to emerge in the digital domain. However, these have not yet been incorporated into a standard methodology and require considerably more development and user prove-in. Thus, simulation within a verification environment is still the most widely used method.

block unless the first answer is "yes," and the second has been answered in a sound manner.

We focus on answering this equivalency question in terms of function, e.g., are the specifications, transistor level implementation, and HDL models equivalent in terms of function? The traditional IP-due diligence checks still need to be conducted to view overall performance. For example, does the IP provider have a proven track record? Do they have silicon of their IP that works? In some ways, performance specifications are easier to verify because one knows what is being measured, and analog designers are aware of verification issues where they exist. The use of extensive simulation or lab measurements is a common method to verify overall performance. Functional verification is difficult because one is looking for what has not been previously checked. The difficult question of how to define "coverage" thus arises.

In analog functional verification, many of the typical errors caught are due to communication issues between IP suppliers and developers and also within the IP development group itself. Common examples of errors include inverted signals, flipped busses, and unaccounted-for dependencies such as in power-up.

Analog Verification Flow

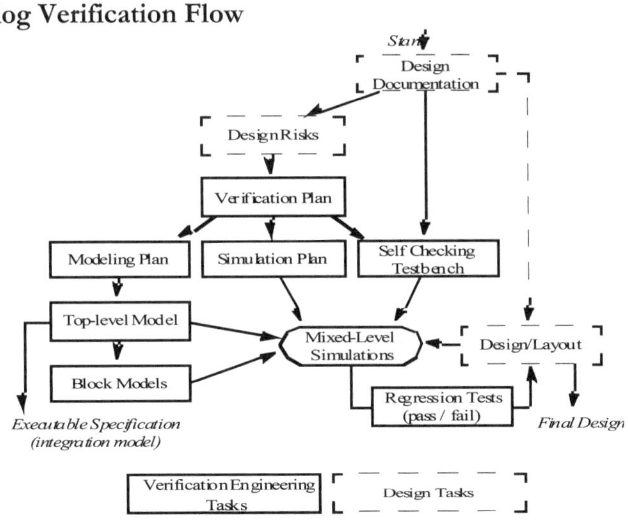

***Figure 4.1:** Analog Verification Flow*

In *Figure 4.1*, we show the analog verification flow [3], and we separate the design and verification tasks. We show all the verification tasks and lump most of the design tasks all together. A key aspect in the flow is the first verification task, which is the creation of the verification plan. This outlines what is most important to the designers to be verified, and it is highly based on their past experiences. The three most critical and time-consuming verification tasks are modeling, testbench development, and debugging. Each is of similar effort, and each requires specific expertise.

As in a typical digital team, we recommend adding an analog verification engineer to the team who is an expert on modeling, testbench development, and debugging for analog blocks. Their sole focus is to find bugs, unlike the designers who have other objectives to meet such as punctuality, finishing the design, and meeting specifications. Their role is shown in the tasks with the solid line boxes.

Model Writing

The first of the three most time-consuming tasks is model writing, which has always been considered a daunting task. Writing complete models for functional verification is a straightforward process, and it consumes much less time than writing models for performance verification. It is made simpler because functional models are easier to create than models for performance, as will be described later. Following the verification plan also simplifies this task since it bounds what has to be modeled. In addition, if the designers follow certain conventions while setting up their design hierarchy, they can significantly reduce the number of models that need to be written.

The key requirements in the model are the allowance of mixed-level simulation (pin accurate), the inclusion of behavior of all bias and control signals, and the same function as IP or blocks. As options, more details can be added for performance verification and assertions can be added for more checking.

Assertions or monitors check, rather than model, block usage for connectivity and function, and it parallels the use of assertions in digital verification. In analog, they are used to check the functionality of often ignored non-signal pins. Common assertions check for expected power/grounds, clock, bias, and analog input/output (I/O), power consumption in various modes, digital signals, and signals required for test. The idea is to flag any errors, rather than model what happens when a block is incorrectly used, which ultimately makes model writing much simpler. Finally, as with most tasks, experience and judgment calls are often required to decide exactly what to include.

The Testbench

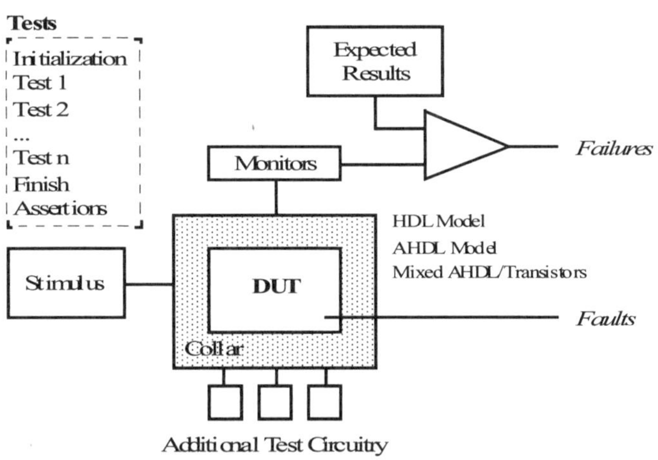

Figure 4.2: *The Testbench*

The second major piece of the puzzle is the testbench [3] as shown in *Figure 4.2*. The strength or weakness of analog verification rests here, and this is the most critical piece to determine success or failure for the verification effort. The testbench has to be rich enough to find all errors in the design, model, documentation, and testbench itself. As shown in *Figure 4.2*, the testbench applies the stimulus to the design, model, and mixed-level representation, and then it checks the results for correctness. The correct results are coded from the specifications, and thus all the specs are checked. The identical testbench is applied to all representations of the design. If they all pass, it means that the specifications, the transistor level representation, the HDL models, and all other representations are equivalent to the extent that the testbench is rich enough to test all pieces.

The testbench consists of both tests and assertions. Tests put the circuit in a certain mode, apply input stimulus, and check the output of the circuit. Tests can be run sequentially or in parallel. Assertions, similar to those in the design model, continuously check for such conditions as setup and hold times.

Finding "bugs" is a very tedious task, as it requires a solid verification plan and reviews. The only approach is to apply a systematic methodology to write the testbench. The verification engineer who writes tests cannot solely rely on designers, and they need instinct and experience to determine when enough verification has been performed. They also need the experience not only to detect design errors, but also to see verification performance. Their sole job must be to find bugs.

Regression Testing and Debugging

Regression testing can begin on the models once they and testbench are in place. This will guarantee that the models match the intended specifications. As analog blocks are designed at the transistor level, regression testing of the analog blocks can begin using mixed-level, mixed-mode simulation. Debugging, here, means debugging the specification, design, models, and testbench.

The next issue is what to do when an error is found. Who does the debugging? The verification engineer is the first line of defense, and they will attempt to debug it by:

- Obtaining a clean testbench with a pure model view
- Verifying that the testbench is okay if there is an error on the schematic
- Diagnosing the schematic
- Diagnosing or contacting the designer when the schematic looks suspicious

Fortunately, functional errors are typically easier to diagnose than performance errors. If it is a design bug, the designer should be notified, and it is their duty to hunt around for more. In theory, the testbench should catch all of the bugs, but the mere presence of one sole bug can often signify there are many more. It is, in other words, a good practice to inspect the region for others.

Analog Summary

Top-down analog verification, as previously described, can be applied whether the designers use a top-down or bottom-up design style. This methodology gives traceability from the specification to the models to the transistors. It verifies that the specification, models, and circuits all match with each other, and it is driven by the verification engineer or the peer to the design lead. When applied correctly, this will answer the key questions of the analog IP buyers.

4.3 Star IP

Introduction

Star IP, or alternatively called stellar IP, attracts a high price for one of many reasons and maintains its value for many years without suffering the "commoditization" that commodity (or 'less-than-stellar') IP undergoes, as will be addressed later. Eventually, most star IP may undergo "commoditization," but this is often a protracted process that takes many years.

Star IP may command and retain a high price because:

- It is based on proprietary definitions of functionality. This is particularly true of instruction-set processors that use a proprietary instruction set, which are the most common form of star IP and the ones that have attracted the most interest. It may take many years for instruction set processors to move into the commodity camp, and this may not occur if the IP protection surrounding the instruction set and architecture are rigorous.
- It is difficult to design, and it thus has a high barrier to entry. Although common memories such as standard single and dual port static random access memory (SRAM) may be argued to be commodity IP, more exotic or higher specification memories, such as ultra low power SRAM,

devices with many operating modes, or high port counts, or various kinds of content-addressable memories (CAMs), may remain in the star IP camp for many years due to the high level of design skill and volume required. Processors will require complete tool chains including compilers, instruction set simulators, profilers, debuggers, assemblers, and often an integrated development environment (IDE), and this represents a considerable barrier to entry.

- It has various kinds of patent protection. Although instruction sets may not be patentable, some kinds of instructions in processors may be just so complex and difficult to implement, that the only way of implementing them may violate patents. Other features of the micro-architecture that interact with the instruction set may also be patented, thus making it extremely difficult for other design teams to design an equivalent star IP block with the same functionality without infringing on the patents.
- It is in a complex, emerging, and "hot" new application area. This is very common in areas such as multimedia applications, where new audio and video CODEC standards seem to pop up every week. The drive to higher and higher definition means that coders and decoders are difficult to design. In addition, they very often use both a combination of programmable processors and dedicated hardware (HW) blocks in RTL, as well as memories, direct memory access (DMA) and similar devices, which together form IP–based subsystems. To be early into the market with an IP block may allow you to command premium pricing; in addition, the relative barriers to entry posed by complex design, along with the needs to write software, possibly license IP from other bodies to create a solution, (as commonly practiced with multimedia CODECs), and homologate the results to get official certification of the resulting IP, all mitigate against a large number of competitors entering both quickly and cheaply. As a result, such standards-driven, but complex, IP may avoid commoditization for a long time, which remain both stellar and premium-priced.

Star IP then falls into three major categories:

- Embedded instruction set architecture (ISA) processors which may be fixed ISA or configurable and extensible processors, where major choices of both micro-architectural features and modifications to the instruction set to add very application specific instructions can be made. In the latter case, a configured, extended processor is often called an application-specific instruction set processor (ASIP) and may of course be discrete packaged parts, fixed ISA processors, or the output of processor configuration and extension based on IP. The 800-lb gorilla of processor IP is ARM which has grown from the mid-1990s to be the lion's share of the embedded processor market. ARM offers fixed-ISA processors and at one point offered an electronic design automation (EDA) processor toolkit for ASIPs, called OptimoDE. Other fixed-ISA processor IP companies include MIPS, Ceva (for digital signal processing [DSPs]), and major semiconductor companies such as Freescale are beginning to offer their designs as IP blocks, rather than just as packaged parts or integrated processors on larger SoC offerings. In addition, configurable extensible processor IP providers include Tensilica and ARC.
- Memories from such companies as Virage Logic and ARM (via its acquisition of Artisan). Specialized memories attract and retain higher prices and are thus more stellar in the star IP firmament. Virage Logic, for example, offers special memories optimized to work with a variety of embedded processor IP, ultra-low power and high performance memories, CAMs, and complete memory subsystems.
- Special media processing blocks for audio and video processing. These may be offered by processor IP companies, for instance, both Tensilica and ARC have a number of audio and video IP blocks that incorporate instruction-set processors configured for the applications along with specialized application software and other blocks. Other companies offer more fixed IP blocks in these areas. Such continuing challenges include staying

up-to-date with the varied audio and video standards, offering configurable solutions that span a wide range of performance, quality and price, or bracketing this range with a sequence of more fixed blocks, are continuing challenges.

There are also some other categories of IP which may be stellar:

- Specialized on-chip IP, for functions such as on-chip tracing, diagnosis, maintenance, and repair, including offerings from Virage Logic, MIPS, and new companies such as DAFCA and Temento. These IP blocks may fall out of the stellar firmament fairly quickly if standardization leads to a rapid commoditization in the area and the entrance of new competitors. Proprietary standards may be the best way to avoid this kind of commoditization.
- Software could form an extra category of star IP especially if it has been ported to a variety of processors, or is written in a standard language without any processor-specific extensions, pragmas, or specialized instructions. It might be delivered either combined with or separately from processors in specific application products such as the media IP blocks previously discussed. If such software implements complex algorithms or takes-up a large amount of effort to define, implement, validate, and homologate, such IP might stay in the stellar cosmos for a considerable amount of time. If it can be easily implemented by a few people on standard workstations, it will be rapidly commoditized in price and become more widely available.
- Sometimes micro-architectural IP that is associated with processors, such as sets of instruction extensions that can be used to configure a processor for specific applications and software, can be delivered as star IP. This is found in the multimedia CODEC area.

Star IP is delivered in very interesting forms, and the most well-known are soft and hard which are also the case for commodity IP, with the former being the most common. However, star, commodity, or analog IP forms part of a possibly elaborate design chain in which IP may change its form several times from specification to soft, possibly to firm, then to hard, and finally is delivered to manufacturing. Let us consider a configurable processor, as one example, offered by a configurable processor IP provider. This piece of IP starts as a specification created by one set of designers within a piece of configuration software, possibly running in client-server mode on the Web. The specification is created by a group of designers who are architecting an SoC, for example. They run the IP provider's configuration tool with the specification as input, and receive back many deliverables including the processor implementation in hardware in a "soft" (synthesizable RTL) form. They might then pass this on to an implementation group, possibly in a separate company who are acting as contractors for the front-end implementation. The front-end implementation team will synthesize the design in its soft form into a target technology and cell library, ending up with a "firm" (cell–based, post-synthesis) netlist. This might be passed to the SoC physical implementation team, that might even be a different contracting company, who stitch many IP blocks in a "firm" form together into the SoC and run physical placement and routing on the whole design. At that point, the IP block, together with others, is in a "hard" (post-layout physical design) form, and it is passed along to a semiconductor manufacturer in the form of mask data files for manufacturing.

At any point in this design chain, the IP might break out of the design chain and enter into a commercial form for sale and delivery as a commercial IP product. The processor IP company in this hypothetical chain offers the IP in the form of access to a configuration tool that delivers it to the design team as a soft entity. Assuming the right kind of commercial agreement, this design team could potentially resell the IP in its firm or hard form, targeted to a particular design process and technology library. The IP block could be combined with others to form a soft, firm, or hardened application-oriented subsystem, either hardware-only or hardware-software. Again, this could be marketed as a commercial product in its own right, strictly layered on the included IP. Stars would aggregate onto stars, forming a veritable supernova that might attract a very high value due to its functionality, performance, and complexity.

Use Models and Deliverables

The deliverables for star IP are not so different at the base from those for commodity IP, but particularly processor IP comes with an entire extra layer of software-oriented deliverables.

The basic deliverables for any soft IP block, whether stellar or commodity, include the RTL code, scripts for synthesis, target results, or data sheets for synthesis runs using standard cell libraries and synthesis tools (performance, power, and area), and possibly power models and testbenches for verification. To turn the synthesized netlist into a placed and routed block, scripts for standard placement and routing tools using industry standard libraries are useful, along with tables of results and constraints and control information for various implementation and verification tools, such as equivalence checkers.

IP providers have major decisions to make as to which particular languages, synthesis, placement and routing, verification and other EDA tools and design flows to support. Many years ago, it was almost mandatory to support both Verilog and VHDL, since Verilog was widely popular in North America, VHDL in Europe, and both languages had their followings in Asia. Despite the relative decline in VHDL popularity, there is enough residual use in Europe and Asia to justify continued provision of VHDL representations and support for VHDL–based flows for some star IP. A more interesting question is the arrival of SystemVerilog and the standing question of when enough users will want to use it for design and verification to justify providing these representations and models as IP deliverables. However, since SystemVerilog is a superset (there are a few exceptions) of Verilog, delivery of such representations will be sufficient to satisfy a SystemVerilog user until some of the new design or synthesis capabilities are required to support the IP.

Similar to the language issues, the issue of which tool flows to support for synthesis, placement, and routing is an evolving one. Cadence and Synopsys both have credible flows and enough users to justify support in many cases, even though the Cadence share of RTL synthesis is still quite small. Magma's layout flow is growing in popularity and some support may be justified. Specialized tools may be more one-offs and not require multiple vendor support.

This is an important issue for IP providers. Each flow that must be supported requires considerable development effort by the provider, as well as working out intricate licensing deals with EDA vendors. Two competing flows are usually enough to provide alternatives to users but IP providers must continually keep a watching brief on the EDA industry and designers to decide when to support new formats, design tools and flows.

Recently there has been a growing interest in system level design or what is now called electronic system level (ESL) design [1]. This ranges from interest in design space exploration via architectural and algorithmic modeling and simulation tools, high-level synthesis via new behavioral synthesis tools, usually driven from C, C++, or SystemC, and software development and verification environment for programmable processors with fast functional simulation models.

For star IP, behavioral synthesis in these new forms is usually not part of the flow, since most star IP have their own specification, synthesis flows, and tools. However, the other ESL tools and flows have led to growing interest and need for IP providers to provide high level simulation models, usually in SystemC, which has become the system level *lingua franca* modeling environment. These may be both cycle accurate models: instruction set simulators, in the case of processors, and fast functional models. The former may be most useful in architectural design space exploration and verification; the latter for software development and verification. In addition, specific integration of IP ESL models into third-party ESL tools using proprietary formats and adapted to specific tool capabilities may be useful. However, it is not yet the case that provision of such models and tool integrations is mandatory for all star IP—but this point may come with further ESL development over the next couple of years.

For processors, it is important that IP providers either deliver themselves or work with third parties to deliver a complete software development environment. This may be based on a commercial, proprietary-integrated development environment (IDE) such as Green Hills or Wind River; or an open-source tool such as Eclipse. Aside from the IDE, optimized compilers, assemblers, debuggers, and profiling tools are all mandatory; although these may come either from the IP provider or a third-party or combination of both. Debug boards, probes and trace hardware, and models to run on emulators or hardware simulation accelerators are all part of the IP design "ecosystem" for programmable processors. Instruction set simulators at least at the cycle-accurate level are mandatory for users who need to profile embedded software targeted to the processor.

Real-time operating system (RTOS) providers are of growing importance for processor IP providers, as the software load being mapped to these processors grows and the need to have both high-level messaging and task control capabilities escalates. IP providers usually do not provide their own RTOS although they may offer simple sets of OS-like functions and capabilities. They usually work with commercial third parties to make their RTOSs available on the processor IP.

Although fixed ISA processor providers may be able to work mostly with third parties for compilers, debuggers and assembler tools, and instruction set simulators, those offering configurable and extensible processors need to invest considerable internal development resources for these tools, to allow them to be automatically tailored to each configuration and new instruction as the design teams explore various alternative micro-architectures.

Development Process

The development process for star IP does not differ significantly from commodity IP, except for IP-specific flows. For example, processor IP requires the development or acquisition of the various special software tools previously discussed. These may be developed by the IP provider—especially important for configurable processor IP—or acquired from open source software tool providers or by collaborating with commercial third parties. We have mentioned Eclipse as a popular open-source IDE that is growing in importance, moving from its Java development roots on to general support of C/C++ software. The gnu tools may also be a basis for compilers, debuggers, and assemblers for various processor IP. When the processor IP is configurable and extensible, then each of these tools must be modified to allow the runtime incorporation of variations on the instruction set architecture. The tool that requires the most work from an IP provider is the instruction set simulator (ISS). Although tools exist that allow the generation of ISSs from architectural description languages, these are not universally used to generate ISSs and may not be able to generate both cycle-accurate and fast functional ISSs which is an increasing requirement nowadays.

As a result of these issues, processor IP companies need a considerable software team in addition to the hardware development team—perhaps of equivalent size; or a large number of external relationships so that commercial third-party tools can be adapted for the compiler, debugger, assembler and IDE roles. Even if the development and porting of such tools is done by third parties, a number of staff within the IP company will be required to manage partner relationships, initiate and review development contracts and agreements, and support the porting and adaptation projects. The IP provider will also require field staff for customer support, training and perhaps dedicated contract development assistance. Although not exact, a good rule of thumb may be that support of processor–based star IP may require about double the number of development and support staff as a hardware only IP.

Development Process

The verification issues surrounding star IP do not necessarily differ from commodity IP, except for the addition of specific IP-related verification steps. Processors, over the years, have particularly attracted special verification steps and tools. Configurable processors have had special verification requirements due to the enormous configuration space [4]. The major processes involved in configurable processor verification are:

- Verification of configured processor
- Verification of customer instruction extensions
- Verification of multiple configurations
- Verification of processor in its system context.

The last verification step is carried out by customers with the IP integrated in their system and involves the kind of architectural and high-level system models previously addressed.

Verifying an individually-configured processor requires the development of an extensive verification environment to verify the processor both at the instruction or architecture level and at the detailed micro-architectural level to verify the pipeline execution and data hazards. Such verification environments can be built as diagnostic programs that take advantage of the fact that the IP is a programmable processor, and be built by assembly instructions rather than at the signal level. In addition, the availability of cycle-accurate ISSs allows the execution of the verification programs to be compared to the same program running on the ISS for state compatibility and identical outputs. Directed and variously randomized diagnostics are possible and useful in this context, and specific processor verification generation tools using the micro-architecture of the processor as input allow test creation that exercise valid, yet unusual, sequences of instructions to exercise unusual combinations of processor state.

Verifying instruction extensions requires a combination of simulation–based methods and emerging formal methods that can be used to verify a specification of a new instruction's semantics versus an implementation of that instruction generated from instruction extension compilation.

Verifying multiple configurations occurs during the development of a major new release of the micro-architecture, with the use of predefined and randomly-generated processor configurations that stress various existing and new features. These are simulated for millions of cycles through the individual configuration verification environment mentioned earlier. The number of configurations that are exercised during the multi-month development of a new micro-architectural release might range from the many hundreds to the many thousands or more. Uncounted trillions or more simulation cycles are used to keep the development on the right track and to prevent regression from occurring. The regression environment can be used after product release with new automatically generated configurations to continue the verification and to explore new regions of the verification space.

Although other kinds of IP may be less complex than fixed instruction-set or configurable processors, they will still require sophisticated verification processes; if configurable, they will require verification over an extensive part of the design space by generating and verifying multiple configurations. Complex multimedia IP, as discussed earlier, that may utilize multiple processors, devices, and extensive software, is an even bigger problem especially with the software load and the difficulty of devising complete corner cases and test scenarios. Here, given enough simulation cycles, the reuse of conformation test sequences and homologation corner cases can be very effective in verifying the correct functioning of the IP in its real world context.

Star IP Conclusion

Star IP commands and retains a high intrinsic value because it provides an economical alternative to in-house development of complex functions with large internal teams that might fail to meet aggressive product development deadlines. The intrinsic value may be bolstered by proprietary instruction sets, unique configuration options, patents, high barriers to entry, high software content, or large amount of functionality required. Star IP will increasingly require embedded processors and software content, and its provision may be done differently and following varied design and business models.

One key issue in star IP is the need to maintain its competitive nature and high value through iterations of the product over time, adding new features and capabilities, and staying ahead of the competition. Another issue is the need to expand the IP market, to constantly seek new uses in new designs, to support the IP for a worldwide community of users, and to see it both used and reused by design teams. This requires a continued stream of ongoing investment in the IP and building an engineering and support team of the right size to match the IP customers' license fees and royalties.

4.4 Commodity IP

Introduction

Commodity IP is, as the term suggests, where IP blocks are commodities. These are typically digital electronic IP blocks which contain insignificant proprietary IP, which perform functions that are readily available from competing vendors. Typically, this IP performs some function involving the translation from one well-defined industry standard format to another one, and while the techniques employed may be proprietary, they do not have the advantage to limit the availability of alternatives of equivalent size and performance. As a result, such IP is readily available from multiple vendors, and while the initialization and control functions may change from one vendor to another, IP functionality across multiple vendors is essentially the same.

Such IP is generally soft, or soft with external hooks, for special hard IP blocks. This is preferable to firm because a single RTL design can be foundry and library-independent. The necessary hard blocks, usually memory, are generally available from the semiconductor vendor chosen by the integrator.

Over a decade ago, the Peripheral Component Interconnect Special Interest Group (PCI-SIG), led by Intel, created the initial PCI specification, which was made available for an annual fee of $1,000 to anyone who wished to implement it. Further, reference test suites that verified the correctness of implementations were later created. As a result, many small start-ups and a number of larger companies began implementing and verifying the accuracy of their own versions of PCI IP blocks. This is perhaps the most visible example of commodity IP, but it was followed by USB, Firewire, Moving Pictures Experts Group (MPEG), and a myriad of other standards–based blocks.

Star IP contains both unique, vendor-specific, proprietary intellectual property, and interfaces for software and control code, which serves to lock in its customer base. By contrast, commodity IP may have either software, such as the 8051 microcontroller, or proprietary intellectual property, such as described earlier, but it does not contain both. The proprietary nature of the 8051 design has long since expired and, as such, it has become a commodity.

In this section, we will explore the economics of development and sale of commodity IP in its open competitive market. We will discuss the evolving marketing techniques used to survive in such a market, and follow up with some observations about the impact such openness has on the long-term viability of the market.

This will be followed by an outline of a typical development flow, elaborating on the areas that are critical to maintain profitability is such a competitive environment, and conclude with some integration related issues.

The Third-Party Commodity IP Market?

The market for third-party IP began out of the need to keep up with Moore's law. The size and complexity of integrated circuits grew the cost of development and even the ability to find enough qualified engineers, and it began limiting the number of design starts. EDA tools were increasing the productivity of engineers, but that increase was not keeping pace with the increasing complexity of IC design. This prompted discussion about the "design gap," namely, the gap between what an engineer can design and what a chip can contain, as shown in *Figure 4.3*.

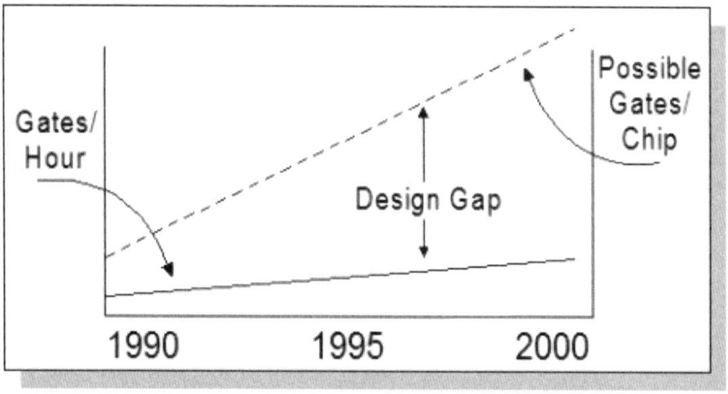

Figure 4.3: *Design Gap*

It was predicted that the inability to close this "design gap" would lead to a decrease in the number of new chips. The Virtual Socket Interface Alliance (VSIA) [7] was formed, around the notion of "design reuse," the building of plug compatible third-party IP, which would vastly enhance the productivity of chip developers by allowing them to architect and integrate new system chips quickly. Such books as *Surviving the SoC Revolution* [2] led this trend by advocating platform–based design reuse. As a result, the third-party IP market began with great expectations, but as can be seen in *Figure 4.4*, design starts continued to decline until just a few years ago. In other words, it has taken the better part of the last decade to realize the dream.

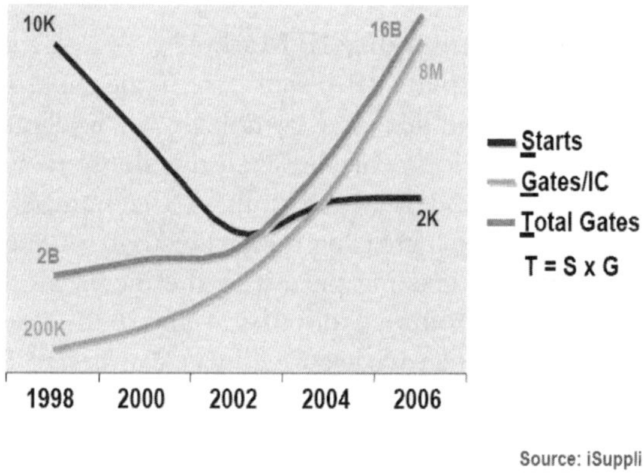

Figure 4.4: *Design Complexity*

Initially, as with any new market, many small companies were created to develop third-party IP, and most of them took advantage of the emerging new standards. Each of these new companies saw a large untapped market with an increasing demand and sought to move quickly to satisfy it.

Time-to-market has always been critical to the electronics industry. Their marketing argument suggests that the maximum sales life of a given product has been eroding over time from over three years decades ago to under 18 months today, and every month a product's launch is delayed is a month of peak revenue lost. In the face of such an economic imperative, almost any price can be justified to obtain the IP early. As a result, the first third-party commodity IP block introduced into the market commanded high prices, but, as is the nature with commodity markets, the price rapidly fell as more competing products enter the market.

The economics of the electronics IP market is similar to all high intellectual content markets. There is a large cost of development coupled with a small to almost zero cost of manufacture. So falling prices should be even less of a problem than for other electronics companies, particularly if as is typical of the IC industry, those falling prices are more than offset by rising sales. Unfortunately, as can be seen in *Figure 4.4*, this market was slow to develop, which meant the falling prices were not adequately offset by the rise in sales.

To further compound these problems, while the cost of manufacture of third-party commodity IP is inherently low, the cost of sales is not. After the initial euphoria of introducing third-party IP, it became increasingly obvious to customers that the IP needed to be both verified and supported. In addition, every customer and vendor had to negotiate liability and other terms of the sale. Such organizations as the Scottish Enterprises Virtual Component Exchange (VCX) and Japan's International Press Telecommunications Council (IPTC) made attempts to reduce such costs by creating standard exchanges, and while they collectively helped standardize the industry each one was an unsuccessful venture. As a result, many of these start-ups could not sufficiently recoup their development costs, and ultimately merged into larger EDA or semiconductor companies, which have increasingly moved from selling individual third-party commodity IP to selling whole libraries of commodity IP blocks, to reduce the cost of sales per IP block.

Nowadays, the further independent development of third-party commodity IP has been hampered by this difficult economic environment for many reasons. First, unlike the general interface standards around which the initial third-party IP market started, most emerging standards today are more specific niches, such as graphics or wireless communications, which means the projected volumes for new commodity IP is likely to be lower than in the past. Second, the cost of individual IP continues to be below the cost necessary to recoup its development. Third, the demand for either customized or customizable software and hardware interfaces has grown as the customers continue to reduce their level of custom design. As a result, managing costs is the key to surviving in the third-party commodity IP marketplace.

Currently, there are a number of other cost management strategies besides selling bundled libraries of commodity IP, such as developing cost-sharing with key customers in exchange for early access, parameterized customer interfaces to minimize the support cost by "shrink wrapping" the IP, and adopting EDA maintenance models to cover the cost of ongoing performance upgrades and software support for initializing and controlling the IP.

On the other hand, subsequent new start-ups and internal development efforts have attempted to avoid the commodity IP marketplace by focusing on generating both proprietary IP and connecting the IP to software or other customer-developed intellectual property, all in order to emulate star IP.

Such new efforts include hardware IP coupled with software diagnostics tools such as provided by DAFCA, IP models coupled with software development tools, and tools for user customization of commodity IP interfaces. For example, Sonics, ARC, and ARM have developed aspects out of these last two as part of their IP and design services.

As this binding continues, commodity IP will ultimately become embedded in new forms of star IP, at least as long as third-party commodity IP continues to be uneconomic to develop in a stand-alone fashion.

Development Flows

In the past, a typical development flow for commodity IP was similar to the block design in a block–based chip design methodology. At a high level it consists of architectural and RTL design, gate level design, verification and implementation steps, though there are significant differences due to the nature of the resulting product.

All of these activities need to be carried out with a minimum effort and cost. Ideally, the most productive EDA tools are used to ensure that engineering costs are minimized which, in turn, makes EDA tools the biggest fixed-cost that third-party soft commodity IP developers have. It is not surprising thus that EDA companies have acquired many of the commodity IP vendors. These EDA companies' IP divisions have to augment their tools with specific IP scripts and flows to further minimize their development and support costs.

Due to the nature of this business, managing commodity IP, unlike other digital design development efforts, must focus on cost containment. It thus needs to be organized more like a manufacturing engineering than research and development (R&D). In the future, successful commodity IP developers will focus on IP that can be developed with maximum productivity and minimum risk. Typically, this will be a process involving recasting, rather than developing, IP which will need to be done by more traditional R&D groups and paid for by key customers. As a result, the development effort outlined earlier will evolve into the reference–based design recasting effort described here:

- The existing standards, along with the original architectural specification, form the basis of the architectural specification. As with reference design, this is augmented to include any additional parameterization, initialization and external control as required by marketing.
- Care is taken in the RTL development phase to tune the existing RTL for synthesis into a generic library, to allow a wide variety of implementations, since most third-party soft IP is shipped in this form.
- The gate level design, verification, and implementation steps are limited to verifying the design. Unless previously done, they are targeted to FPGAs to test with the verification test suites in a timely manner. In some cases, they are targeted to existing foundries, being implemented in a single-chip form in a multi-design prototyping run, particularly when there is a need to foundry qualify the design. Gate level design, which includes synthesis, place and route, and back-annotated timing, is also done to

verify the performance of the design. Key parameters from the original process are used and extracted to provide the customers with "firm" directives to meet their timing requirements.

- Besides running the existing verification test suites for the standards during verification, a complete set of verification tests are ideally constructed and automatically run to adequately test the software and processor interfaces, as well as the range of parameterization and customization provided to the customer in the final IP.
- Finally, unlike internally developed IP, third-party IP developers must extensively document their IP in order to allow integrators, who are unfamiliar with the IP, to use this with minimum support. For such an engineering-limited effort, this can become one of the biggest efforts in low-cost commodity IP development, unless it is at least partially automated. Wherever possible, both marketing and engineering documentation must be extracted from the actual design information.

Support of the resulting product is largely determined by the nature of the customer's integration and IP usage, which is the subject of the next section.

Integration Issues

The issues regarding commodity IP integration may be divided into two groups: the concerns of the vendor and the customer. This section will deal with both as they relate to integration of commodity IP into FPGA, ASSP and application-specific integrated circuit (ASIC) designs by large integrated device manufactures (IDMs) and end users.

The vendor is particularly concerned with minimizing costs and the risk of losses, such as the cost of support, liability, and theft risks.

The cost of support is primarily the cost of ensuring a customer gets the proper function and performance from the integrated IP. This can vary widely depending on the application, interfaces, type of IP, and the target technology. In addition to thorough verification of their IP, vendors can further minimize their ongoing support costs by augmenting their user documentation with reference designs, guard-banding the specifications of their IP, and tuning their IP for consistent implementations in the customer's targeted technology. IDMs frequently required IP to be qualified in their processes prior to purchase, and they usually only cover part of these costs. However, these high up-front qualification costs can be offset by large bulk sales agreements. Such foundries like IDMs require qualification for hard IP, but leave customer support to the third-party IP provider. End user support costs are generally related to problems in integration and implementation of the IP, which are best solved by the vendor providing adequate documentation, appropriate reference designs, and ideally implementation scripts and notes for the target technology.

The liability risk is generally contractually limited to the cost of the IP, but that may be augmented by insurance.

The risk of theft is reduced by various forms of IP protection such as tagging, watermarking, and encryption. Tagging provides a way for honest customers to properly track their IP usage. Watermarking merely provides a way to determine if the pirated IP is the vendor's, and encryption ensures the IP distribution is limited to those who purchased it. No system of IP protection is foolproof. Tagging soft IP requires tools which support the transfer of the tagging information through the implementation process. Watermarking requires access modes to extract the watermarks to be preserved through the implementation process, and encryption does not protect the resulting decrypted data. However, such measures do reduce the risk of theft roughly in proportion to the cost of implementing them. The specific implementation techniques vary depending on the targeted technology. Tagging and encryption may be a substitution for watermarking in FPGA-targeted IP, but it is limited by the facilities available in the FPGA. Encryption does not extend to the fabricated IP in ASICs and ASSPs, and tagging is usually not applied, which leaves watermarking as the only method for determining the source of the physical IP.

Customer concerns are primarily related to the ease of use and risk of redesign, both related to the cost of IP integration. These concerns are largely covered by the same measures discussed in the vendor's cost of support. Yet customers, who are creating ASSPs or ASICs, also want access to the third-party IP support engineers to ensure their particular integration problems are addressed in a timely manner, because of the large cost of prototype re-spins.

In conclusion, while the key to minimizing integration issues is proper up-front, verification, specification, and documentation, there will always be some need, however small, for live engineering support of third-party commodity IP.

Conclusion

In the end, similar to the smaller arithmetic functions and basic cell libraries before them, commodity IP will become absorbed into the libraries and tools needed to generate future ICs. They will either become embedded in the tools as generated functions, or will be added to the basic libraries as parameterized elements; regardless of their new form, they will fade into the rising fixed cost of developing new IC designs.

The commodity IP market will most likely follow the EDA tools model. Future development and verification of commodity IP will probably be done through joint ventures between key customers and the IP developers. Ongoing support will be covered by a combination of specialized integration services, tools, developers, and semiconductor qualification groups, as part of the necessary overhead of doing business, and end users will need to add specialized engineers within their development groups to support this added complexity.

As with development, the most likely economic model for the future sales of such commodity IP will be similar to EDA tools, namely fixed prices plus annual maintenance fees, or annual subscriptions, and as with other maturing markets, cost-cutting and sustaining engineering will take precedence over R&D engineering.

4.5 Emerging Issues

Introduction

Just as IP blocks are the components of IC designs, the emerging IC design issues apply to them as well. With the relentless march to ever-finer dimensions in IC processing, DFM has become an emerging issue. Similarly, the finer dimensions have increased power density which makes low-power design an emerging issue when coupled with the rise of wireless applications. Lastly, the ever-increasing complexity of ICs has led to emerging needs for and provision of architectural, system-level tools, and models, which place further requirements on third-party IP.

These emerging issues, as they apply to the development and use of third-party IP blocks, will be described, along with possible solutions and predictions for future trends in IP development and use.

The Need for Lower Power: Techniques and Issues

The need for low-power IP blocks will place additional burdens on the developers of IP. There are a number of techniques currently available to minimize power usage, in addition and separate from low power IC processes. Some of these include the following:

- Gated and multiple clock frequencies—These involve using clock controls to either slow down or turn off resources that are not being used.
- Asynchronous design—This involves either data generated clocks or data timed clocks. In addition to using less power, asynchronous design also smoothes out the demand for power, but is much less predictable than synchronous design techniques and consequently harder to debug.
- Orthogonal design—This involves adding extra terms to minimize the switching of gates that do not contribute to the result of any local computation.
- Data movement minimization—This involves architecting the design to stage data based on where and how often it is needed. Cache designs for central processing units (CPUs) have long since been optimized to minimize misses for

statistically average computation, but more specific optimization may be done for specialized processing where the frequency of data usage is better defined.

While any number of these techniques may be employed in the hard IP design, they all pose problems either in the design or transfer of the traditional deliverables for soft IP. Clocks may be the easiest to deal with, but regardless, additional testing is needed to ensure the IP continues to operate over the intended variety of clocks and frequencies. Asynchronous design requires special tools and implementation, thus making it difficult to fit into traditional synchronous design implementations. Orthogonal design requires detailed RTL and special synthesis scripts to ensure the preservation of the seemingly redundant orthogonal terms during synthesis optimization. Finally, while data movement minimization may not require special implementation, it will likely require significant additional, and seemingly unnecessary, external resources and interfaces. Furthermore, it may have limited results unless it is applied to the entire system during integration.

As a result, low-power IP blocks have, and will continue to require, special development and processing, and thus will be sold as separate IP, be they hard or soft. An alternative delivery mode is to use generators for configurable and extensible IP blocks that provide trade-offs between power, performance, and cost (gate count) so IP consumers can make choices that optimize their particular design. This has been successfully used with star IP such as configurable processors. In the processor domain, in particular, the specific instruction set used can have a very large impact on energy consumption and peak power requirements for a specific algorithm. Thus, the use of extensible processor technology, where specific application-oriented instructions can be added to a processor targeted at a specific application, is a viable and now well-proven mechanism to, at times, dramatically reduce overall energy consumption [6].

DFM and How It Relates or Will Relate to IP Development and Usage

DFM deals with detecting and correcting manufacturing yield-related issues. As the dimensions in IC design fell below the wavelength of light used to pattern the IC designs in current IC processes, many problems associated with IC design reproducibility began to arise. These problems, having started as second-order effects, have grown to the point where some designs cannot be manufactured in the current IC processes without some DFM engineering. The two areas of DFM are detection and correction in both manufacturing and design issues.

Most of the manufacturing-related DFM tools consist of collection, management, and analysis of the manufacturing process and test data with regard to an IC design. The analysis uses statistical methods to isolate yield-related detractors within the design. Although this is slightly irrelevant to IP design, the traditional black-box approach to third-party IP will ultimately limit the failure-isolation ability of these tools. As such, independent third-party IP design companies will need to make their internal designs available to the semiconductor companies that use such tools and create isolation-related test suites to use in detecting and removing design-related yield detractors. Or, they can alternatively supply the engineering support necessary to analyze their sections of customer designs to the semiconductor company.

DFM for design issues primarily consists of optical correction of mask level data. These programs consist of simulating the distortion in optical projection methods used in an IC process, analyzing the resulting distortions of mask level data to find cases that violate the IC processing design rules, and fixing the mask level data to correct the inherent distortion. Being a newly emerging field of EDA, these tools are still rather limited in their abilities. For example, while most tools on the market can analyze single layer errors, only a very few can handle registration errors between layers or make the proper trade-off between opposing distortions across multiple corners of the optical projection process. In addition, while these tools have improved the design manufacturing yield in deep sub-micron IC processes, some of the problems cannot be corrected at the mask level. Instead, prior to generating the mask level patterns, they require implementation-aware tools to avoid critical dimension distortions.

Hard IP will ultimately require such processing to ensure that it does not detract from the IC yield that uses the IP, but soft IP will also eventually need implementation scripts or directives which have been tuned by DFM analysis. Alternatively, the building block (basic cells and macros) will be DFM-corrected, and the EDA tools will become sufficiently DFM aware, thus limiting the work needed to ensure correct implementation of soft IP.

Architectural IP

Much of the star IP and some of the commodity IP that exists nowadays require architectural-level models, especially if the IP executes software or a specialized control code. Only very little star and commodity IP is now provided to users with architectural models, and their very nature does not guarantee the timing relationship of all events, which under certain implementations may pose problems to the customer's integrated design.

In the near future, customers will expect architectural models to be provided along with the RTL, and as a result the third-party IP developers will need to also include all required event ordering constraints to their designs and documentation, to ensure correct customer integration of their IP.

Programmable star IP, particularly processors, are increasingly offering two levels of instruction-set simulation models for use in design space exploration, detailed architectural verification, and software development and verification. These are cycle-accurate and fast functional models; cycle-accurate may perhaps run 100 to 1000 times faster than RTL models, and fast functional models execute a further 10 to 100 times faster because they are instruction-accurate, and thus do not model internal processor pipelines and state. These models may come as stand-alone C/C++ models or integrated with SystemC, or both, and they may also be offered within third-party ESL tools and integrated development environments for software. All of this can also be called architectural IP.

References

[1] Brian Bailey, Grant Martin, and Andrew Piziali, *ESL Design and Verification: A Prescription for Electronic System Level Methodology,* Elsevier-Morgan Kaufmann, 2007

[2] Henry Chang, Larry Cooke, Merrill Hunt, Grant Martin, Andrew McNelly, and Lee Todd, *Surviving the SpC Revolution: A Guide to Platform–Based Design*, Kluwer/Springer, 1999.

[3] Henry Chang and Ken Kundert, "Verification of Complex Analog Integrated Circuits," *Proceedings of the IEEE Custom Integrated Circuits Conference*, 2006.

[4] Dhanendra Jani, Chris Benson, Ashish Dixit and Grant Martin, Chapter 18, "Functional Verification of Configurable Embedded Processors", in Brian Bailey (editor), *The Functional Verification of Electronic Systems: An Overview from Various Points of View*, International Engineering Consortium, 2005.

[5] Grant Martin and Frank Schirrmeister, "A Design Chain for Embedded Systems," *IEEE Computer*, March 2002, pp. 100-103.

[6] Grant Martin and Steve Leibson, Chapter 8, "Commercial Configurable Processors and the MESCAL Approach," in *Building ASIPs: The MESCAL Methodology*, edited by Matthias Gries and Kurt Keutzer, Springer, 2005.

[7] Larry Rosenberg, "Virtual Socket Interface Alliance," 2003 *SoCTEC* Conference presentation.

Chapter 5

IP Protection

Saverio Fazzari, Associate, Booz Allen Hamilton

5.1 Introduction

Intellectual property (IP) is the distinguishing feature for every company in the world, and this allows companies to set themselves apart from the rest of their competitors. Coca-Cola, with their secret formula, is able to distinguish itself across the globe from all other soda companies. The IP is the differentiator between the various customer choices, and it becomes more valuable as companies are successful since it sets up goals and increases the potential barriers to new market entrants. A consulting organization, for example, has the intellectual capital of its people which clients look to employ. Finding and defining the worth of an IP portfolio becomes increasingly more valuable to companies nowadays. Wall Street investors are currently trying to establish a method to quantify this IP, especially since it is critical in defining the true value of a company.

In a modern electronic product development process, there are functional blocks which are used in the design. A design is made of sub-blocks that represent different functions or operators, and this has enabled the creation of a more precise engineering definition of IP. In the engineering sector, IP is the source code, or block, which is made available to use in many designs. This new environment is fragmented due to the numerous partners and companies that interact to produce the final product. The IP must be protected from loss during these transitions between companies because there is a high risk that the value might diminuate.

Modern electronic design processes have changed to meet these new environments where design teams must succeed. The market requirements dictate that designs must work with existing standards of available technology. The product developed might contain twenty or more different functional blocks, and these could originate from many different sources. Internal groups may have these blocks already or they may need to be built. More often, there is an external source from which the block is acquired, such as the mobile phone. Its own functionality has advanced over the years and, as a result, phones can now feature games, video, music, and can be used to take pictures or make phone calls. The size of the device is however shrinking, but in order to achieve these product requirements, a company needs prompt access to emerging IP. The number of different blocks needed could be significant, and the complexity of the projects means that companies can no longer do it by themselves.

These complex projects are undertaken by many different companies in order to help manage the risk involved. The electronics marketplace has evolved dramatically over the last 20 years and, at the same time, companies have begun to recognize the value in the various segments of their product and are looking for ways to exploit that value.

5.2 Why Protect IP?

The market evolution has provided multiple choices for customers, and product difference, whether real or perceived, allow companies to demand more for their offering. Customers are willing to pay more when allowed to choose from a wide selection; the difference may lie in, for instance, a new capability absent among the other competitors' products, such as a TV embedded in a cell phone. There may not be any functional differences in the offering, for it may be associated with the company's reputation for quality, reliability, and other predominant factors. For example, an IP core from a large established company may be more valuable since it needs to uphold its reputation, it is more mature and likely to be present during the course of the project, and the risk of usage may be lower. The IP can be useful to both customers and partners in terms of joint development. Companies are attempting to unlock more value from the IP that they have or can obtain, and this IP can be shared with other companies in return for profit.

This introduces a number of different business models. Now, companies are looking at new potential markets with a higher level of sophistication than ever before, as they are selling to different customers, locations, and competitors. When the financial analysis is completed to determine a company's success, the development cost for the next-generation product is an important factor. Companies that pursue smaller or different markets must look at diverse ways to successfully build a given product.

Globalization has created various opportunities worldwide. For example, the capabilities of a cell phone depend on regional markets. South Korea, which has a strong 3G network, can support the vision of the next-generation cell network. This allows all sorts of additional capabilities, such as gaming and video over the cell phone, to be successfully sold. Whereas, in the United States, the network coverage is not complete or comprehensive enough to make this possible. Although the United States likes to be considered a very advanced nation, its own failure to adopt standards, or even following them under duress by the government, has resulted in a less capable technology base.

However, the development strategies had to evolve to handle all these new potential market products that have particular capabilities and release window. Companies could not afford to build everything themselves because of excessive costs. The risk was too high, and the turn-around time would be unnecessarily long. What if the product missed the release date? Even worse, what if the product is released and there is a defect forcing additional expense to correct problems? The history of computer products is filled with similar situations. The various video game systems that were released after the Christmas season caused a substantial drop in revenue, and customers later faced the troublesome Intel bug in the Pentium processor. Design teams are now exploring various alternatives to aid in the design of the product in order to manage the cost and risk.

For a typical systems-on-chip (SoC) product, there may be many external sources for an IP block such as a universal serial bus (USB) interface core. In the past, such information, along with developed products, was developed internally by companies. A large integrated device manufacturer (IDM) developing a new electronic device, for instance, was able to achieve everything such as internally creating the IP and delivering the final design. These solutions were considered both stable and secure since they were developed by one vendor. As a result, companies were intent on using the system from the single supplier. It became crucial to address communication if the system needed to work with other solutions from different companies. Customers were looking for more capabilities that required interaction between different devices, which consequently led to the creation of various standards to promote interoperability between systems. The goal was that, as the requirements changed, customers could pass information back and forth. As the markets grew and became more dynamic, companies recognized the value of using partners to carry out some of the development work. This allows the spreading of the risk and potentially raising the likelihood of a return.

In the modern electronics world, companies design a product by themselves, and then turn to an application-specific integrated circuit (ASIC) company for fabrication. The companies are doing more of the development work internally and going to foundries such as the Taiwan Semiconductor Manufacturing Company (TSMC) for cost reasons, which is generally referred to as the fables semiconductor model. Nowadays, designs involve multiple companies co-operating to design the SoC. Market evolution that is served by these designs has created more partners in the development process. Companies need to use the available resources worldwide to create products at the right cost, and thus leverage technology and people.

The urgent need to use other people in this process brings about a new approach to the various IP blocks. These new development chains change the view of how the IP must be used and watched.

5.3 History of IP Protection

IP protection has always been a critical requirement, as people always look to secure their own information and to prevent unlawful usage especially in their interactions with each other. There has been a historical level of trust associated with the amount and quality of shared information. A king did not want information about his kingdom's status to reach his opponents due to both commercial and military requirements. Armies do not want their opponents to know or be able to predict their strategy. While information had to be shared between different groups, the goal was to protect this information so it could be used in an engagement with the enemy.

Cryptography was used to pass secret information between different groups of people. The Egyptians had their hieroglyphics which could be changed for a higher level of protection and still allow effective communication. There are many other methods that were used to share information between people. The common goal was always to prevent its dissemination from being exposed too soon. Protecting information was easier when only few people could read or decipher codes. History is filled with different stories about methods previously used by civilizations to create and protect information, and these cases often involved a protection mechanism known only to a few select people. This method carried the risk, however, that if discovered and interpreted by the enemies, it could easily be exposed without the original author's knowledge. This occurred in World War II when the British managed to crack the code for the Enigma cipher[1].

As written forms of communication became more common, IP protection became the means to limit access to certain communications. The language was protected as messages were transmitted back and forth, which was immediately applied in the military sense. As a specific example, it took armies a few days to properly position themselves in the battlefield and then they spent relatively short periods of time for fighting. Capturing messages about the enemies' plans could dramatically change the course of events on the battlefield. Even Julius Caesar introduced methods to protect his written orders to his generals in the field for the upcoming battles. He created a simple shift cipher to protect his messages as seen in the following example:

---

[1] Report on Enigma decipherment, 1 November 1939, http://www.turing.org.uk/sources/nov39.html

The message: Attack at dawn
With a simple 3 letter shift in which A become D and B became E.
The translated message: Dwwdgn dw gdzq

This information would be transferred by messenger to his armies, but enemies would need to know how to decipher this if they needed. It was irrelevant to Caesar, a long time afterwards, that the message could be decoded, since it would be too late for his opponents to benefit from this new information. This simple approach of IP protection allowed information to be securely delivered and had a significant impact on the overall process of communication.

History is full of diverse approaches to protect various types of IP. Countries created codes to share important information with their representatives stationed in external locations. Diplomats and generals could pass data back and forth with their leaders, but the opposing countries could tap in and try to break the various data codes. The protection mechanisms became increasingly more sophisticated, especially as different devices were used to convert text with advanced protection schemes in order to create unbreakable codes. People became adept at breaking and changing the various available codes, and the goal was to break a code and not let the enemy know that their data had been discovered.

Some famous cases occurred during the world wars of the twentieth century. The key-protected IP were the code messages which various countries did not want to expose to their enemies. Germany sent a telegram to Mexico during World War I promising them the return of territory lost to the US in the Mexican-American war 75 years earlier. However, the British were able to find and decode this message called the Zimmerman telegram and later use it as an effective propaganda tool.

In World War II, the allies were able to decipher many of the different codes used by both Germany and Japan, but did not make them aware that their codes had been broken. The allies were then forced to determine if the value of the information was sufficient to risk Germany knowing that their codes had been broken. Many lives were actually lost in order to protect this piece of knowledge.

5.4 EDA and IP Protection

Modern design is a sophisticated process with multiple steps. A typical SoC design, according to many studies, is in the multi-million gate size, and these are systems in cell phones, videogames, and other devices. These devices have demanding requirements for performance, size, and availability, and the need for increased productivity has led to the development of new tools that will create a quick and efficient design process. Design teams must create, integrate, and develop gates more quickly, and there is high pressure to bring about a successful product the first time.

In the typical design flow, numerous combinations of tools may be used, including both commercial and internal tools and scripts. Companies interact with different design teams internally and externally, all of whom may be using different tools, such as synthesis tools. One team might have made a language choice that is different from the other team. There are different tools from each of the major electronic design automation (EDA) vendors, in both digital design and verification, that might be used in a given company or their IP suppliers. In this design environment which contains many different tools and processes, the various blocks of IP for the project must be integrated and verified. As design information for IP is shared, there may also be different forms of constraints, along with other input files, that require protection. Here are some of the different views that may be provided:

- A hardware description language (HDL) model for integration into a register transfer level (RTL) model for synthesis
- A behavioral view that allows for faster verification
- A layout view for better prediction of performance.
- Various constraint files to work with the available EDA tools

Other tools in the flow have formats that are typically open to a certain degree. Some might have proprietary views as well for performance reasons. The design team might not be able to afford to build the IP block and would like to get some IP from an external source that could allow them to spend that development and integration time more effectively elsewhere. This led to the growth of IP suppliers who can provide these blocks. The IP providers want to furnish information that allows their customers to be successful with their IP, yet at the same time, they want to maintain the security of the information.

An interesting part of the problem is how various tools support IP protection. Verification tools can secure the data so a design will be tested against such a model. A synthesis tool experiences a small problem in cogently addressing IP protection which must be integrated into a complete design and be exposed for debug purposes.

5.5 Different Technical IP Protection Methods

There are different ways involved in IP protection, which include securing the IP and its value. Each IP block has a different value in a complete system, so one should not uniformly apply the same strategy to all IP blocks. For example, if the consumer is buying commodity IP, then the level of protection may not need to be extremely high. However, if the IP is a unique block, then all efforts need to be taken to prevent the source information from being available. This is important not only to the IP provider, but also to the user since seeing its usage by other people without proper licensing will make their product less competitive in the market. One should note that not all methods are effective for each IP type, especially since each approach includes technical and legal means, and both types have value and inherent risks. This section will focus on some of the technical approaches that are available today.

There are different delivery formats for IP, as mentioned previously, which a vendor may want to secure. These delivery formats are text, compiled models, and combinations.

Security will take two different approaches. The first is more passive, and here, the goal is to identify whether the IP is as expected. The authentication approaches are designed to make sure that the IP can be identified and validated correctly. These have many applications and they are, as examples, used to determine the royalty rates that a customer might pay, used by organizations as authentication techniques to combat counterfeiting problems, and used to prove theft in a court of law.

An efficient form of detection mechanism is needed to authenticate the IP which involves, at the basic level, adding some form of tags so it can be identified later in the process. The stronger option is watermarking where a tag is embedded into the IP and has a level of protection that can not be easily removed. The issue here is that it requires some degree of trust between the IP provider and user. The tags can be removed or potentially modified, so many IP providers are concerned about their value in providing solid protection. It is used more often as a tracking mechanism to identify the owner of the IP.

The second approach is to limit access to the IP by protecting its own visibility. This prevents users from removing the IP value through reverse engineering or sharing the concepts. People want to avoid company secrets being shared. Interestingly enough, however, there are several different reasons for usage which are irrelevant to the IP developer and the customer relationship. For example, a user wants to allow a contractor to perform validation on their design. They may use some form of IP protection to allow them early access without having to wait for any legal approval. Yet, the problem here is how to make the IP available for use, while disallowing the user from stealing it. The IP needs to work in the desired EDA tool in a secure manner. However, many IP vendors are concerned about the security tools that are required to handle the IP.

The more active approach is IP access prevention. The IP is usually in text format, so encryption is a common form of prevention. Data is obfuscated by using various ways to secure the information. The IP data must be protected at two different points, and transferring the IP between a user and provider is the first case. The IP provider will use encryption to convert the plain text using a key. This IP can be transferred to the user who will use the key, or some alternate combination, to decode the IP and use it successfully. The more complicated approach is to protect the IP inside a tool during usage. In this case, the encrypted IP must be read and used by the tool, but not be made visible to the user. Software tools must be able handle the IP, and not release the data.

The second set of protection methods includes compilation. In this case, the model is imbedded directly into the tool itself. The tool contains the model with instructions on how to link it to other blocks. The connection is often achieved through a programming interface, and this technique is actually used by many IP providers. They could take a processor core and create a verification model that could be linked to a customer environment to test their solution. This allowed the customer to check their solution yet have no visibility into the IP.

A number of positive and negative differences appear in comparing encryption and compilation. Compilation techniques are powerful since they are embedded into the machine code and ensure the behavior of the block. However, one would have to decompile the model to determine the IP.

5.6 Legal IP Protection Issues

The legal foundation is an important element of protection, since there are currently numerous legal techniques to protect companies' IP. There are many relevant legal issues, such as encryption, that relate to the technical approaches.

The ability to encrypt data has been of great concern to governments across the world. Earlier historical examples highlighted the value of IP protection which led to the governmental pursuit to control technology access. Encryption capability was later banned by a number of countries for many years, and the United States limited access to cryptography technology. It was placed on a classified munitions list, and it was considered critical to national security. This list contained such issues as nuclear weapons technology as well as guided missiles. At the time, encryption technology was viewed as munitions which were governed by the various laws for arms trafficking.

The growth of global business led to some changes in governmental thinking. Companies were demanding access to new methods in order to protect their data. For example, the growth of banking required the use of techniques to confirm user identification in a bank machine. An urgent need later arose to easily encrypt the IP, and companies then needed to protect the shared IP with different partners throughout the world.

In response to this globalization, over thirty-three countries met in the Netherlands to discuss the security concerns associated with such items as encryption. The Wassenaar Arrangement[2] arose as a result of this meeting which established international rules on technological use such as encryption for businesses. The export restrictions still remained for seven countries, then considered terrorist nations, which included Cuba, Libya, North Korea, Iraq, Iran, Sudan, and Syria. The rules considered encryption as two types related to the ability to have access to the protected code. Businesses now could authenticate passwords for business transactions. The second type was the ability to encrypt and decrypt the IP which, however, still had some restrictions in place. The primary limitation was that the weak encryption strength allowed. This enabled governments to read messages.

This provided a global basis for companies to share data with encryption. Other countries, such as Israel, Russia, and France, still had requirements that were placed on encryption technology. The laws are now changing, however, as the global business community seeks to make this type of technology easily available; at the same time, it becomes more difficult for governments and international bodies to regulate the protection scheme.

5.7 Different IP Types

There are different types of IP employed in the typical design flow, and within these, there are various techniques available for protection which highlight the different file formats and tools that may be used with them. It is important to choose the best technique to protect the IP yet still make it available to the end user.

---

[2] http://www.wassenaar.org/

Soft IP

The term soft IP refers to the IP block that is not implemented in a specific technology. This soft IP package contains the model, a test suite, and constraints for implementation. The test suite has a series of tests to demonstrate the correct functional behavior of the design IP block. A constraints file is used to achieve the desired performance with the appropriate technology implementation tool.

Protecting the soft IP is a particularly tricky problem. The model is typically provided in text format, and it is integrated into the rest of the end user's design. The user needs to be able to embed the IP, validate it in their RTL model, and must synthesize it to achieve the desired result. During this process, they need access to the internal details of the model in order to ensure the effectiveness of the debug process. One needs to be able to see into the model to trace paths and determine where a problem might exist.

In order to provide enhanced debugging and performance, the software tools need to see how the IP is created and integrated in the overall design. One has the problem of a block which, for power and timing requirements, may need to be linked with other pieces to ensure an effective result. Now, the IP is spread across different parts of the design and needs to be visible for debugging and testing.

This problem is more critical in the standard cell community as one needs to be able to see the individual pieces. A field-programmable gate array (FPGA), for example, is easier to protect since it has higher level blocks. Today, trust is the number one form of protection for soft IP and, where protection is inserted, it is usually to point an honest user to a violation of usage restrictions. They are actually not intended to stop a thief from stealing your intellectual property.

Hard IP

Hard IP provides a pre-developed block in the particular technology and needs to be integrated into the rest of the design. In this case, because the block layout will not change, it is easier to provide the models for integration. Only verification occurs during the design process. The protection of the hard IP is easier to achieve since it is provided as a black box that can be linked into the design and test environment. The IP provider guarantees performance characteristics that can be tested, but they cannot be changed by the user. It is relatively easy to identify a company's intellectual property just by looking at a chip under a microscope, and as such, usage violations can be immediately spotted. If a company attempts to modify the physical IP form to prohibit its visibility, then they have a considerable problem since they have no source views to help them make this transformation. Upon making manual changes, the re-verification costs may well exceed the license costs for the block.

Abstract IP

In abstract IP, protection was a very critical factor, especially since various organizations needed the technology to allow IP sharing with their partners. Abstract IP was used by partners to make sure that their design would work successfully in the environment needed. These models are required to validate and test the desired functionality at the system level. Of course, such models can be at different levels of abstraction and for different purposes, for example, it could be a transistor model provided by the foundry used for validating performance. Many companies will provide abstract IP models to partners before silicon is ready to enable the full chip verification to work successfully. The important point here is that the model does not need to represent the actual solution form in any way and thus may be freely given away since it reveals nothing about the implementation.

Other IP

In this class, there are rule decks used by different tools to ensure correct behavior of a block in a design flow. One might provide a design rule checking file that must run successfully before a design is released. The IP provider wants the user only to see the results but not to be able to modify the IP.

5.8 Challenges for Successful IP Protection

Designer Challenges

Many problems are faced by the designers using the various protection methods successfully while still trying to maintain the needed design productivity. The first question that is always asked is: Why do we need it? Don't our current techniques address this today? Many companies rely on a variety of legal and trusted approaches. The problem, however, is that there are many more cases of IP theft since it is almost impossible to recover the IP once it is made available to the public. Many companies rely on creating newer and different models which force the potential thieves to take additional time to attack the IP. So the question is: what level of protection is really needed in the flow?

Rules are needed for protection to work successfully, but it will ultimately add time to the development process. One of the concerns is that this time can impact a schedule, and in order for a designer to be successful, they will require full access to the IP to debug any problem. The cost of a failed block would pale in comparison to the loss in profit if a project is delayed due to an IP problem. Users want to share IP, but they are uncertain if it can be protected.

This problem, as mentioned previously, is critical as one looks at the various supply chains being created. Global users are collaborating with various organizations which may have less stringent legal protections for the IP. The techniques need to be in place to protect the IP, yet they cannot guarantee its own stability.

The development team needs to look at these key issues to determine how to use the technology:

- How will the IP be used by the designer?
- How will the IP be used by the design tools?
- What degree of protection is needed in the IP?
- What level of security is required to make the IP transferable?
- Does it make sense to encrypt a particular model that is used in the design?

One needs to look at the effectiveness of the protection in the different technologies. The strength of the protection is dependent on the tool usage. For example, with encryption, key management becomes a big issue as separate keys become generated on an individual basis for each IP block. This is important since it limits access to any particular IP block.

Tool Developer Challenges

Nowadays, the lack of standard methods for encryption can place a large burden on tool developers. As a result, IP developers have partnered with major tool providers and have implemented proprietary mechanism. These partnerships often come about due to customer demands that allow encrypted IP to become accessible to the tools in the designers development flow. However, these ad-hoc schemes end up being a drain on the IP and tool developer's resources and often leave the user having to juggle multiple versions of the same IP block for different parts of the flow. This might also lead to additional problems if proper source control is not used by the IP developer or user since there are now more files that can become potentially incompatible.

Encryption can also have an impact on the tools themselves. They must be able to read an encrypted source and ensure that they do not enable back-door access to the tool information once it has been decrypted. This is not an insignificant problem, and it is taken seriously by other industries. Let us consider the IBM cell processor initially used in the Sony PlayStation 3 which has a single PowerPC control processor and eight special purpose digital signal processors (DSPs). The system is architected such that if one of these DSP engines is decrypting and processing a video stream and then re-encrypted before sent out of another port of the device, then, under these circumstances, it is not possible for the PowerPC core to gain access to the memory in which the un-protected version is kept.

Simulators, synthesis engines, and other tools that work on encrypted models do not have architectural isolation, and it may require considerable effort to ensure that the tool will enforce the required security. Software companies can and even do get this wrong. Let us consider the recent case with the breaking of the protection scheme employed by the new high definition-digital video disks (HD-DVDs). While the security system has not yet been fully compromised, careless developers of one or more HD-DVD players made it possible for the decryption keys to be extracted for certain disks, thus stripping those movies of any kind of protection.

5.9 Future of IP Protection

Nowadays, there is an increasing demand for IP protection, and companies would like to be able to use various methods in the entire process. The problem lies in the need for the techniques to work in all tools and in all flows. IP providers want to guarantee that their models are not negatively impacted. A stronger emphasis is needed to create standards to implement various protection methods. Tool developers will not be open to protection schemes if it costs them significant development time or if it slows their tools down. At the same time, users will react negatively if it impairs their ability to get the job done. The industry also has to be very careful that the adoption of any scheme that relies on keys does not preclude new tool or IP companies from entering the market.

Chapter 6

Software

Richard Vlamynck, Verification Engineer, Mentor Graphics Corp.

6.1 Introduction

When talking about intellectual property (IP), one first tends to think of semiconductor IP, but software is actually one of the ways in which commodity IP can be turned into a stellar resource. This chapter will discuss the potential, if not considerable, value that software and hardware it runs on can ultimately acquire.

6.2 Software Considerations for Hardware IP

What are software device drivers? Does software play any significant role in the marketing, sales, or delivery of IP? Does software play any significant role in the operation of a system? After designing and building the first Cray1 supercomputer, Seymour Cray was reported to have said, *"The hardware was the easy part..."*

When programmable logic blocks were smaller and simpler, one could first design, tape out a chip, and then deliver a prototype to software engineers who were given the task to create the software that actually made the chip valuable to the end customer. For example, the documentation supplied with a universal asynchronous receiver/transmitter (UART) was usually all that was needed to write a software driver for that chip. This is no longer the case now that IP such as universal serial bus (USB) controller soft cores, reconfigurable video blocks and multi-core system-on-chips (SoCs) have arrived.

Nowadays, the situation occurs time and time again where an SoC is delivered to the software group long after the hardware group has moved on to the next design. The hardware team is typically unavailable to assist in understanding and debugging the interface between the hardware and the software. This means that the time to market gets pushed out and the quality of results suffers.

Additional difficulties with the SoCs include the problem of visibility into the chip internals. This is exaggerated with multicore SoCs and one of the reasons why on-chip IP that increases visibility or aids in debugging is becoming a new and growing breed of IP

The interface of the hardware IP to the rest of the system is via software drivers. Software drivers are a part of the virtual machine (VM) that belong in the hardware domain. Therefore, if one is to deliver a complete IP product, the following rules must be established and followed:

- Software drivers must be a part of the hardware IP deliverables.
- The functional specification of the software drivers must be included in the functional specification of the hardware IP.
- For in-house SoC designs, functionally-verified software drivers must be specified as deliverables of the tape-out which should occur as part of the register transfer level (RTL) verification cycle.

The costs associated with being late to market have already been mentioned, and this implies that the possibility of delivering IP starts when the software for that specific IP is finished. This signifies that, with IP, there is an even bigger incentive to move software development and verification further up in the flow and can, in some cases, mean that the hardware development flow must be modified to make this possible.

While drivers may be considered to be the lowest level of software that connects to the hardware, it does not imply that only the drivers should be developed and verified up-front. In many situations, additional layers of software can help to make the IP an even more valuable resource, and thus this needs to become one of the significant trade-offs that are made in planning a new IP block. While this chapter focuses on the driver level of software, one should be aware that many of the same arguments can also be made for higher levels of software.

6.3 The Fundamental Problems of Software Drivers for Hardware IP

What software is needed to support all the different end customers?

The examples in this chapter will assume that the software is being created for a configurable hardware core that is intended to support USB functionality on some application-specific integrated circuit (ASIC). This hypothetical USB core will be used to provide some generic guidance and some plain examples of the problems faced when creating software to support IP. This chapter will only address certain issues of relevance, and it will not entail an exhaustive study of all related topics and answers.

In some cases, the USB software example can be translated to support some other hardware IP but, in other cases, this is not always possible. This is because some hardware subsystems bring some presumptions, specifications and other "baggage" to the table.

If a generic software device driver for a hardware block could be created and if a given driver was adaptable to all the ways the end customer could use the IP, the following considerations could be accounted for:

- Central processing unit (CPU) locking primitives' differences
- CPU derivatives executable binaries differences
- CPU interrupt request (IRQ) handling differences
- Address and data bus differences
- Endianess differences
- Compiler tool-chain differences
- Application binary interface (ABI) differences
- Real-time operating system (RTOS) Differences

- Hardware implementation dependent timing differences
- Module level verification support
- System level verification support
- Timing variability caused through variable voltage on the chip and variable frequency of the processor(s)

The fundamental problems tend to indicate that a universal device driver is not feasible unless it is delivered as one component in a more comprehensive "platform." This also shows some of the costs that must be considered when deciding on the amount of delivered software. As we move up the application layers, it becomes generally more difficult to create generic software and thus it would probably make more sense to consider tighter forms of business relationships.

Software Abstraction Layers

A main ingredient of portability for software drivers is the use of abstraction layers. The two most common abstraction layers are hardware abstraction layer (HAL) and operating system abstraction layer (OSAL). In the higher abstraction layers, the software could go all the way up to a full reference implementation for a system platform company which would ultimately enable a customer to start with a fully working solution, such as a cell phone. On the other hand, the developer would be able to extend this platform to add in new capabilities such as an MP3 audio player or a camera capability.

Software abstraction layers provide the benefit of black box interfaces which hide implementation details from the above and below layers. The advantage is that functionality can be "plugged in" or "swapped out" as needed by each system implementation. The disadvantage of abstraction layers is that they add to code bloat and have a negative impact on performance. The added complexity of abstraction layers can also lead to maintenance and debugging headaches. *Figure 6.1* shows one view of the possible abstraction layers as developed by the Virtual Socket Interface Alliance (VSIA).

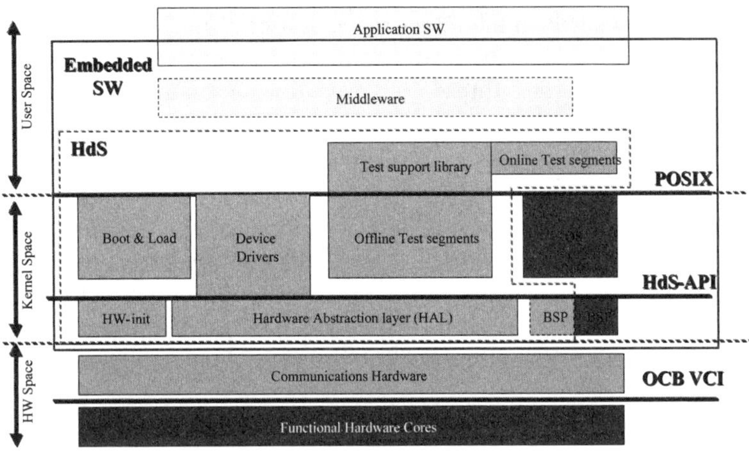

Figure 6.1: *Software Abstraction Layers*

The HAL is the application programming interface (API) provided by the device driver that is used by other software modules in the system, such as the operating system. This forms part of what is often called hardware dependent software (HdS), which also includes such software as boot loaders, diagnostics, and others that are highly dependent on the actual hardware implementation.

The industry taxonomy adds that: *The HAL consists of three sub-layers: the access shielding, register shielding, and functional shielding layers. The access shielding layer contains the mapping between the variable names and the addresses of the memory locations, and macros enable the use of the HAL in simulations. The register shielding layer enables access to the memory locations using read and write functions. The functional shielding layer allows higher layers to use a device, without knowing all the details of the device implementation.*

While not all systems may strictly provide these three layers, it does demonstrate the kinds of function and abstraction that are provided by this software layer.

The OSAL is the set of system level services needed to support the device driver. The definition of system level services in this context should be reduced to the minimum necessary to deploy the driver.

This can usually be a highly restricted subset of the system services that are delivered with most commercial grade RTOS's such as creating a separate thread of execution, synchronization and communication primitives, and acquiring and releasing memory.

A separate thread of execution is usually called a task or a thread by the RTOS vendor. The ability to have multiple threads of execution brings many advantages and some challenges. One of the advantages is the ability to logically partition the software into separate work units. One of the challenges, meanwhile, involves the communication and synchronization that exist among the various threads.

A synchronization primitive is needed when there are multiple threads of execution in a system. The most common form of synchronization primitive is a called semaphore, and the most common form of communication primitive is the queue. A very fast non-blocking primitive is needed to acquire and release chunks of memory for driver usage. Frequently, a driver needs to acquire memory for incoming data, and then put a pointer to that memory into a queue.

Design Requirements of Software Drivers

Register and device mapping of individual hardware modules into the system level memory map is problematic, but it is crucial to the success or failure of a "system agnostic" device driver. One will find that hardware devices are usually represented to the software as sets of registers. The registers can be written and read on certain alignments, such as a 1-byte, 2-byte, or 4-byte alignment.

A single hardware register can be represented by an object that is the properly sized, signed, and aligned integer type. Therefore, register sets must be represented by C or C++ structures with the proper size, sign, and alignment for each of the structure members.

The structures must be "mappable" or "re-mappable" depending on the system configuration. The American National Standards Institute (ANSI)-C standard does not allow you to declare a variable at a hardwired memory address. This is easy enough to get around, since you can declare a pointer to a register and assign the pointer the address of the hardware. Therefore, when you de-reference the pointer, you are addressing the register.

For the sake of argument, let us consider the hypothetical USB controller core used for example in this chapter. One of the register sets is the control and status register set (CSR) for an endpoint. The register set, which consists of five 16-bit and five 8-bit registers are shown in *Figure 6.2*.

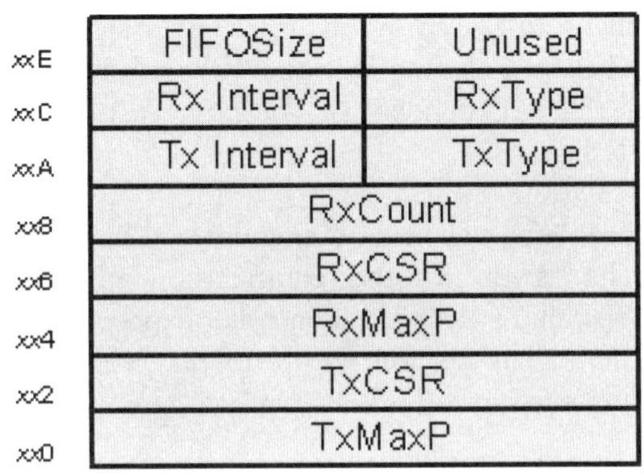

Figure 6.2: *Register Map*

In order to effectively program this set of registers, the driver writer will need to use a structure and a pointer to a structure. Now, let us suppose that during system initialization, the register set ended up at address 0x8000 in the system memory map. In order to set TxMaxP equal to 512, the driver writer will need to use the following, if not a similar, C-like pseudo-code:

```
typedef struct MYCORE_CSR {
        ushort  TxMaxP;
        ushort  TxCSR;
        ushort  RxMaxP;
        ushort  RxCSR;
        ushort  RxCount;
        uchar   TxInterval;
        uchar   TxType;
        uchar   RxInterval;
        uchar   RxType;
```

```
uchar   FifoSize;
} MYCORE_CSR_0;

MYCORE_CSR *myCSR_p;
myCSR_p = 0x8000;

myCSR_p->TxMaxP = 512;
```

Another consideration in software design is the layout of data structures which can be laid out in either big-endian or little-endian byte order. They can also be laid out with or without padding in memory, and this is dependent on the targeted processor and the compiler tool-chain used.

One of the problems that will be encountered in our hypothetical USB controller is during the enumeration phase of a USB peripheral. Typically, a USB host controller expects to unpack serial data in a little-endian fashion and with no padding in the data structure that it is trying to re-constitute.

```
typedef PACKED struct _STD_CONFIG_DSCR {
 UINT8  bLength;
 UINT8  bDescriptorType;
 UINT16 wTotalLength;
 UINT8  bNumInterfaces;
 UINT8  bConfigurationValue;
 UINT8  iConfiguration;
 }
```

If an ASIC with a big-endian processor was created, there will be serious challenges to face when we move the 2 byte entity wTotalLength from system memory into the USB first in first out (FIFO) for transmission to the host. When the data gets to the host, the big and little end of the 2-byte word will be swapped which will eventually make the enumeration fail.

The following example illustrates the need for member alignment within a data structure. In this example, we have a data structure with four elements and a one byte entity followed by a 2-byte entity, another byte sized element, and a 4-byte sized element.

```
typedef PACKED struct _FOOBAR {
 UINT8  Element1;
 UINT16 Element2;
 UINT8  Element3;
 UINT32 Element4;
 }
```

The problem here is that without the "packed" keyword, the compiler tool-chain will probably insert one byte of "padding" between Element1 and Element2, especially since most microprocessors make efficient use of the data bus when data is aligned on "natural boundaries." The natural boundary for an element size of two bytes is an address with the lower order address bit set to zero. The compiler tool chain will probably insert three bytes of padding between Element3 and Element4 since the compiler will probably try to get the address of Element4 to occur at an address where the lowest two address bits are both zero.

In this case, let us presume that we are dealing with some communication protocol and that we are going to send data from one machine to another, similar to the enumeration case in the previous example. We do not want the compiler to add the "padding," so we introduce the keyword "packed" to eliminate the unused bytes between the actual members of the structure. All ANSI-C compilers are presumed to have control features, pragmas, and invocation switches to deal with these kinds of problems.

Implementation Requirements of Software Drivers

Drivers should be written in C or C++ for portability, and they must follow certain coding standards and conventions. Therefore, within each company or IP provider, coding standards must be established, published, and followed. The advantage in defining and following coding conventions is shown during the maintenance and enhancement phases of software projects.

The establishment of programming conventions is an important part of design reuse. The standards for software source code include, but are not limited to, formatting, naming, and coding conventions.

In addition to self-generated internal coding standards, there are some industry standards, such as the well-known is Motor Industry Software Reliability Association (MISRA)-C where MISRA-C is a subset of the C language. It is a collection of rules that define the subset of the C language such that it is less error-prone and more suitable for safety critical systems. Safety critical systems are those that are used in motor vehicle electronics, avionics, medical systems, and defense systems.

The following example reflects the least amount of coding practice that should be used. A commercial effort should develop and enforce its own set of rules that are tailored to the companies or the projects software product line intentions.

Source Code: Example of Formatting Conventions

- Indentation—Four spaces, with tabs used whenever possible. Some editors' default settings may need to be changed to view the code properly.
- Bracing style—Open and closing braces on a new line.
- Comment style—Comments are in C (not C++) style, and are indented with the code.

Source Code: Example of Naming Conventions

- Namespace pollution protection—All global symbols, including preprocessor macros, begin with MYCORE_ or IPD_. MYCORE_ is for things intended for customer use and/or customization. IPD_ is reserved for things intended to be purely internal.
- General naming rule—All names are descriptive rather than abbreviations.
- Structs and enums—These are typedef-ed and are named per the global symbols convention. This eliminates the need to use the extra keyword everywhere.
- Variables and struct/union members—These use Hungarian notation. This is consistent with the USB specifications and eliminates ambiguity when looking at something whose declaration is far away.
- Constants—They are all uppercase, whether preprocessor macros or enum values, with underscores between words.

- Non-constant identifiers—All other identifier names following the prefix use mixed case (even for acronyms) and no underscores.
- Filenames—For compatibility with old compilers on some platforms, source code filenames follow the 8.3 format which entail no more than 8 characters with no more than a 3-character extension. As a result, all files have a comment header stating their purpose and including their revision. All interface headers and top-level source filenames begin with mu_, platform-specific customization files begin with plat_, and driver files are named according to the device and class they represent.

Source Code: Example of Coding Conventions

- Equality comparisons—Any constant appears on the left-hand side and prevents accidental assignment.
- Compound statements—All for, if and else, do, while sub-statements use braces, even if they are a single line. This prevents accidental misplacement of statements during maintenance.

6.4 Binaries versus Source Code

Hardware IP runs in the realm of timed clock cycles, handshaking protocols, and signals while software drivers run in the context of microprocessor execution. The execution context of a program or a driver running on a microprocessor is untimed by its very nature. The software running on the microprocessor can experience timing differences which are caused by RTOS preemption, interrupt latencies, interrupt masking, and instruction cache misses. These differences are inherently built into a microprocessor system just by its own nature. When, for example, IP is produced for an ASIC, the issues of voltage and frequency scaling for power control are also introduced.

The higher layers of software stacks and protocols can be delivered as binaries (*.a or *.dll), but drivers that connect directly to the hardware must be delivered as source code.

Software exists largely thanks to flexibility and, as a result, customers usually have some expectations in this area. Each customer's threshold is different, so suiting everyone is not feasibly possible.

However, certain forms have become a standard in the industry, and it seems prudent to take at least these as minimal goals. For example, software can be delivered as source code or binaries and even if source code is available, it can be cumbersome for customers to integrate into their build environments (one area of tremendous differences in practice). Therefore, it can be a great service to customers to provide baseline binaries for certain platforms and configurations.

There are several standard types of binaries for usage. There are static libraries (*.a) that, aside from being a part of the customers' application, have all symbols resolved and need to be linked with the product of the customers' build. There are also dynamically-linked libraries (*.so) that are not included in the executable, but they have to be available to the system at runtime.

For example, in our hypothetical USB software, a static or dynamic library is the appropriate type of binary for the protocol stack. Therefore, callback interfaces are designed to use function pointers which allows a library to be built and delivered to a customer. Further, by designing software builds as such, customers are able to deliver binaries to their customers. Yet, legal constraints are sometimes involved. For example, it is possible to deliver a proprietary subsystem that works with a general public license (GPL) Linux–based system by following careful rules in loading and running the proprietary software subsystem.

6.5 Software Verification and Debug

The verification and debugging of software drivers suggests a debugging process consisting of four phases: testing, stabilization, localization, and correction. The testing of a driver includes the detection of software bugs such as:

- Lexical errors—These will be caught by the compiler tools and will create a problem for the compiler to identify precisely where the error occurred. The concept of lexical proximity is used by the verification engineer to track down the exact error. Lexical errors are often on the line directly above the one the compiler is complaining about.
- Syntax errors—These will typically be caught by the compiler tools.
- Execution errors—These are the "tricks, traps, and gotcha's" of the C-language and the runtime environment on the embedded systems where this type of IP is typically deployed. The origins of this issue will be addressed in the remaining sections this chapter.
- Intent errors—One of the greatest issues for delivering software for IP soft cores is centered around the ability to debug errors of intent in the software. A rigorous and methodical verification plan developed independently by the verification team is the most straightforward means to catch these bugs.

The problem with debugging ANSI-C is related to the generalization of the virtual machine that is provided by ANSI-C. These errors include pointer bugs and the stack, out-of-range subscripts, variations, uninitialized pointers, and pointers that overwrite code.

Debugging execution errors on a running system:

Observability of control flow and the instrumentation of data read/write can affect the timing of the driver, thereby masking the error that the engineer is trying to observe. Often, you will find a "printf" or a simpler "hexout" type of function that prints values to a console or saves them in a memory buffer.

The following pseudo code illustrates the technique:

```
int set_reg( int new_value) {
int old_value;

old_value = hw_reg;
```

```
hw_reg = new_value;

if (debug == 1)
        printf("previous value was:" %d, old_value);
else
{
}
return (old_value)
```

The problem introduced by the previous approach is in the amount of time required to perform the write to the console or the data write to the buffer. The amount of time may only be milliseconds or microseconds, but it does change the overall timing of the system. In addition, several locations in memory will have been modified which can hide or modify other kinds of errors, such as those caused by reading uninitialized memory.

Therefore, when debugging a live system, a better approach would be to use Joint Test Action Group (JTAG) trace modules and the tools built into a high level source code debugger to get the control and data flow information through unobtrusive means. This involves setting soft breakpoints and watchpoints that collect data without having a big impact on the system level timing. However, the disadvantage of these systems is that when the captured data is transferred out of the chip, the dynamic operation of the system must then be suspended. This can cause serious problems with dynamic logic such that it, or other parts of the system, may lose its value, and may prohibit the generation of regular interrupts. Further, in some cases, the first test chips returned from wafer-fab may have broken scan chains, in which case the software developer must have a fall back plan.

Therefore, the best approach is to perform the hardware integration before the ASIC is built. This entails the use of a hardware emulator, a hardware software co-verification environment or a system-level virtual prototype. The advantage of the co-verification environment is that the software team can create the exact boot code and driver code that will run on the actual ASIC. However, each of these options has a number of advantages and disadvantages:

- Hardware emulators—These allow large quantities of software to be run and are usually cycle accurate. In many cases, actual silicon devices can be used for the processors to ensure 100 percent accuracy. It is usually possible to run the emulator in-circuit with appropriate speed gaskets meaning that long and complex testbenches do not have to be created, and the system can be subjected to live traffic streams. The problem is that hardware emulation is not going to be available until all of the RTL code has been created and debugged, which defeats some of the objectives stated at the beginning of this chapter. In addition, emulators are expensive and, as a result, it is usually not possible for every software developer to have access to one.
- Hardware/software co-simulation—Using the Mentor Seamless co-verification environment, for example, it is possible to create and verify the entire software load for an ASIC before the chip is even built. The use of co-verification speeds up the execution rate by substituting the processor model for a much higher performance instruction set simulator (ISS). However, while the ISS runs 1,000 times faster than the rest of the logic, simulation performance can still be slow because of the need to simulate the rest of the hardware at the RTL level. The other potential problem with this solution is that the ISS is a model of the processor and may have bugs or behaviors that are different from the actual RTL. Some of the RTL models may be replaced with higher abstraction models to speed up the simulation. The same problem applies here in that these models may not be 100 percent accurate. However, with the lower cost of this virtual environment and the improved debug capabilities that it provides, they are used by a large number of low-level software developers.
- Virtual prototypes—Similar to the co-verification environment, these rely on software models of the hardware, but they are generally all at a higher level of abstraction. The models also do not attempt to be completely cycle accurate and because of the relaxation of this constraint often run 100 or a 1,000 times faster than

co-simulation. There have been some reported cases where the virtual prototypes run faster than the real-world systems which only demonstrate that they can be used in-circuit with the right input/output (I/O) circuitry.

All execution environments certainly present some level of trade-off between cost, performance, and accuracy, and in many cases more than one option will be employed during the development of an IP block.

6.6 Conclusion

This chapter has explored some of the issues related to software delivery that is customized to support hardware IP. It has also addressed the considerable risk associated with waiting until an ASIC is built to verify that the software drivers are functioning correctly.

Software drivers are a part of the VM that belongs in the hardware domain. Perhaps hardware IP will never be delivered in a truly "shrink-wrapped" configuration, but it is possible to move closer to the ideal. Many IP customers are looking for a "platform"-type of solution that integrates the maximum amount of IP functionality into the cheapest, densest, and most complete package. In this case, software drivers are a crucial and essential element of the hardware IP deliverables. The software also adds significant value to the IP offering and allows the product to be differentiated from the competition.

References and Recommended Reading

[1] Brian Bailey, Grant Martin, and Thomas Anderson, *Taxonomies for the Development and Verification of Digital Systems,* Springer, 2005.

[2] Daniel Bovet and Marco Cesati, "Posix APIs and System Calls" and "I/O Architecture and Device Drivers" in *Understanding the Linux Kernel: Third Edition*, O'Reilly Media, Inc., 2005.

[3] Michael Keating and Pierre Bricaud, *Reuse Methodology Manual for a System-on-a-Chip Designs: Second Edition,* Springer, 1999.

[4] Jean Labrosse, Microtec, Mentor Graphics Paper #4350: "Adopting Programming Conventions."

[5] www.misra.org.uk.

[6] Chris Rowen, "Typical Software runtime and development structure," figures 2-3, pp. 42, and "The Software Development Development Environment," pp. 226 in *Engineering the Complex SoC: Fast, Flexible Design with Configurable Processors*, Prentice Hall, 2004.

[7] Richard Vlamynck, "Visualizing Programming Models in Embedded Systems," and "Event Handling in Embedded Systems," pp. 242-249 in *Embedded Software: The Works*, Colin Walls, Newnes, 2005.

Chapter 7

Verification IP

Sean Smith, Chief Verification Architect, Denali Software

7.1 What is Verification IP?

Although terminology has changed over time, engineers have been utilizing the functionality of verification IP (VIP) for the past two decades. In fact, about twenty years ago, the bus function model (BFM) was popular, then that term was often shortened to just "model" and then we proceeded to start hearing terms like "soft model," "hard model," "source model," and even proprietary terms like e-verification component (eVC) from Verisity, now Cadence Design Systems. Recently, the industry has adopted the term VIP, probably due to the more broadly accepted systems-on-chip (SoC) methodology that centers on the use of design IP.

So, what is VIP? Formally, VIP is defined as a packaged component intended to help its users perform functional verification on some form of electronic circuit. In reality, the largest application of VIP is more commonly a component designed to be used in a digital logic simulator to generate stimulus and check a standard protocol. Here, the value is that development teams can leverage a complete, pre-packaged verification component that provides modeling, protocol checking, and other verification functions, without requiring the team to invest in the necessary domain knowledge to produce such a component. This value is magnified by the increasing complexity of interface protocols (e.g., peripheral component interconnect [PCI] express, serial advanced technology attachment [SATA], universal serial bus [USB], etc.) which are defined in "paper" documents that now regularly exceed one-thousand pages. In this sense, VIP can be considered as the encapsulation of domain knowledge for the protocol in a package utilized as part of an overall functional verification strategy, including a test bench, test harness, or test fixture to create a virtual simulation of a hardware system. See *Figure 7.1.*

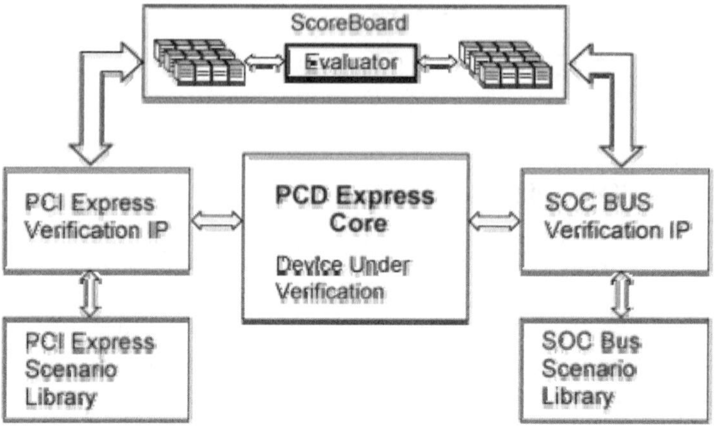

Figure 7.1: *Block Diagram of a Typical Functional Verification Test Bench*

It is also worthwhile to note the recent emergence of VIP for use in formal verification environments, as well as analog verification environments.

Analog IP

Analog VIP, commonly referred to as "pre-built test benches" in the analog world, is very similar in concept to digital VIP, but the next levels of details are quite different.

Analog VIP works with a circuit or spice simulator to make sure that a given design under test (DUT) is compliant with an analog specification. Some examples of checks at this level related to opening an eye, include jitter, pre- and de-emphasis ratios, common mode voltages, leakage currents, and phase margins. Generally speaking, the design must be set up differently for various tests, so it is nearly impossible to attach a test bench to a design and have it perform all the tests. In addition, there is no standard interface supported by the spice simulators as there is with digital logic simulators so the VIP cannot exchange information with the simulator and configure itself dynamically. Instead, it has to rely on a two step process:

1. Set up the test by including test vectors and conditions in the netlist and run simulation
2. Once the simulation results are available, process the output waveforms for checking compliance with the specification

This situation will somewhat change as the use of mixed digital and analog signal simulators increases when those simulators converse with an attached test bench using standard programming language interfaces [1].

Formal/Assertion VIP

Formal/assertion VIP is another specialized form of VIP, which is gradually being accepted in some verification processes and the commercial markets. Formal VIP is typically a collection of assertions for a specific protocol or topology that has been optimized for use in formal verification tools, such as a model checker. On the other hand, assertion IP creates the VIP through a series of assertions, as does the formal VIP, but it may not be optimized for a formal verification process. Instead, it is designed to be used in both dynamic (simulation) and static (model checking) verification tools. Formal VIP may often take advantage of abstraction in the specification, e.g., symmetry reduction, data independence, non-determinism, over approximation, etc. Formal VIP will thus not necessarily work as simulation–based VIP due to these abstractions.

7.2 Benefits of IP

Overall, the ultimate goal in using any IP is to reduce the project cycle time and improve overall product quality. "Considering that verification consumes 50 percent to 80 percent of the total development effort, verification reuse brings tremendous benefits to the verification team," said Steve Ye, a senior engineer in the IP reuse design and development group at Agere Systems [2]. With the sheer volume of IP that goes into a modern application-specific integrated circuit (ASIC) or SoC, it is not feasible for development teams to create all IP, whether for design, verification, or software from scratch. This is especially true for design and verification IP targeted at standard interfaces which are used in most, if not all, chip designs. The value of VIP providers is clear in this case; unlike any of the individual design companies, VIP vendors can afford to make significantly greater investments in domain knowledge and the associated VIP infrastructures, and then amortize that cost across a large number of customers. Beyond these high level goals, what are the specific benefits associated with deploying VIP? Let us explore some of these in more detail.

Reduced Time-to-First Test

The first and most obvious benefit in using VIP comes from not having to write the VIP from scratch. This saved time means an engineer can simply plug in a VIP, configure it, and start writing and running tests. The result is a large-time savings at the beginning of a project, which enables the verification process to begin much earlier. The amount of time saved here depends on the complexity of the protocol and VIP completeness. Let us consider a recent protocol like PCI express (PCIe) where the provided value here is self-evident. "We have over 20 man years of development effort invested in developing our PCI Express verification IP," said Jiurong Chen, vice president of software engineering at Denali Software.

Reduced Time-to-Test Complete

Another, less obvious, benefit associated with VIP is reduced time to testing complete. This is not only due to the time savings that occurred in writing the model, but also includes efforts associated with developing stimulus libraries and functional coverage models. Modern VIP often includes robust libraries of canned stimulus or test cases that can be applied out of the box or modified by the end users. These test cases raise the abstraction level of test creation to make it easier and quicker to create complex test cases than the traditional approaches of writing tests a transaction at a time. In order to accomplish this, many modern VIP products leverage complex constraint solvers to create directed random stimulus. Since this stimulus is created in a directed random fashion, it becomes important to objectively measure our verification completeness using objective techniques like functional coverage [3]. Many modern VIP products include functional coverage metrics which help the user determine objectively what they have tested. These aspects of stimulus libraries and coverage provide more significant benefits than just the base model in reducing the time it takes to get testing completed. In some cases, with modern industry standard protocols like PCIe, the standards bodies sometimes offer check lists to help define what basic compliance means for these protocols. These check lists can be automated by VIP in the form of stimulus libraries, assertions or checkers, and functional coverage metrics. The end result is nearly a turnkey operation for verifying baseline compliance with the protocol specification. Although these attempts at compliance only cover a very small percentage of a comprehensive compliance suite or test plan, they do offer real value to customers in identifying basic compliance criteria. It is useful to note that verifying compliance is very different from verifying interoperability with other compliant devices. This becomes obvious when considering that protocol specifications are used to define the complete set of features and functionality for a particular standard, and that device designs that use the protocol almost always use some unique sub-set of the overall standard.

Improved Quality by Broad Usage

Since either internally- or externally-created IP is designed to be reusable, its users greatly benefit from this reuse by having the VIP tested by different users, devices, and platforms. This becomes a critical factor for commercial VIP where the entire user base can have a significant impact on increasing the quality and robustness of the VIP product over time. One must realize that any given design will only exercise some subset of the protocols and VIPs state space. Each customer will only exercise some small subset of this state space takes time and a variety of users and applications to thoroughly exercise VIP. This type of robustness simply cannot be engineered into a new model or one created in house for a specific project especially since it requires widespread usage over time. Let us consider the Denali PureSpec™ PCIe VIP. Denali officials indicate that this particular VIP product has been used in over 150 PCIe designs, in applications ranging from desktop computing, graphics, networking, telephony, and consumer mobile. The richness gained by such broad usage provides a very tangible benefit to teams looking to deploy VIP for PCIe design verification.

In conclusion, it is easy to see that high quality VIP can provide substantial productivity gains, especially for large ASICs and SoCs designed around industry standard protocols.

7.3 Make versus Buy

The practice of creating a verification environment from scratch for a project is nearly always an impractical option for today's designs. Almost all modern interfaces have evolved to use complex, multi-layer protocols that are defined in documents that regularly exceed one-thousand pages. Developing VIP requires complete, accurate, and industry-consistent interpretation of the specification into an efficient verification implementation, which then needs to be exhaustively tested, maintained, and supported. When this is combined with continuously shrinking schedules and a high cost of a re-spin or product recall, it makes VIP development prohibitive for nearly everyone, except for commercial VIP vendors. Thus, assuming that the need for VIP has been established, the next important question remains as how to acquire the VIP. There are two basic options here:

1. Roll-your-own VIP (make)—Users harvest code from previous applications or write it from scratch taking into account goals for reusability and future needs.
2. Commercial VIP (buy)—Users can leverage pre-existing VIP from a variety of vendors that were written from the ground up to reuse in a variety of environments, methodologies and languages. These tend to have rich feature sets, are optimized for performance, and often conform to standard application programming interfaces (APIs) which make learning and deploying a variety of protocols much easier. It is not unusual for modern verification environments to be composed of heterogeneous mixtures of the earlier addressed options. This is usually the result of commercial VIP not being available for all protocols used by a design, especially proprietary interfaces unique to a specific design. Unlike five to ten years ago, virtually all common and many leading edge protocols now have well supported and tested VIP available.

"It's a worrying state of affairs when the verification IP is almost as complex as the design IP itself, but that's the stage we've reached," said Clive Maxfield, a noted author and columnist for EE Times [4] This is an obvious point when you consider that VIP must check for all possible violations of the protocol specification, while design IP instance is developed within a defined subset of the specification. This is perhaps an understatement as commercial VIP is often more complex than the design it is associated with. Despite this observation, some teams still develop their own VIP. Teams who choose this route often discover that the effort required to create their own VIP is much larger than they originally anticipated. There are a number of common challenges that appear repeatedly in this process.

A deep understanding of the specification is required. When you create your own VIP, it is essential, first of all, to completely understand the specification in order to know what to check for, how to check it, and how to create stimulus. Today's industry standard protocols are very large and complex. Let us consider the PCIe specification. After summing up all the pieces of the specification, we arrive at an excess of 2,000 pages of material that one might digest comprehensively to create a quality VIP. Its time aspect will scare most potential authors away. Another often ignored issue is that the human effort of interpreting and translating the paper specification into VIP must be consistent with the interpretations of other developers in the industry. Despite the enormous efforts of industry standards organizations, paper specifications still contain ambiguity that requires interpretation.

- *Harvesting code is seldom easy.* Harvesting code used on previous projects can be a costly endeavor especially if the original author is not available, or if the code is poorly structured or poorly documented. This code will need to be re-verified if it is modified. Practical experience has shown in many cases that it takes more time to harvest the code than to re-write from scratch.
- *Support is never free.* Once a VIP has been crafted, the ownership of support, fixing its bugs, and synchronizing the model with the ever changing revisions of the protocol specifications must be carefully scheduled and managed.

Experience has shown that support can add 20 percent to 30 percent of the initial development cost. Even worse, is the issue that if the model is not well supported future use of VIP will be restricted, thus eliminating much of the perceived value of writing it in house.

- *Developing VIP is rarely strategic.* As with IP and other non-critical aspects of our engineering work, it is useful to ask such questions as "what is my core business?" and "where does my company add value?" Given the state of SoC design today and the existence of a commercial VIP market, it is hard to imagine any engineering manager concluding that they are aiding the company's profitability or providing better products to their customers by writing their own VIP.
- *VIP quality is very expensive.* VIP needs to be tested and then applied by teams in real-world settings on a variety of projects and platforms before it can be truly robust. Those considering writing their own VIP should ask themselves, "do I really want to be my own beta customer?" While some VIP vendors have created valuable internal testing environments, the consequences of trusting a completely unproven VIP model can be disastrous to a project or platform.
- *Supporting modern verification methods is crucial.* The verification world has evolved tremendously in the last 15 years. Today's best of breed verification environments use electronic system level (ESL), directed random generation, functional coverage, and assertions to achieve better results. Creating VIP based on these best of breed techniques requires deep knowledge and investment in the techniques and the underlying infrastructure required to support them.

Let us apply some of this to a real world example and consider the PCIe protocol which is very popular in modern chip designs. Its specification is more than 2,000 pages of text. Comprehending its purpose and behavior alone is a gigantic task, independent of actually creating an implementation for design or verification. From a verification perspective, it is not the sheer volume of the specification that is a daunting task, but the understanding and implementation of a VIP that we would use in constructing a verification environment. Let us consider Denali's PureSpec PCIe VIP. According to company officials, there are nearly 3,000 assertions or checkers contained in this VIP. Over 1,300 static checks are checked at time 0 in the simulator to validate the overall architecture and topology that was described to the model is valid. In addition, there are an additional 1,500 dynamic checks in this VIP to ensure the protocol is being obeyed during simulation. Overall, they estimate they have invested in excess of 15 years in creating this VIP. Clearly, this is not a trivial effort and most design and verification teams could undertake these tasks without introducing serious risk into their schedule. This cost alone should justify to most, that rolling your own would be a bad idea, but that is just the initial implementation effort and not inclusive of the overall effort required to produce a proven and trusted PCIe VIP.

7.4 Evolution of Verification IP

VIP has evolved significantly in the past decade. In this section, we will review some of the history of VIP and examine where it is today.

Simple BFMs and Checkers

In the early days of register transfer language (RTL) simulation, VIP was a much simpler component than it is today. Most VIP was developed in-house, and there were not many commercial VIP providers. These early VIPs generally consisted of just simple models to generate stimulus at the transaction level. Stimulus generation was handled solely in a directed fashion, and support for error injection was very limited or non-existent. From a checking standpoint some models would include cursory protocol checking, but generally the amount of checking included in the models of the day were minimalist and inadequate. The concept of using assertion languages, libraries or functional coverage had not gained widespread acceptance yet. These early models often fit into one of three categories.

1. Source model—A model written and delivered in some simulator specific language like Verilog or VHDL. The real source language may have been some other language, and

the Verilog and VHDL may have been machine generated. Furthermore, many of these source models may have been encrypted or obfuscated. Obfuscation is the process of taking source code, changing the structure and the variable names to long, or highly unreadable tokens such that a simulator can still run it. However, it is very difficult to cogently understand its structure and operation.

2. Object model—An object model is a written in some language other than the Verilog or VHDL, most often C, and was delivered as object code that would be linked in with a simulator through one of its programming language interfaces (PLI). These models were not that common early on because Verilog simulators did not universally support PLI 1.0 and VHDL simulators lacked a common interface like PLI 1.0 which made investing in these PLIs costly and risky from a business perspective. Object models typically were controlled, and stimulus was injected by some proprietary language used to control the functionality. In the case of microprocessor models, the stimulus generally provided by compiling object code for the processor in question and then converting it to a popular format like ELF, SREC, etc.
3. Hardware model—The hardware (HW) model was a model that leveraged an existing ASIC/central processing unit (CPU) along with some custom hardware to create a simulatable model of a device. These models functioned very differently than all the software models which have been discussed so far. A HW model would use a real chip and get the stimulus from a simulation environment, replay those simulation vectors against the device, and then monitor its output. This output would be driven into the simulator for use by the test environment. There were a number of issues with the other model types that lead to the creation of HW modeling technology:

    a) Creating VIP for large chips would take a long time, and overall performance would be very low.
    b) The market for certain specialized chips was so small it did not warrant investment in creating software VIP.

c) At that time, engineers, did not feel it was feasible to generate VIP for something as complex as a microprocessor or DSP and capture its behavior accurately at the cycle level.

HW models successfully filled these gaps, and provided general purpose platforms to generate stimulus to the chips and to measure responses (or, mini chip testers). As a result, users could purchase or build their own adapter boards for parts that were unavailable in a vendor's library. Since these models used real silicon devices, there was never any doubt about the accuracy of their behavior.

Despite having a few advantages, there were a number of drawbacks to the HW model as well:

d) No visibility into the model, so debug was difficult and time consuming.
e) HW models did not scale well into the concept of a simulation farm or multiple users. Generally, you would need N HW models for N users or N simulation seats thus making it very costly.
f) Some devices could not be stepped along at very slow clock rates which created a problem that every time the simulator would interact with a device, it would reset the device and play an entire pattern just to get one clock cycle of output. This led to performance problems and cost issues associated with large amounts of memory required to buffer all the patterns.

HW models played a significant role in the 1990s, and although they are still used nowadays, their popularity has diminished.

Advent of the eVC

In the late 1990s, a company called Verisity Design (now part of Cadence Design systems) began selling VIP as eVC, named after Verisity's own proprietary 'e' hardware verification language in which the VIP was written. This gradually raised the standard for VIP functionality and increased the benefit provided to the verification engineer. Other VIP providers eventually followed suit and, in the end, resulted in a large change for the better in features, functionality, and quality of the VIP available on the market.

The advent of the eVC brought about the following important technical advancements:

- Directed random stimulus generation—Up until this point, VIP depended on directed stimulus generation which is time-consuming to write and lacking in its ability to efficiently and effectively find corner case bugs. Random generation entails generating random valid or invalid transactions atomically. These techniques have been in use for some time, but they lacked a key aspect to make the technology very useable. That technology would be a constraint solver. The addition of the constraint solver and the ability to write constraints dawned the era of directed random test generation. In the directed random approach, users can easily steer a random stream to different corners cases and easily create massive amounts of stimulus with very little effort compared with traditional approaches.
- Functional coverage—This is an objective metric designed to help verification engineers answer the question: "are we done testing?" By including functional coverage models for a protocol with the VIP, this allowed the engineers conducting verification to easily measure how much they had accomplished and what remained to be verified. These functional coverage metrics, despite being inherently useful, are starting points to conduct more comprehensive functional coverage metrics at the device level.
- Sequences and stimulus libraries—Sequences are collections of transactions that are of interest to a protocol or verification engineer. A CPU cache hit, CPU cache miss, or two red packets followed by a blue packet, are a few examples of sequences. Raising the abstraction level to generate stimulus above the transaction level adds more value to the VIP by allowing its user to quickly and easily author complex stimulus. This concept was further developed, and libraries of stimulus specific to a protocol were provided that leveraged directed random techniques and functional coverage. These facilitated higher quality

testing in reduced time compared to traditional approaches.

One can refer to [5] for more information on how these techniques have had a great positive impact on the verification.

Modern VIP

VIP architecture continues to evolve from the concepts and the functionality originally introduced by the eVC. Modern VIP development is now being largely driven by dedicated vendors who specialize in VIP products, and may not necessarily participate in the electronic design automation (EDA) market for related simulation tools, testbench languages, etc. The adoption of commercial VIP is also facilitated by the industry's migration to SystemVerilog, and to a lesser extent, standardization efforts for the 'e' language. This is a significant change from the previous generation of VIP offerings, where the primary mission of the VIP development was to support a particular EDA tool or enable a particular testbench language. In these cases, the overall VIP quality was subject to risk from a lack of investment in a secondary product from these companies, and from the lack of widespread industry use since the VIP was targeted at a single, and often proprietary verification environment. Today, nearly all providers have attempted to take best-of-breed concepts from many areas and apply them to modern VIP. Some of these useful improvements are described in the following:

1. Language agnostic interface—Modern VIP is often written in C/C++ and then linked into the simulator using various PLI's which enjoy broad support today. Providers often wrap the VIP in the native languages used by the end user like Verilog, SystemVerilog, VHDL, OpenVera, and *e* to name a few. The end result is VIP that can be used in many different languages and verification methodologies that adapts itself to the language being used. The associated quality advantage is that one VIP source can be deployed across the entire industry, spanning multiple verification environments, flows, and languages.

2. Passive mode operation—Traditionally, VIP was only used in an active mode where it was generating stimulus. Today, modern VIPs can be used both actively and passively.
3. Statically and dynamically configuration—Models were traditionally configured via settings or configuration files that were processed at simulation invocation. Modern VIP has many more features for dynamic configuration which are especially useful when combined with directed random generation techniques.
4. With the rise of transaction level models being used for architectural exploration, performance analysis, and early software development, the concepts of having common VIP from the transaction-level to the pin-level to the gate-level has become more desirable. The industry has responded largely with modern VIP that offers both transaction and pin-level interfaces.
5. Debugging modern protocols has become even more complex and the traditional methods of text log files and signal–based waveform debugging are being strained. Modern serial protocols which utilize complex encoders and scrambling algorithms further complicate the debugging of these protocols. Modern VIP is addressing this by providing transaction level debug and providing higher levels of integration into the simulation debug environments.
6. Functional coverage has become a widely accepted technology in functional verification environments. Modern VIP is embracing this by measuring protocol specific coverage and then providing ways to export or integrate this coverage information into coverage analysis and management software provided as part of a hardware verification language (HVL) or as part of the digital simulator.
7. Technical support has always been part of the VIP business, but it has taken a long time for support features to be integrated into vendor product offerings. These features usually involve capturing the protocol and VIP topology, as well as strategically capturing stimulus from critical interfaces. This is extremely valuable since it allows users to submit support requests without sending their

VIP providers their DUTs and their test benches, which often complicates getting technical support from a legal/licensing and technical standpoint.

7.5 Architecture of Modern VIP

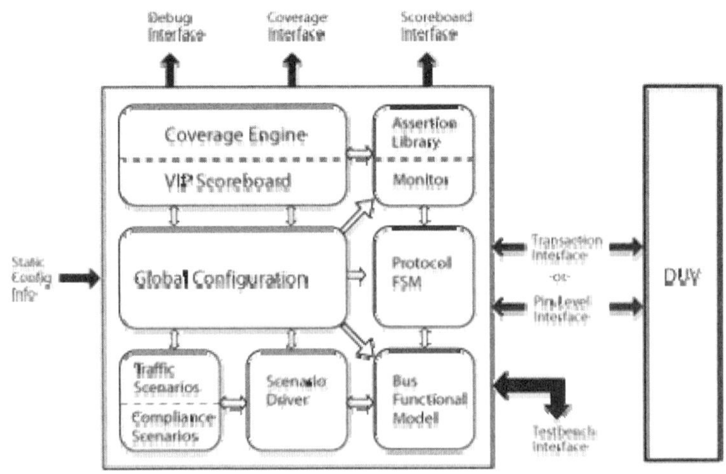

Figure 7.2: VIP Components (source: Denali Software, Inc.)

As seen in *Figure 7.2*, modern VIP has evolved significantly from early models in terms of features, functionality, and the underlying architecture. We will now explore some of the components that differentiate modern VIP from its ancestors.

Protocol FSM

Many VIP authors have discovered that there is often a lot of overlap between passive and active components or those components that just generate stimulus and check results. Commonly, large pieces of functionality must be used in both stimulus generation and checking. This functionality was often duplicated between the active and passive components and resulted in performance degradations and potentially lower quality. Modern architecture has a common protocol-finite state machine (FSM) that is responsible for managing the state of the protocol. This logic is then shared by checkers and stimulus generators for increased performance and robustness.

Bus Functional Model

The BFM is the most traditional part of modern VIP. Its purpose is to take transaction level information and convert it into pin level information. The BFM is an active component for injecting stimulus. In modern VIP, the BFM does little more than convert transactions to pin events. The more complex stimulus generation features are handled at a higher level by the scenario driver.

Protocol Monitor

The protocol monitor is another traditional aspect of VIP. Its function is to convert pin events into transactions and check the validity and sequences of transactions, along other protocol rules. Only temporal checking is done in the monitor and the data-checking is implemented by some higher level scoreboard. The protocol monitor is a purely passive component.

Scenario Driver

The scenario driver is the primary vehicle to create stimulus in modern VIP. It allows stimulus generation to be raised past the transaction-level and consider sequences of transactions generated procedurally, declaratively, or using some combination of both. Traditionally, there has always been a battle between procedural, directed, and constrained random test camps. Scenarios make this argument obsolete by allowing the best of both worlds. They can be comprised of random and directed components, and the entire scenario can be used as a random component or as a building block to create more complex scenarios. For example, let us consider the verification of a CPU. If we write a random instruction stream generator and atomically generate transactions, then there is a low probability of generating any non-infinite loop. Further, there are even lower chances of generating a nested non-infinite loop. Now, what if we create the scenario called FOR_LOOP that is similar to the following assembly pseudo-code?

```
INIT:  LOAD X, 0
START:  INCR X
.
.
.
Some_random_#_of_instructions_or_scenarios
```

.

.

```
    CMP X, LOOP_VALUE
    BNE START
```

Now we have a building block called LOOP that can be used as part of our random generation. This creates a tree of stimulus, as seen in *Figure 7.3*, which includes atomic transactions and scenarios.

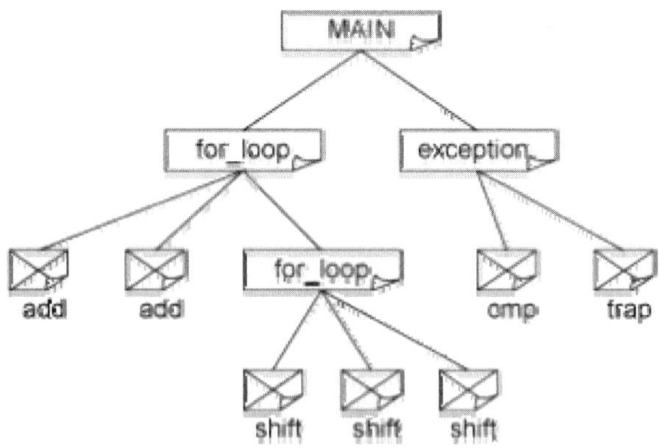

Figure 7.3: *Scenario Tree Structure Example for CPU*

Scenario Library

Modern VIP should come with libraries of scenarios that help automate protocol verification and provide the end-user with building blocks for creating more complex tests. VIP that includes non-modular and non-reusable stimulus is far less valuable than VIP that includes a library of traffic scenarios. Scenarios should cover basic usage scenarios, corner cases, and stress tests that are tailored toward the protocol in question.

Global Configuration Information

Global configuration is the means to configure the model. At first glance, this may not seem too important, but considering the overwhelming amount of configuration information and checking in a modern protocol, it is an important part of the VIP. Modern VIP often provides numerous ways to configure models ranging from graphical user interface (GUI), text files, and APIs that can be used to dynamically configure the VIP. The best-of-breed VIP provides all the configurability means needed to suit a variety of different environments and test methodologies.

Coverage Metrics

The coverage metrics is a block of the VIP that tracks coverage information for the VIP. Coverage information can be simple like whether an assertion fired during a test, or perhaps something more complex, such as measuring transaction or state machine coverage of some protocol aspect. Modern VIP includes coverage measurement and the ability to merge and aggregate coverage data or some means of interfacing with another coverage engine provided by an HVL or the simulator platform the VIP is used with. Executable test plans and compliance suites are very dependent on these coverage metrics for measuring completeness in an objective fashion.

Scoreboard Evaluator

The scoreboard is the VIP component responsible for high-level checking which includes data checking as well as higher-level temporal checking. It is likewise responsible for making sure that transactions are not dropped either unexpectedly or as scheduled. Modern VIP should provide the means to enable and disable certain aspects of score boarding as well as comprehensive hooks so end users can write additional checking or coverage metrics, or even modify the internal state of the scoreboard to deal with special cases in their testing.

7.6 Future VIP Directions

After having surveyed the current state of VIP, we can now look at the future directions for VIP. Current research trends imply that we will see several directions for VIP in the future. The first of several directions will be ESL-centric models. The second major direction of VIP evolution is the attempt to try to eliminate the human feedback loop between functional coverage and stimulus generation. New technologies are emerging to generate constrained random stimulus dynamically based on the functional coverage desired. This eliminates some of the human analysis in today's coverage-driven verification methodologies. For the purpose of this discussion, we will refer to this as smart coverage-driven VIP.

ESL VIP

Nowadays, ESL is becoming more common in mainstream designs to evaluate product architectures, measure performance, and enable early software development. A new class of VIP is, in other words, emerging for ESL. These ESL-centric VIPs are authored with performance in mind and are capable of running orders of magnitude faster than traditional VIP. They leverage new architectures for VIP design to enable increased execution speeds needed for software (SW) development. There are trade-offs in accuracy and one is unlikely to see complex stimulus generation or functional coverage measurement as part of these models. This is due to its significant impact on the resulting model's performance and on the system it is used to simulate. The current wave of fast instruction set simulators (ISS) can be considered part of this class.

Smart Coverage Driven VIP

On the other end of the spectrum, instead of optimizing for speed, there are efforts to optimize for test efficiency. In today's best-of-breed coverage driven verification methodologies, verification engineers leverage constrained random stimulus to exercise their DUTs. Coverage is then measured, and the verification engineer must analyze the coverage, adjust constrains, or write new tests to plug the coverage holes. This feedback loop is the source of much effort in today's coverage-driven verification (CDV) methodologies. New technologies, allowing intelligent constraint modification, are now emerging based on coverage to automate hitting coverage holes that eliminate the manual feedback loop of today's methodologies. This technology holds big promises in improving verification productivity.

Executable Verification Plans

The verification plan is a cornerstone of modern functional verification. Its purpose is to describe the goals of verification from a subjective and objective standpoint. Code and functional coverage were some of the first objective techniques to measure verification completeness. This concept is now evolving further with new languages designed to make engineers able to formally express entire test plans. This allows easy information collection, reporting, and provides an objective way to not only control the daily aspects of verification, but also an automated objective means to measure verification progress. With today's common industry standard protocols becoming increasingly complex, it is becoming crucial that protocols define test plans objectively in order to ensure interoperability. As protocols start to provide this level of detail, we can expect VIP to provide even richer stimulus generation and verification metrics to simplify and automate the test procedures for industry standard protocols.

How HVLs and Verification Methodologies are Affecting VIP
In the last decade, with the proliferation of verification languages like openVera, *e*, and various C++ libraries, along with the recent arrival of languages like SystemVerilog, there is an extensive array of choices that affect both VIP producers and consumers. To further complicate the problem, a plethora of the verification methodologies, both by the EDA community such as e-reuse methodology (eRM), advanced verification methodology (AVM), reference verification methodology (RVM), verification methodology manual (VMM), and object role modeling (ORM), have been created to further enrich different HVLs and their usefulness in real world application. In general, HVLs provide a more robust and feature-rich foundation to create VIP each with its own specific bells and whistles.

General purpose languages, like C/C++, provide fewer features, but provide opportunities for greater portability across different simulations and testbench languages, at the expense of some tool and language-specific bells and whistles. The smart money, in the last decade, has relied on C/C++ to develop VIP due to the laws of economics and the opportunity cost of using proprietary languages. SystemVerilog, emerging as a standard, can hopefully allow the VIP industry to leverage more advanced features and still produce a product that can serve the entire market without ties to a specific flow, toolset, or methodology. Unfortunately, SystemVerilog has so far failed to provide any tangible benefit in the real world due to the following two large problems:

1. All the vendors' implementations are missing large sets of features from the SystemVerilog language reference manual (LRM) and many of the features that are implemented simply are not robust enough to use.

2. The introduction of new proprietary verification methodologies on top of SystemVerilog like VMM and AVM provide valuable features and abstractions that verification engineers will benefit from, but each are proprietary or at a minimum not interoperable across different platforms to enable wide scale use.

7.7 Checklists for Evaluating VIP

A detailed check list for criteria to consider when choosing VIP or a VIP vendor is provided here. This list includes criteria gathered from our own professional experience, as well as from other experts in the semiconductor and EDA industries. Michael Horne of Cadence Design Systems contributed to this area and to the following checklists.

Authoring Process

Knowing how a VIP was created tells you a lot about the quality and maturity of the VIP, and the commitment of the vendor to delivering high quality products.

- Documentation—Specifications and development plans used in the VIP development process can give you a clear understanding of the vendor's attention to detail and completeness. Vendors should be able to provide copies or excerpts of their development documentation under non-disclosure agreement (NDA).
- Requirements specification—This document defines at a high-level the key features of the VIP, which portions of the standard that are implemented, and what usage features are supported such as languages, environments, etc. This spec is sometimes combined with the functional specification.
- Functional specification—This document details every aspect of the VIP functionality, including a listing of every command, parameter, and error message, and this is sometimes combined with the requirements specification.
- Verification plan—This explicitly identifies how each of the items listed in the functional specification are tested.

- User's manual—This provides detailed information on VIP usage including a detailed listing of commands, parameters, etc. The user's manual should also document example usage in all supported languages.

Development Methods
How a vendor develops its VIP directly affects the VIP quality. The development methodology can also show commitment to the VIP area. Leading VIP vendors have significant investments in infrastructure thus allowing them to produce high quality VIP with less effort.

- *Did the technology evolve or was it designed with intent?* Modeling expertise is a critical issue, and VIP should be designed with a platform concept to ensure simplicity of support and ease of updating. VIPs that evolve from inferior base technology could be abandoned by the vendor when support costs become unwieldy.
- *Was the VIP harvested?* VIP cobbled together from code blocks used on past projects or consulting engagements typically provide inferior functionality and robustness.
- *In what language is the VIP implemented?* VIP created from test bench languages typically provide the best feature set and are the most powerful. VIPs, created in languages like C, are inherently more portable and reusable than those constructed in test bench languages. Both approaches have their merits, and hybrid models that leverage the strengths of both represent the current state of the art.
- *Do they use revision control and bug tracking software?* The use of software development tools helps to make the vendor more efficient and able to address bugs – resulting in shorter development and model support time.
- *Do they perform portability testing?* The VIP should be tested across all supported simulators, operating systems, and languages for consistent operation.

Project Staffing
How a vendor staffs VIP development provides insight into the long-term commitment of the developer.

- *Where is the VIP developed?* The physical location of the developers, whether they are all in one place, and their time zone distance from your team can affect problem resolution time.
- *Is development performed in-house or outsourced?* Outsourced development can imply a lack of commitment by the vendor and introduce delays in resolving model problems.
- *How many engineers are dedicated to VIP development?* Serious VIP vendors invest substantial resources in VIP development for generating new models and evolving model features.
- *How many application engineers does the developer have and are they located geographically in areas that can benefit the user?* Leading VIP vendors have experts in specific industry standard protocols around the globe to assist their customers.
- *Protocol expertise.* Is the vendor a member of the standards body and do they participate in that forum? Leading VIP vendors are members of the standards body, consortium, or forum for a particular protocol. Do they actively participate or even lead committees? Beware! There is a surprising amount of VIP developed by organizations without proper access to the specification and the resources required resolve ambiguities and tackle the difficult corners cases in modern protocols.

Functionality and Portability

Not all VIP are created equal, since functionality and portability differences can dramatically impact your total coding effort and future flexibility.

- *How much of the protocol is supported?* Protocol depth support can vary greatly between models. Most modern protocols have optional features or extensions that may be important to your application. One has to make sure that the VIP vendor supports these optional sections or extensions to the specification.
- *Is full constrained random test supported?* Extensive support for random test methodologies ensures maximum usage flexibility and model evolution.

- *Is it partitioned logically?* Models that are partitioned into blocks that reflect how the protocol is typically used provide maximum usage flexibility. For example, can a certain layer of model be used by itself? Or can the model work both actively and passively?
- *How many languages and platforms are supported?* VIP restricted to use in a single language or platform limits your ability to migrate usage to other languages and platforms in the future.
- *Are functionality, usage model and results consistent across all languages?* VIP that supports multiple languages should have a consistent usage model and provide identical operation and results.
- *Is it compliant with emerging methodology standards?* VIP that support emerging standard methodology guidelines like AVM, eRM, RVM, and VMM save verification implementation time for end users.
- *Are all VIP available in a common library?* Mature VIP vendors tend to include all of their models in a single library for customer ease of use and financial value.
- *Do they deliver a full VIP package?* A complete VIP release will include a fully documented user's manual with examples in many languages.

Verification

Ensuring that a VIP conforms to the published specification is both crucial and non-trivial. VIP users need solid confidence that the VIP will perform to spec, or they simply will not use it.

- *Is code coverage measured on the VIP?* Vendors should be able to meet modest coverage goals of more than 95 percent line coverage with clear reasoning for the remaining 5 percent.
- *Do they verify against a golden reference?* Some standards bodies provide golden reference models that can be used to validate the VIPs core functionality to spec.
- *Do they perform portability testing?* Is the VIP tested across all supported simulators, operating systems, and languages for consistent operation?

Maturity

The stability and trustworthiness of a VIP will be a direct function of how many times the VIP has been used in the real world.

- *Has the VIP been used on a taped-out design?* While some vendors with new VIP offerings will provide early access to their models, mature VIP on the market will have a proven track record of five or more successful tape-outs.
- *Are customer references readily available?* One should ideally obtain at least three solid references from the vendor, call them, confirm how it was used, problems found, responsiveness to support requests, etc. One can find other references by checking the Verification Guild [6], DeepChip [7], ESL forum [8], Specman User Group [9] or Usenet community forums [10].

Vendor Capability

Since most vendors prefer to license their VIP using time–based models, one must confirm that a vendor is financially stable, and committed to delivering quality VIP and capable of promptly supporting the customer for the duration of the license period.

- *How long has the vendor been in the business?* What is their primary focus? VIP vendors who have focused for some time in this space, built up a real core competence in VIP, are more likely to deliver mature product and be more responsive.
- *How often are models updated?* A regular release schedule of updates is critical for newer protocols as they evolve over time. Leading providers even provide timely updates to existing protocols as they improve the feature set and robustness of a model. Frequent releases do not necessarily indicate an immature model.
- *How many engineers are dedicated to VIP support?* The vendor should dedicate a reasonable number of full-time staff members in supporting customers.
- *What technologies does the VIP include to simplify support?* Best of Breed VIP can capture a customer scenario in a trace file without needing the testcase, test bench or DUT source

code. This is a big time saver when questions arise and they always will.

- *Where are the support staff located?* Confirm the location of the support staff and evaluate whether time-zone differences will be a problem for you.
- *What is their service record like?* Try and obtain metrics from a vendor regarding issue response time, average time to resolution, etc.

7.8 Summary

VIP has become a necessary and valuable element of electronic system design, and similar to design IP, VIP is at the very heart of modern SoC methodologies. Also like design IP, the discipline of creating VIP has evolved from internal development, to mainstream EDA vendors, and now includes commercial vendors dedicated to VIP product development. Unlike design IP, the role of VIP can be expanded beyond verifying interface functionality; where complex traffic generation and coverage analysis can add significant value in chip-level verification and performance analysis.

In only a short time, we have seen the status quo of VIP grow from simple BFM functionality to include more comprehensive verification functions such as directed random stimulus generation, functional coverage, scenarios libraries, and advanced debug. While RTL level simulation is still the primary workhorse for verification, we are also beginning to see meaningful adoption of formal tools and the associated VIP, in order to augment this effort. VIP is an exciting and rapidly-evolving technology that is playing an increasingly critical role in SoC design, and it will be exciting to see that pace of development increase and where it will lead.

References

[1] Sandipan Bhanot of Knowlent Corporation, a commercial analog VIP provider. www.knowlent.com

[2] Steve Ye, "Best Practices for a Reusable Verification Environment," *EE Times*, July 12, 2004.

[3] Andrew Piziali, *Functional Verification Coverage Measurement and Analysis*, Springer ,2005.

[4] Clive Maxfield, "Max Bytes" editor's column, *EE Times*, September 16, 2004.

[5] Janick Bergeron "Chapter 3" in *Writing Testbenches: Functional Verification of HDL Models: Third Edition*, Springer, 2006.

[6] www.verificationguild.com

[7] www.deepchip.com

[8] www.ElectronicSystemLevel.com/phpBB/index.php

[9] http://tech.groups.yahoo.com/group/specman/

[10] Some useful EDA related usenet groups include: comp.lang.verilog and comp.land.vhdl.

Chapter 8

Verification of IP

Leena Singh, Senior Staff Manager and Engineer, Qualcomm, Inc.

8.1 Introduction

Nowadays, time-to-market pressures and design complexity enforces the usage of third-party intellectual property (IP) in designs. This is the case in the design and verification worlds, but problems are much more complex in the latter. The IP offered by vendors covers the entire range of functionality which provides a vast range of available configurations and features. To deliver the IP to a specific customer, it may be altered and packaged to suit unique functionality and requirements. The users, upon receipt of this IP, verify it again in order to check the IP behavior in their own system and to cover the specific customized scenarios. The key objectives are to ensure that the IP quality is met and to verify IP integration.

This chapter discusses the industry trends in IP verification and makes recommendations on how to employ the verification efforts to both promptly achieve the functional coverage goals and reduce the bugs remaining in the final system. In addition, we will discuss reusability aspects from any set of exiting verification IPs (VIPs) to verify a specific IP in a system. As seen in *Figure 8.1*, it is possible to achieve the objectives of fully-covered and verified IP in a system by creating a more robust verification environment which makes use of the latest trends, technologies, and proper third-party VIP reuse.

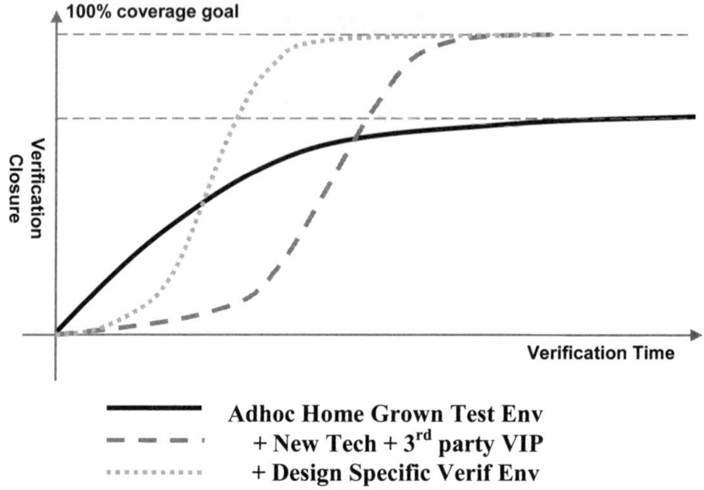

Figure 8.1: *Verification Approach Comparison Graphs*

The dark solid line in *Figure 8.1* represents home grown ad-hoc test environment being used for IP verification. At some point in verification, only through this approach, it becomes very time-consuming to reach the final goal of 100 percent coverage. The dark gray dashed line shows the adoption of newer technologies, trends, and some reusability of third-party VIP which helps you get to the final goal in a reasonable time. The light gray dotted line shows that to get to the final coverage goal much faster—along with the use of newer technologies and reusability of third-party VIPs, one also needs a dedicated well-planned verification environment that gives insight and visibility into the design.

We will use an example of a double data rate-x (DDRx)–based memory system verification environment using specific design IP from the memory controller. In a DDRx–based memory system, as shown in *Figure 8.2,* there are different aspects and challenges in putting together the whole verification environment using existing design IPs.

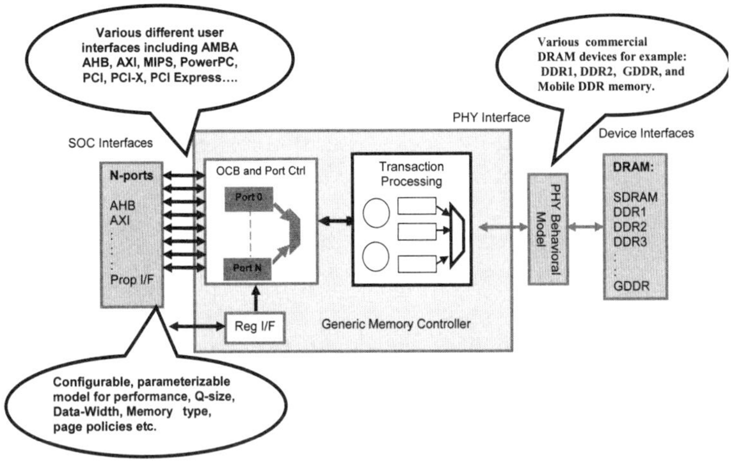

Figure 8.2: *DDRx System with Memory Controller Design IP*

This example also shows a generic memory controller as design IP that supports many different configurations. The transactions from different ports and interfaces are routed to the device based on page policy, queue management, and finite state machine (FSM) implementation in the transaction manager. The design IP, once integrated in the specific customer memory system, has to be verified for its standalone quality as well as within the complete system. One should note that verifying a block and verifying something provided as an IP is completely different. The IP user has to re-verify that a particular configuration is fully functional, even though the same block might have been fabricated in silicon for different a customer but who uses a different configuration.

8.2 Challenges in Verifying IP and Industry Trends

IP verification can involve building a complete verification environment or re-using third-party VIP, or some required components from an existing set of VIP, along with specific customized verification components. There is, at times, also a tendency to re-use existing verification testbenches from the IP vendor itself, but this depends on the reusability of the IP vendor's verification environment. This section discusses the existing industry trends followed to verify an IP and the challenges involved. The different ways of verifying an IP including creating a new verification environment, reusing existing VIP or parts of the vendor's testbench will also be addressed.

The design IP sign-off includes verification related items needed to begin planning for verification of the IP which include:

- Checklist—A detailed list of various checks that will be performed for the design IP. This is an exhaustive list of checks extracted from the functional specification of the design that covers the protocol-specific compliance checklist or any internal design specific checks. The actual checks can be implemented as assertions, end-to-end checkers, reference checkers, or formal proofs. Acquiring this checklist from the vendor gives assurance to the IP user for compliance with the functionality and allows planning for specific IP integration and system level checks. This document should describe the following aspects on each check: name, description, can be further categorized into protocol, end-2-end, reference, end of simulation, etc., and comments stating if implementation is assertion or end-to-end check code etc. If it is assertion – name of the assertion.
- Coverage numbers—The design IP provider should also provide a coverage plan and numbers obtained to the IP user. The coverage number includes both functional coverage against planned coverage and the code coverage (line, expression, FSM and toggle coverage). The IP user will use these to add any missing coverage for specific customized requirements and also to include the coverage model in their system level simulation environment. This document should describe the following:
    - Name of the coverage item
    - Sampling point at transaction boundary, every cycle, or some specific event
    - Some implementation specific comments such as the covergroup name, cover property etc.

- Can be categorized based on each interface and end-to-end coverage
- Testlist and coverage plan—The vendor will need this to add more directed tests for customized functionality or adding random tests for system level environment.

Testlist documents relevant to the example design IP (some excerpts for example are provided) is as show in *Example 8.1*. One should note that it has references to the checker list and coverage list document for specific checks or coverage items:

<table>
<tr><td colspan="6">This is a test list document for the memory controller design IP. The verification plan is discussed in a separate document. It is a living document to add more tests based on coverage and track the test status with regression. The same testlist document is to be used for simulation, formal, or FPGA-based verification.</td></tr>
<tr><td colspan="6">MC design IP description: DDR SDRAM memory controller (MC) is highly configurable memory controller which supports all flavors of industry standard DDR SDRAMs. Its superior architecture and high configurability provide high performance, smaller footprint, low latency and low power solution for various types of application, including consumer, compute and communication.</td></tr>
<tr><td colspan="6">Coverage info: Each test lists the mc functionality it represents and how it is covered.
Running Tests: All the tests listed here are based on constrained random generation parameters that have been setup and also will be run for different device/system configurations as listed on sheet2.</td></tr>
<tr><td colspan="6">Checking strategy: All tests will be automatically checked using assertions or end-to-end response checking. Separate documents will discuss the assertions and checklist. This document will map the test to what checking is being used.</td></tr>
<tr><td colspan="6">Description of fields:
Category/scenario: Main feature/functionality category to be tested
Scenario: Corresponding scenario that test uses from scenario generation
What to test: Actual testname file that exists
Description/functionality: What the test intends to do
Coverage/goal: Map to coverage item in the coverage list document e.g 3.1.2
Checking: Map to checklist in the check list document e.g. 5.1.2; there has to be at least one entry per check in checklist
Priority: Priority of implementing and running the test - related to which feature is more important
Test status: PASS, FAIL, Not Yet Implemented, Not Yet Run, Is the Coverage goal met. Will be automatically updated by regression runs
Test coverage: After regression run, how much %age coverage test contributed to. This column is also automatically updated by regression</td></tr>
<tr><th>Test Name
Scenario Name</th><th>Scenario</th><th>Description/Functionality Tested</th><th>Coverage Criteria</th><th>Checking</th><th>Test Status</th></tr>
<tr><td colspan="6">Category: Basic Read Write Tests</td></tr>
</table>

basic_csrRdWr	**REG_SINGLE_SCEN**	Basic Read Write CSR tests. Will send in single/multiple combinations of read/write on PBUS if.	1.2.1	1.1.2	Pass
basic_memRdWr	**GI_SINGLE_SCEN**	Basic Read Write Mem Tests. Will send in single/multiple combinations of read/write with variable length packets on Baseline if.	1.2.4	1.1.3	Pass
basic_memRdWr_csrRdWR	**Use above two scenarios**	Mem, CSR Read. Write combination. Will send in single/multiple combinations of read/write with variable length packets on Baseline if and on PBUS.			Pass
Category: Address Mapping					
mc_mem_col_interleaved	**GI_ADDR_INTERLEAVED_SCEN**	Multiple read writes with Address interleaved.			Pass
lo_dev_lo_bank_interleave		Low-order device/low-order bank. Long run random test.			Pass
lo_dev_mo_bank_interleave		Low-order device/mid-order bank interleaving.			Pass
psuedo_random_mem_access		Mem accesses are random, no dependency between requests. For random long tests. Algorithm chosen for address mapping makes no diff here.			Pass
Category: MRS EMRS					
MRS_EMRS_CAS_latency	**GI_MRS_EMRS**	CAS latency in MRS/EMRS: Test the effect of change in MRS/EMRS during normal operation by re-programming CAS			Pass

		latency values			
Category: Ordering					
arbitration_algo	**GI_ORDER_SCEN**	Program specific arb policy, send in traffic; dynamically change it followed by more traffic. Round-robin, strict, no-order			Pass
ref_order		Send in referesh with rd/wr, change arbitration policy, resend.			Pass
mrs_order		combine MRS with above			Pass

Example 8.1: *Test List Document for Example Design IP*

The key point in verifying design IP is that before undertaking any of the verification efforts, one should obtain all the essential verification metrics, as outlined in *Example 8.1,* from the IP vendor and both review and identify the missing information.

Challenges

There are a number of challenges in developing a verification environment and adopting an existing one in terms of the functionality it can support, languages and methodologies that can be used, and the simulators or other tools it can run on. In the verification environment, one can acquire reusable VIP components, but one would still need to create a custom environment to define the proprietary interfaces, as well as control and configure the whole system as shown in *Figure 8.3.*

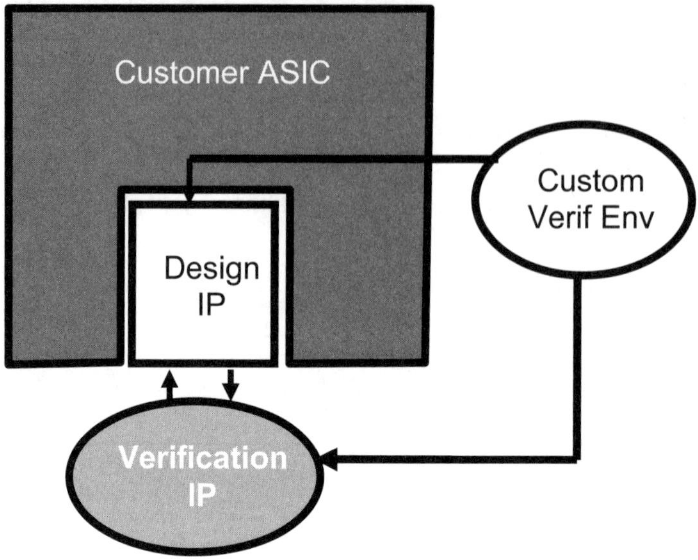

Figure 8.3: *Design and Verification IP*

Some of the challenges include the fact that:

- It is difficult to create a plan based on existing information from the IP vendors since there is no standard way of exchanging specific documents like check list, coverage plans, compliance check list, test plan, etc., which can result in missing information and multiple iterations. It takes considerable time and communication to understand customized formats and extract the right information.
- The verification components for the standard interfaces can be acquired from third-party VIP vendors, but it is unlikely that all of the interfaces for the whole system will be available from the same VIP vendor. This leads to integration challenges between different VIP vendor environments.
- The verification components, such as bus functional models (BFMs), monitors, checkers, and generators, for proprietary interfaces have to be developed. These have to exist along with other reusable VIP acquired, appropriate language and tools set has to be identified.

- The integration of design IP in the system has to be tested thoroughly which requires good insight into some of the internal interfaces. There are generally not enough hooks in the design IP for white-box testing.
- White box testing will require being able to insert monitors at internal interfaces that are available for access.
- Pre-existing testbench components may be available in languages other than the one preferred by the user.
- There are many complexities of realistic verification environments composed of multiple languages. Even using single Verilog–based components written for one simulator needs extensive testing to prove the same behavior in different simulator. There can be difference in clock cycle interpretations and the component has to workaround any of the simulator differences.
- There might be methodology-related bottlenecks in reusing existing components in your custom verification environment and, in this case, extra effort is required to build appropriate wrappers and synchronization processes.
- The system has to be configured for generation scenarios to direct test cases or random cases for specific coverage goals.
- When different VIPs are used along with a proprietary verification environment, one has to deal with licensing issues for all the different languages used. There might be some features in a specific language that are enabled by one vendor tool but not supported by others. Appropriate workarounds have to exist.
- A well-defined architecture must have global control over configuring, enabling, or driving various interfaces or blocks.

Figure 8.2 shows an acquired memory controller IP that has to be verified for all different configurations a user may want to support in a specific system environment. IP users acquire the memory controller for all ranges of configurations and features required by their own customers—which is usually a subset of ranges the original IP was developed and tested for by the IP vendor. For example in a user's environment, the IP must support different customer requirements as follows:

- DDR/DDR2/DDR3/GDDR/Mobile DDR
- Various different channel width support: 4 to 256
- Packet size support range: 128B to 64KB
- Number of banks supported: 1 to 8
- Number of ports supported: 1 to 64
- Data width: 32b, 64b, 128b or 256b
- Number of devices : 1 – 16
- Page policies : Open/Closed along with associated variants for timeouts, look-ahead logic
- Arbitration policy and support: WRR, RR, R>W, W>R.
- Application interface: Packet–based/advanced high-performance bus (AHB)/advanced eXtensible interface (AXI), etc.

This requirement poses multiple design specific challenges:

- Serial path
    - Initialization
    - Link training, recovery, etc.
    - Deskewing, elasticity buffer, etc.
    - Logical physical layer, much of data link layer
    - Formal does not buy much over simulation
- Concurrent path
    - Multiple open banks, pipelined command sequencing
    - Arbitration between different banks and ports
    - Request reordering, address hazards, coherency requirements
    - Command sequencer (transaction manager), multi-agent controller, on-chip/off-chip interconnect protocols
    - Complex power management and refresh tracking

Every customer will have different configurations and/or combinations so there is a new verification problem for every system.

The previous list of attributes can cause configuration explosion in terms of possible different combinations that have to be verified. Besides, the planning for verifying this IP requires VIPs for all standard protocol interfaces such as AXI, AHB, peripheral component interconnect (PCI)-X, PCI-Express, MIPs, DDRx interfaces checks, and proprietary configuration drivers along with stimulus generation environment with global configuration control.

8.3 Creating a Verification Environment

Main Components of a Verification Environment

The main components for any verification environment include:

- BFMs (or drivers)
    - To drive transactions to signal level protocol
- Monitors
    - Data and transaction collection
    - Events for interesting signal activities
    - Scoreboard hooks
    - Performance monitors
- Coverage
    - Coverage collectors
- Checkers
    - Reference models
    - Block-level and end-to-end checking
    - Interface protocol checkers
- Scenario generation
    - Scenario library
    - Constraint and coverage–based stimulus
    - Scenario generators
- Assertions
- Formal properties
- Performance models
- Verification documents
    - Test plan
    - Coverage plan
    - Check list
    - Verification metrics

Most of the required components can be obtained from third-party verification IP vendors, but there is still the challenge of availability and complete integration. Third-party VIP deliverables generally only include BFMs (drivers for specific standard protocols), monitors (written in standard languages) and protocol checkers (written in standard assertion languages). Such documents as an existing check list, coverage plan, verification metrics, and sets of formal proofs, may be obtained from certain design IP vendors. The other important components related to integration, stimulus generation, and global control necessary to put together the whole testbench has to be developed in-house.

Recommended Architecture

The overall effort of dealing with the complexities of putting together a verification environment can be greatly reduced by "sticking" to a well-defined architecture. Almost all of the existing known methodologies (e-reuse methodology [eRM], verification methodology manual [VMM], advanced verification methodology [AVM], etc.) or reusable verification environments follow a similar architecture; there is a further requirement of standardizing some essential components for communication and integration purpose across different languages and platforms.

An architecture for a verification environment related to the memory controller example design IP is shown in *Figure 8.4*. Every IP user needs to develop a similar verification environment to verify the design IP that is acquired from the vendor. There are existing third-party VIPs that are re-used to stimulate standard interfaces, but, the verification environment has to be developed for user specific proprietary interfaces, along with global stimulus generation, testing, and configuration control environment.

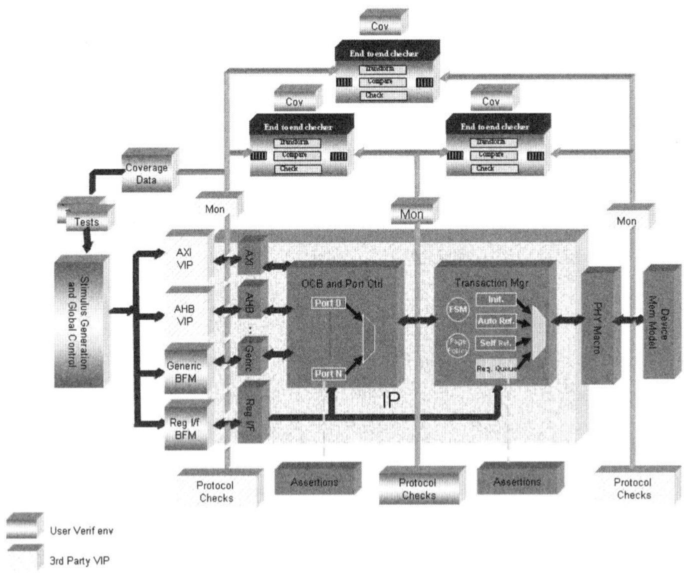

Figure 8.4: *Verification Environment Architecture*

The verification architecture utilizes available third-party VIPs for standard interfaces; one should note that the main testbench components to be developed (shown as the boxes with the lighter middle) are for proprietary interfaces, stimulus generation setup, and global control. The VIPs acquired (shown as the white boxes) can be in different verification languages with different global controls that have to be integrated in the overall verification environment. Stimulus generator infrastructure should handle and co-ordinate stimulus generation to different interfaces. The following list details the components for memory controller IP:

- Third-party VIPs—AXI, AHB, and PCI are standard protocols for which VIP can be bought from a number of third-party VIP providers. Some of these VIPs come with well-defined graphical user interfaces (GUIs) to configure them and send in relevant stimulus from a set of important scenarios already defined in the VIP generation library. Others will only provide the drivers, monitors, or protocol checkers. These VIPs can be in *e*/SV/Verilog/VHDL or any proprietary language and are mostly well tested for support for different simulators.

- BFMs—Generic BFM (for proprietary generic interfaces as shown in *Figure 8.4*) for customer's application-specific integrated circuit (ASIC) interface or register interface have to be written. These BFMs should be written using modular or reusable approach supporting plug-n-play in different configurations.
- Checkers/scoreboards—For user's specific configurations, checkers have to be created which can utilize the checkers or scoreboard provided by the VIP. In the example, end-to-end checkers that connect to the monitors at the interfaces require transformation, comparison, and checks.
- Coverage collectors—Mostly the coverage is defined in acquired VIP for specific protocol and interface. Users need to write their own coverage collectors for end-to-end checks.
- Coverage for proprietary interfaces—DDR2 channel coverage can be acquired but coverage for generic interface and some end-to-end coverage scenarios have to be written e.g. how many packets on port-N came from the AXI interface and were sent to bank 2 of the DRAM device.
- One should assume, for *Figure 8.4,* that some white-box testing visibility for the design IP is provided by the vendor at the input interface of the transaction manager.
- Stimulus generation—This needs to provide co-ordination and control of different VIPs and also write specific generation scenarios for proprietary interface.
- Tests—These have to be written, and they can be directed or constrained random tests. There has to be global control to setup and enable different components and configurations.

The main requirement from all different components and VIPs is reusability, plug-n-play, and the capability to communicate and connect to each other through well-defined interfaces. The interface communication has to be independent of the actual functionality implemented. The different components can be further categorized into different layers as shown in *Figure 8.5*.

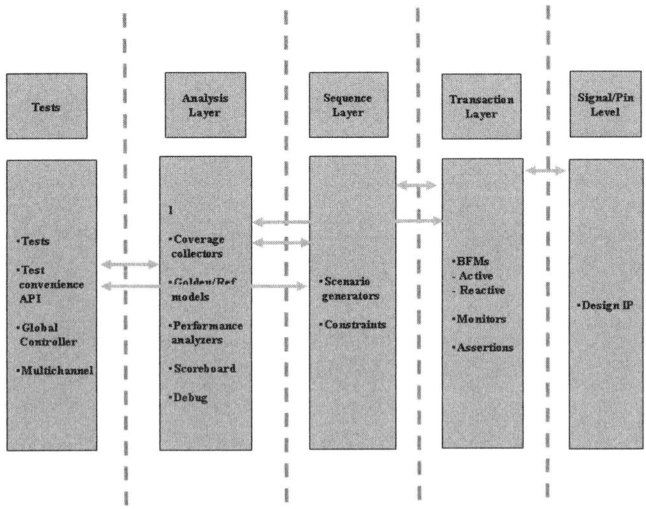

Figure 8.5: *Categorization of Verification Components in Layers*

The design IP is described at the signal/pin level, and the components that connect to the design IP fall in the transaction layer. The BFMs and drivers, at this level, convert the transactions from the higher layers to a signal–based protocol. Monitors then collect the data from these signals and pass them upward to higher level components. Assertions are written at the signal interface level plus any checks or cover points related to that signal activity.

Next, the sequence layer deals with all the scenario generation or constraint writing. For example, based on the valid constraints, generators decide a random scenario that yields valid traffic for interleaving column address for different banks and packet sizes based on specified bandwidth with varied interesting patterns of packet generation. In addition, the generator defines constraints for generating random valid sequence of commands (read, write, active, precharge, refresh, etc.) that can be sent to the device.

The analysis layer has mainly passive components that are observe and analyze the traffic, such as coverage collectors, performance analysers, debugging support end-to-end checkers, etc. Finally, the tests layer deals with global control, multi-channel, and any convenience application programming interface (API) for test writers.

This provides guidelines in terms of the communication between different components. For example the transaction layer components communicate with the design under test (DUT) on one side and one or two higher layers in the hierarchy on the other side. They do not communicate with the tests directly or, that is to say, tests do not have any "out of module" reference or signal reference at the DUT pin interface. Tests deal with scenarios and sequence of transactions making them more reusable and concise.

One should note that the communication between each of the components is using the standard channel interface implemented in the base class library. For example, the generator sends the transaction to a BFM using a blocking put (transaction) channel interface. Once there is distinction in terms of the interface and the channels, it is easier to adapt and integrate to other environments, VIPs, or components written using a different methodology. An adaptor can be written for transferring information from one interface to another. This will become clear from the implementation examples. No direct calls to procedural interfaces or signals between different components are recommended.

8.4 Example Implementation

In this section, we will look into some of the implementation examples for the environment as seen in *Figure 8.4* based on the layered approach in *Figure 8.5* and the recommended architecture. We will use SystemVerilog as the verification language in the examples.

As mentioned previously, the essential requirements for components are to use a standard communications channel and reusability. For communication channels, we can assume using the same interfaces as SystemC transaction level modeling (TLM) standard channels. The implementation of TLM in SystemVerilog can be assumed to be home grown or provided by some vendor's methodology class library or VIP environment.

Besides the channel implementation, a class library is required that provides the methodology implementation that is utilized by the user verification environment. All the methodologies will require at least the following base class implementations:

- Transaction base class—This is the base class for the transactions defined in the verification environment which can have information on transaction properties such as ID, initiator, target, and virtual tasks for printing, recording, packing, un-packing, comparing, or copying. All the transactions that are transported across different components inherit from these. In our implementation example we will call this base class a user_transaction.
- Component base class—All the verification components inherit from this base class. This provides information on component properties like name, ID, connectivity, and synchronization tasks. There can also be virtual tasks defined for starting, stopping, and suspending the component. In addition, all of the following components that inherit from this can have their own base class that is specific to particular system functionality or general functionality e.g., checker base class has an implementation of functionality such as compare and transform that can be overridden. In our implementation examples, we will call this base class the use_verification_component and such base classes, as BFM, monitor, checker, and coverage, can be derived from it.
- Reporting utility—This is required to provide a common format for reporting/logging and verbosity control for debugging. The main functionality is the control on verbosity of the messages and the action associated with each message; this will be called user_report_message in our implementation example.
- Verification environment base class—The environment base class is required to introduce the concept of phases for simulation run. All the components are constructed,

configured, connected, and executed based on the functionality provided in the environment base class. The top level environment inherits from this class and establishes connectivity and phases for simulations runs. We will call this base class user_verification_environment in our implementation example.

This class library, similar to channel implementation, can be taken from any vendor-provided library, open source library, or can be completely home-grown based on the requirements. If a vendor specific class library is chosen, it is always recommended to layer it under the user required class library to give the ability to switch between different implementations from different vendors seamlessly and add in and select only specific functionality.

In the following sub-sections, we will look at example code for some important verification components in each of the different layers. The example code provides relevant excerpts from an actual implementation, not the entire code, and assumes that the reader has familiarity with SystemVerilog.

Transactions

The data item (transaction) flows across different components, and the transaction carries the physical fields (those actually get driven on the interface) as well as meta fields (control and decision-related). There can be different transactions in the verification environment implementation, and *Example 8.2* is related to transactions at a generic BFM interface. The specific standard interface transactions can be derived from this base transaction. One should note that this is inherited from the *user_transaction* which is provided as the base class in the library.

The transaction fields here should be self-explanatory. Note that besides the main data transaction flow, there is also a configuration transaction that carries information about the components configuration and parameters for test scenarios.

```
typedef enum {SCEN_INIT, SCEN_RAND, SCEN_SINGLE} scenario_tr_t;
class scenario_request extends user_transaction;

        scenario_tr_t scen_type;
        rand int count;
        rand int sim_time;
        //Target BFM, if there are multiple targets
        tlm_blocking_put_if#(generic_request) bfm_target;
         //other input scenario knobs
         ..........................................
         //Define constraints on scenario fields
endclass
```

```
typedef enum {READ, WRITE, AREF, PRE, MRS} trans_cmd;
typedef enum {ERROR, SUCCESS} trans_response;
class generic_request extends user_transaction;
        rand trans_cmd cmd;
        rand trans_response rsp;
        rand bit[ID_WIDTH-1 : 0] req_id;
        rand bit[ADDR_WIDTH-1 : 0] addr;
        rand bit[DATA_WIDTH-1 : 0] payload_data;
        rand bit[ID_WIDTH-1 : 0] payload_id

        function print();.......
        endfunction

        constraint payload_size {payload_data <=256,  payload_data >= 0}

        constraint addr_cl {if(DEV_CFG_BANK == 0)
            addr[ROW+COL+BA_WIDTH - 1 : ROW+COL] <= 3;
            else
            addr[ROW+COL+BA_WIDTH - 1 : ROW+COL] <= 7;
            }
        ..........................................
endclass
```

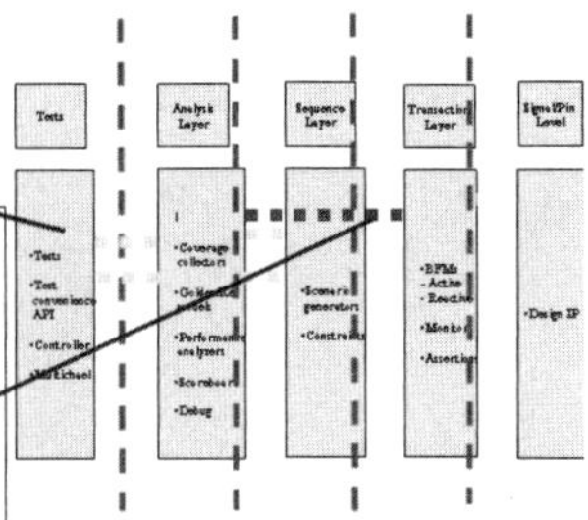

Example 8.2: *Transactions Code Excerpts*

Transaction Layer Component

The transaction layer components deal with the pin interface, and the pin interface layer is again abstracted using the "interface" definition. The design IP pin interface is described in the "interface" block. This interface is passed as a port list to the transaction layer components or connected as a virtual interface. It keeps the transaction layer components independent of the pin level details, and the connection is established in the top level connectivity file. One should be aware that the generic_if is declared as a virtual interface in the BFM definition, whose pin directions are defined by the modports.

An example template of baseline BFM implementation is also provided in the diagram. The main points that should be noted are the connectivity to the design IP through a virtual interface, usage of standard channel interface (tlm_get_if) and run task. The "run" task implements the main protocol: it gets the transactions through the standard channel interface (req_port.get [generic_request]), and it drives the pins. It also composes the generic_response transaction and makes it available to all connected components. The component is inherited from *user_verification_compent* base class.

The structure and flow shown in *Example 8.3* is for a simpler non-pipelined protocol. For a pipelined interface, the main task "run" can fork off two parallel tasks to handle request and data. The communication between two pipelines requires a TLM first in/first out (FIFO) whose depth is equivalent to the amount of request commands that can be queued.

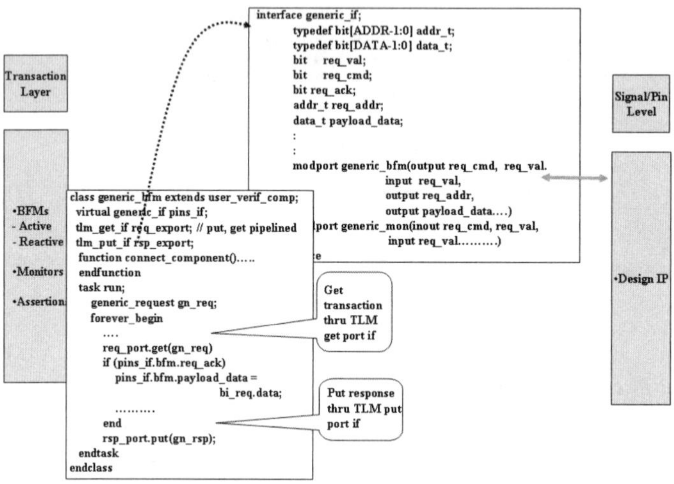

Example 8.3: *Generic Interface and BFM Code Excerpts*

Assertions

Assertions are written as a separate module and are bound to the corresponding design through bind statement. Bind keeps the interface assertions independent of the design. It is recommended that module and port connectivity (instead of interface) be used to define assertions since it provides reusability of the same assertions and files for formal verification. Formal has restrictions in terms of SystemVerilog language construct supported, so it is best to choose the common denominator and spend effort writing assertions only once that work in both simulation and formal environments.

```
module generic_protocol_chk(
 //generic interface signals - don't use interface
);
 // Any verilog code in synthesizable form
  ......................
  generate
    for(k=0; k<1; k++)  begin
      property gi_push_req_q_empty_at_reset_check @(posedge clk)
          $rose(reset_n) |-> !req_sts[k];
      endproperty

      gi_push_req_q_empty_at_reset_check_label:
         assert property(gi_push_req_q_empty_at_reset_check)
            else
               user_report_message("error", "Req queue not empty after reset");
      ...................
    end
  endgenerate
   ............
endmodule

bind generic_rtl_src generic_protocol_chk generic_protocol_chk0(//ports);
```

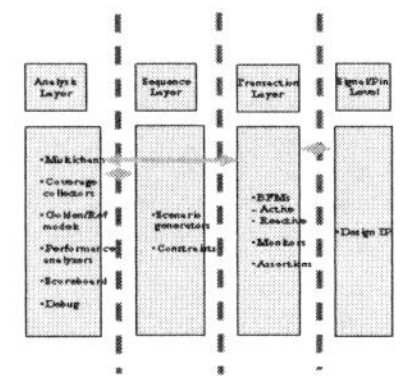

Example 8.4: *Assertion Code Excerpts*

Sequence Layer

The sequence layer implements the generation scenarios for different interfaces, and it is also where constraints for different kind of scenarios are built. Let us note that this component derives from the *base_scen_gen* class that takes care of all the connectivity related and other common functionalities. For a generic interface, this generator implements a few interesting combinations of transaction sequences with constraints. For example, if the generator is connected to 64 ports (meaning 64 instantiations of one BFM for the driver interface), the *rand_ports_full* scenario can randomly distribute the transactions on all 64 ports. Tests can choose between one and a combination of these scenarios, further constraining these if required.

base_scen_gen defines connectivity and all common tasks . It also runs a thread that starts virtual task generate_scenarios defined in each of the scenarios.

```
class GENERIC_RAND_SCEN extends base_scen_gen;
  function new(.........)
  endfunction
//virtual task generate_scenario called in base "run"
task generate_scenario(scenario_request rq);
    .......
    generic_req my_req;
    for (i=0; I<rq.count; i++) begin
        my_req = new();
        assert(my_req.randomize() with {req_id == i; payload_id == req_id;});
        rq.bfm_target.put(my_req);
    end
  endtask
endclass

class GENERIC_SINGLE_SCEN extends base_scen_gen;
//virtual task generate_scenario called in base "run"
task generate_scenario(scenario_request rq);
    .......
    generic_req my_req;
    for (i=0; I<rq.count; i++) begin
        my_req = new();
        my_req.tr_type = SCEN_SINGLE;
        my_req.count = 1;
        rq.bfm_target.put(my_req);
    end
  endtask
endclass
```

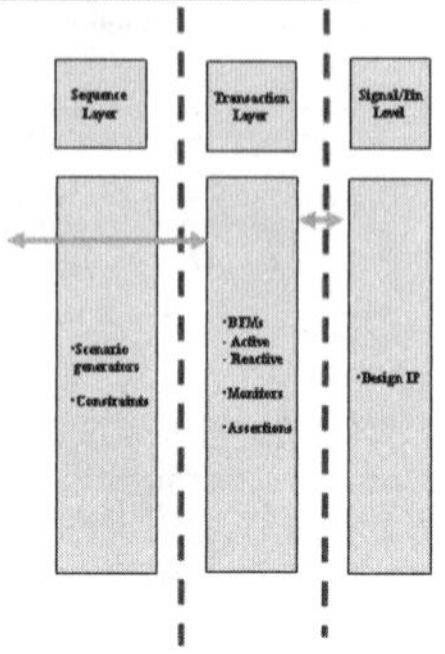

Send concurrently to n instances of BFMs or any combination, Connectivity in base_scen_gen

Example 8.5: *Scenario Code Excerpts*

The example scenarios shown in the example code extend from the base scenario, get the information on parameters/knobs for generating sequence of transactions from the *scenario_request* transaction, and create specific sequence of transactions on the required interfaces.

Coverage

For each interface, there is a coverage collector attached to the monitor or interface and these implement the corresponding coverage items. Run thread gets the transaction from the FIFO and samples the covergroups/coverpoints that are defined for the interface. This is the example of data/transaction related coverage. The information on current coverage numbers can be fed back to the generation.

Covergae base class inherits from user_verification_component, defines common tasks and connectivity

```
class generic_coverage extends user_coverage_base;
 ............
 local tlm_fifo #(generic_request) req_fifo;
 local bit is_covered;
 local bit [31:0] pl;
 local bit cmd_typ;
 covergroup payload_cov1;
  payload : coverpoint pl
  {
    bins low = {[0:14]};
    bins high = {[15:31]};
  }
  typ: coverpoint cmd_typ;
    pXt: cross payload, cmd_typ;
 endgroup
......................
task run;
   generic_request t;
   forever_begin
     req_fifo.get(t);
     pl = t.payload_data;
     cmd_typ = t.cmd;
     payload_cov1.sample;
     payload_cov2.sample;
     if( payload_cov1.get_inst_coverage > 20 ) begin
     is_covered = 1;
  end
 endtask
......
endclass
```

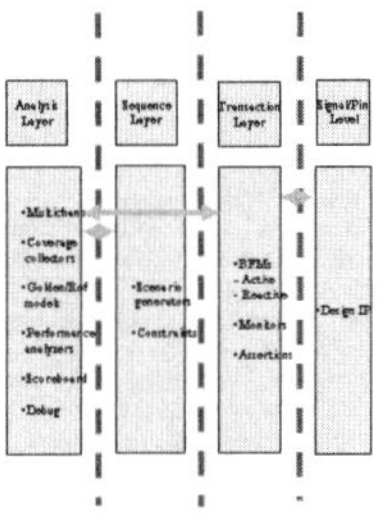

Example 8.6: Coverage Code Excerpts

The signal activity-related coverage can be best written using the cover property. While planning for the coverage items in a coverage document, the categorization should be done based on which coverage item is implemented as "cover property" or "covergroup". There might be some covergroups that are dependent on some complex sequence of signal transitions, which requires combination on sampling event generated by "sequence" written in the "virtual interface" along with the covergroups as defined earlier.

Test Layer

The test layer is where tests are implemented, and the tests can pick any combination of scenarios. If the test does not pick any scenario, a default scenario can be run which randomly picks all the defined scenarios for a chosen simulation time. Tests can either populate or send the transactions or they can pick scenarios from the scenario layer; there are usually also convenience functions such as an API.

Tests decide on the connectivity to the generator so that it can be dynamically modified based on the requirements. Each test inherits from the base test component, picks the scenarios and generators it wants to connect to, and sends in the stimulus as implemented in *run_test* task. Tests can further constrain the scenarios as shown.

```
class my_test extends user_test_comp_base;
 GENERIC_RANDOM_SCEN sc1;
 REGIF_RANDOM_SCEN sc2;
 ......
 task run_test;
    scenario_request mc_req;
    user_report_message("test", "Starting Test........");
   //Select random scenario to send 10 random transactions
    mc_req = new;
    mc_req.scen_type = SCEN_INIT;
    mc_req.bfm_target = bfm.req_export;
    assert( mc_req.randomize() with {count == 10;};
     sc1.mc_scen_req_exp.put(mc_req);
   //Select another random scenario
     assert(mc_req.randomize());
     sc2.mc_scen_req_exp.put(mc_req);
 endtask
```

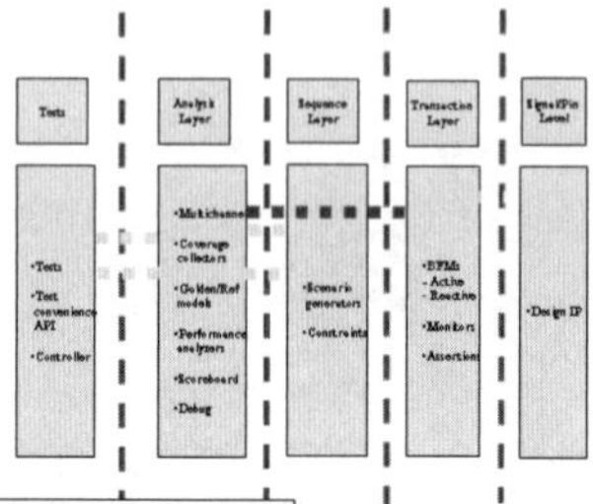

OR

```
task run_test;
      .......
      randcase
            10: sc2.mc_scen_req_exp.put(mc_req);
            40: sc1.mc_scen_req_exp.put(mc_req);
      endcase
endclass
```

Example 8.7: *Test Code Excerpts*

Top-Level Environment Connectivity

The top-level environment is simply the connectivity of the different verification components and the design through the interface. One should note that the top "env" inherits from user_verification_environment.

```
class env extends user_verification_environment
  my_test t;
  user_report_formatter rf;
  function new(virtual generic_if g_if);
    rf = new;
    set_report_formatter(rf);
    t = new(g_if,"top");
    user_verbosity_level(5);
  endfunction
  task execute;
    t.execute(); //Executes the test
  endtask
endclass
```

env brings in concept of phases

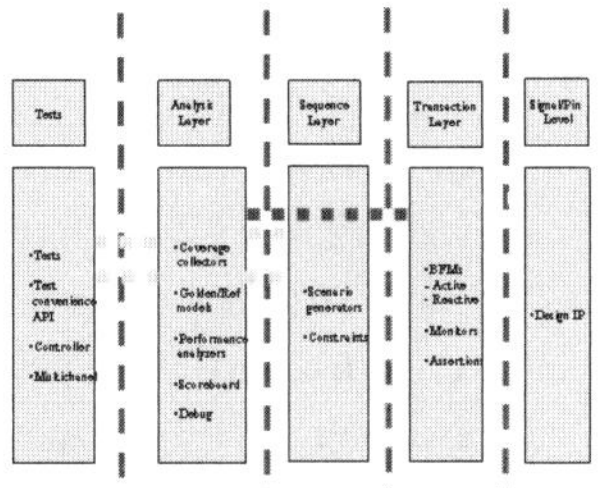


```
program mc_pgm_top(interface b_if);
  import user_pkg::*;
  import ip_pkg::*;
  import tb_pkg::*;

  env mc_env = new(b_if);
  initial begin
    mc_env.do_test(); //calls diff phases
    $finish;
  end
endprogram
```

```
module mc_tb_top;
  // Host interface instantiation.
  mc_generic_if generic_if();
  // MC to DDR2-PHY interface instantiation
  mc_ddr2_if ddr2_if(clk);
  // Design Instantiation.
  DDR2_mc_top mc_top (generic_if, ddr2_if );
  // testbench program
  mc_pgm_top pgm_top ( generic_if );
endmodule
```

Example 8.8: *Top-Level Connectivity Code Excerpts*

8.5 Following Standards

As discussed previously, there are many complexities involved in the verification of design IP. The industry wide practice of following standards while developing verification IPs is essential to boosting verification productivity, and quality enhancement, and to overcome challenges in integrating and building a system verification environment. Following standards is equally important when defining a methodology and architecture for IP verification. There are various dimensions where standardization can be applied, and the most important of these are the standards for exchanging information between different components and VIPs. Having one standard language and methodology of implementation can also help, but it is difficult to stick to one standard in terms of language, methodology, and techniques that can be followed. The choice of language and methodology is mostly based on the developer's preferred language and system requirements.

The main benefits of following standards are described here:

- Ease of integration—If all the VIPs acquired by the user for standard protocols or from the design IP vendor follow similar standards, the planning for the verification environment becomes much easier. This sets an expectation of what APIs will be provided and how they have to be connected together. In addition, users can independently develop their own additional components following the same standards.
- Seamless plug-n-play capability—The same, exact communication mechanism has to be followed in order to achieve this capability. If different components provide various forms of communications, such as signal–based, transaction–based, direct procedural calls, event–based, it becomes cumbersome to integrate new components and devise the ways to synchronize and transfer information between them.
- Documentation reliability and exchange—A standard way of exchanging documentation eliminates any chance of miscommunication or missing information. For better reliability, a standard checklist defining all the items for IP exchange is then required.
- Ease of defining global control—Global configuration and control for a fully-integrated environment is easier to implement if all the components are based on some standard. GUIs can be used by tool vendors or developed by the user that enables global control and co-ordination.
- Ease of extension of behavior—Extending the behavior of existing verification components is always required. Different vendors provide varied techniques for this which involves either callbacks or hooks in the components. Another approach is called aspect oriented programming.
- Categorization—Following a standard class library allows the whole environment to be categorized which improves modularization and scalability. The default behavior for each component category in each of its phases can be implemented and shared in the form of a common class library.

Some of the practices that are now being followed are listed here:

- TLM Interfaces (www.systemc.org)—Standard transaction level model interfaces (from Open SystemC Initiative [OSCI]) define the standard way of communication between different components. These can be followed irrespective of the language chosen for implementation. A large number of vendor methodologies have also started to incorporate these in their implementation. All the examples in this chapter follow the same standards, and multi-language support can be provided by the same interfaces.
- SPIRIT Consortium (www.spiritconsortium.org) provides specifications to help standardize a single representation of design meta-data, and it also provides a complete API for meta-data exchange and database querying. The verification environment configuration description, instance views, monitor insertion, connection point to the design IP, and register configuration, can be captured in SPIRIT-enabled meta-data. This can be used as a standard way to exchange design/verification meta-data that is language-neutral.
- VSIA QIP Metric (now transferred to the IEEE) provides a vendor qualification worksheet to get the quality IP scores that serve as a good set of guidelines. It has a list of metrics to access verification quality for the purpose of VIP development and integration.

It is important to further refine the existing standards and define new ones in order to improve the verification reuse strategy, thus allowing a smoother integration process and minimizing the effort involved in the verification of acquired design IPs. Once these processes are standardized, it is easier to tackle the multi-language, simulator-related complexities. The language transformation, global control, and integration challenges can be tackled easily by the user or tool vendors if the previously mentioned standards and practices are in place and are strictly followed. For example, conversion of a component with rich functionality from one language to another can be quite cumbersome, but transferring information between components written in different languages can be part of a standard TLM interfaces and tool support provided by simulators. The use of SPIRIT standards for defining global configuration allows an independent GUI–based control.

References

[1] Advanced Verification Methodology Cookbook (AVM), Version 2.0, Mentor Graphics.
[2] eReuse Methodology (eRM) Developer Manual, Cadence Designs Systems.
[3] Leena Singh, "Verification IP Reuse: Crossing the Language Barrier," Efrat Shneydor DVCon 2006.
[4] Leena Singh "Building Verification IP for Reusability: Based on SystemVerilog and TLM Standards," DesignCon 2007.
[5] Leena Singh, Leonard Drucker, and Neyaz Khan, *Advanced Verification Techniques: A SystemC–Based Approach for SuccessfulTapeout*, Springer, 2004.
[6] Alistair Bruce, M. M. Kamal Hashmi, Andrew Nightingale, Steve Beavis, Nizar Romdhane, Christopher K. Lennard, "Maintaining Consistency between SystemC and RTL System Designs," Design Automation Conference, 2006.

Chapter 9

IP Quality

Kathy Werner, IP Strategy and Business Manager, Freescale Semiconductor; President, VSI Alliance

9.1 Introduction

In a perfect IP reuse world, one would simply connect a reusable IP block, validate the interface with the rest of the system, and successfully produce fully functional silicon. Unfortunately, this rarely happens in practice, and mismatched expectations for deliverables, versioning issues, poorly documented interfaces and functions—not to mention improper functionality—are all common issues encountered when reusing IP. The pressing question is: how does one assess IP quality and identify the potential reuse pitfalls before it is too late?

Clearly, not all IP is created equal. The actual process of successfully evaluating and integrating the IP to maximize its reuse benefits is as important as the decision to reuse. However, as anyone who has looked for IP has noted, there are many in-house and third-party IP suppliers in the marketplace, and more are entering the fray every day. Functionality simply needs to be the first selection criteria, but when two or more similar IPs are available, the selection criterion quickly becomes more difficult. In the past, IP selection may have been solely based on the purchase price; however, with the rising costs of respins, the risks associated with integrating IP through this decision-making process far outweigh any cost savings based solely on the purchase price. There are other approaches that are used to evaluate IP that meet with varying degrees of success.

The most common method used to evaluate reusable IP has been solely through internal review. This is a resource-intensive process which often requires several months of effort. More often that not, legal council needs to be included in the process for third party IP to generate or evaluate non-disclosure agreements (NDAs) which adds to the total evaluation time. Another method used in the past was the OpenMORE[1] spreadsheet, based on the reuse methodology manual (RMM).[2] Unfortunately, while this method was publicly available as a first step toward measuring and communicating IP quality, actual usage showed that much of the metric was open to interpretation and thus limited its own success. Finally, some companies have developed internal reuse standards for their own development efforts and subsequently attempted to get their vendors to deliver IP that conform to those proprietary standards. Unfortunately, this approach tends to be very company-specific, and it generally addresses a much broader space than could reasonably be expected for third-party compliance. These standards are truly geared towards enabling internal reuse, are typically more complex than needed and tend to obscure, through sheer volume, the important quality reuse measures.

Using the previously described approaches in IP selection resulted in great variation in the actual experience of using the IP depending on the provider, end application consumer, and even individual integrator. None of these methods provide a consistent mechanism to communicate an IP's suitability to purpose, particularly between commercial IP providers and end customers.

9.2 Internal IP Reuse Quality

Many chip designers will affirm that they have been reusing code for years. If one probes a little deeper, one will see that this reuse consists of taking a similar block and using it as a starting point or example for the next-generation design. While using existing designs to get a running start may be an admirable practice, it is hardly reuse. The original design is tweaked, enhanced, and it is the rare designer who can resist putting their own spin on the implementation. Some of the code may be recycled into something

---

[1] OpenMore opens more SoC IP doors – EETimes 05/15/2000. http://www.eetimes.com/story/OEG20000515S0039

[2] Michael Keating, Pierre Bricaud, *Reuse Methodology Manual for System-On-A-Chip Designs*. Springer 1998

recognizable from the original design, but the time spent in implementation and particularly in verification, not to mention the risks in fabricating what is essentially a new IP block, generally work against the goals and benefits of true reuse.

Even minor changes to a design can dramatically increase the risk when integrating the design into a systems-on-chip (SoC), especially when made by someone other than the original designer. Seemingly innocuous modifications may have an impact on an unintended portion of the design, and will most likely not be caught unless a comprehensive verification suite is completely run. With verification estimated at three times the effort of design, these activities can seriously impact final deliverable schedules. The ultimate goal of design reuse is to integrate a "known good" piece of IP and verify the interfaces in the SoC, thereby reducing the risk and improving the time to product and revenue. With this said, how do you know that a design is good?

Internal Reuse Practices

Simplistically, good designs are a product of good design practices from IP conception through customer delivery and support. Considerations must be made by the design team during development to ensure downstream tool compatibility, general IP robustness, as well as what the end user needs to ensure successful integration and reuse of the IP without any changes.

To attain black-box reuse, the IP must first start with detailed system requirements, from which a solid design specification is derived. The system requirements should be cross-referenced to the design and verification specifications to ensure that all requirements are correctly implemented and validated.

It is important to consider the potential system targets when specifying and designing IP for multiple black-box reuses, so the marketing and architecture teams must have a close linkage through the requirement and specification phases. Marketing requirements for similar functionality from several end customers need to be balanced. Tradeoffs may need to be made for size, cost, and functionality as high performance and feature rich parts may not be desirable for cost sensitive applications. While the ultimate goal would be to have a single part address all end customer needs, the reality may not be quite as straightforward. Configurability and building block design styles are two approaches to use in this situation.

Configurable IPs are quite useful when different processors and/or bus architectures must be supported. Bus widths, bus architectures, memory types and sizes, and even architectural implementations such as interrupt controllers can be configurable. The portability and the savings realized when integrating the IP in multiple systems more than make up for the additional overhead of the initial development and verification effort. For example, it is quite easy to use text macros or generics to specify bus widths as opposed to using a hard coded value. This allows the block to be retargeted by simply changing the value in a single location as opposed to doing an error prone global search and replace.

```
'define DATA_WORD 8
'define DATA_LONG (4 * 'DATA_WORD)

instead of:

'define DATA_WORD 8
'define DATA_LONG 32
```

Figure 9.1: *Verilog Text Macros Usage*

Designing IP using a building block approach allows unwanted functionality to easily be removed from the design. Basically, this approach breaks the design up into small, logical groupings which can be removed or modified much more readily than when interspersed with additional functions. Depending on the design, these types of blocks may include small functions such as interrupt controllers, bus interface units, universal asynchronous receiver/transmitters (UARTs), or, for a highly integrated block, more complex functions such as universal serial bus (USB) or processor blocks. These functions can easily be disabled using `ifdef directives in Verilog or configuration files in VHDL for cost-sensitive or reduced functionality end applications.

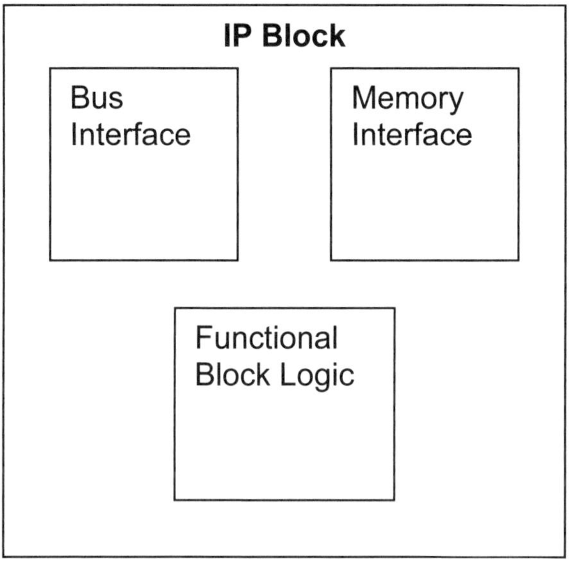

Figure 9.2: *Block Partitioning Example*

The constructs and coding styles used also contribute to IP reusability. For example, the use of combinational feedback loops will make the use of formal verification tools and cycle–based simulators much more difficult. Obviously, the handling of clocks and resets can have a huge impact when the IP is integrated into the SoC. The IP state must be predictable during and after reset, including the direction of bi-directional pins and the state of logic in multiple clock domains. Any required clock gating or generated clocks should be separately partitioned to ease the SoC design for testability (DFT) strategy and implementation. Legal code constructs may result in varying simulation results based on the tool and even the design platform used.

The following illustrates legal and common constructs that may be interpreted differently from the user's intent:

```
wire [63:0] wire64bit;
wire [ 7:0] wire8bit;

// assigns integer (32-bit) to 8-bit net
assign wire8bit = 1;

// assigns integer (32-bit) to 64-bit net
assign wire64bit = 1;

// assigns unsized number to 64-bit net
assign wire64bit = 'b1;

// assigns 8-bit expression to 64-bit net
assign wire64bit = wire8bit;

// bit-wise AND of 8-bit and 64-bit nets
assign wire64bit = wire64bit & wire8bit;
```

Figure 9.3: *Code Example of Potential Simulation Errors*

Verification is arguably the single most important element to ensure that IP can be reused in a black-box fashion. Formal verification methodology, coverage metrics, and DFT must be considered to ensure verification completeness. An additional goal is to facilitate the reuse of the verification environment delivered by the IP provider by the end user as much as possible in the integration and chip level verification of the end product.

The verification environment comprises testbenches, models, scripts, and other deliverables and best practices should be followed for their development and implementation. For example, clock generation must not result in rounding or truncating. This is an obvious concept, but one that is disregarded fairly regularly in practice. Many of the reuse concepts of importance in the development effort may not be readily visible to the end users. They are primarily interested in how to actually integrate and use the IP without needing the original designer or having in-depth knowledge of the design. While generating documentation is not generally high on the list of activities on which engineers want to work, detailed integration, hardware and programmer reference manuals, as well as errata or release notes are invaluable.

High-quality integration manuals should contain all the information that the end users need to know and understand in order to implement the IP in their SoC. This should include information such as interface descriptions with functionality, polarity, and timing, instantiation instructions that detail any special considerations or additional logic required, naming conventions that were followed, and a listing of deliverables provided in the IP packet, along with any applicable information, such as the file size and description.

A separate integration manual or a section of the hardware reference manual should describe the IP functionality including all of its operating modes. Timing diagrams illustrating the key signal and functional relationships is vitally important and should also indicate any technology dependencies. Finally, an architectural overview is helpful to illustrate the basic IP structure.

For IP that contain a programmable instruction set, a programmers' reference manual must also be available which should contain all the specified information on how to program the device, including instruction sets, sequencing, and mode set-up. Any configurable control registers must also be defined, and their usage must be specified. A complete register map that lists all of the user-accessible registers including the bit fields and their usage must also be available.

The release of errata notes is last among the documentation requirements. This document should indicate any differences or updates from previous released IP versions, as well as list any known, unresolved, or outstanding issues with the IP. If available, a solution or means for the user to workaround the issue should also be detailed.

Internal Reuse Certification

The end user of an internally-developed reusable IP must have a high level of confidence in the suitability, functionality, and quality of the IP to resist the temptation to tweak or otherwise recycle the design. Objective means to communicate and measure the reusability of IP must be implemented to result in reduced integration and SoC verification time, and attain the benefits of black box reuse. Several factors must be considered to attain confidence in reusing a piece of IP.

A universally accepted and adopted set of reuse standards must be implemented in-house to support these practices and a means to quantify the conformance to these standards must be implemented. Conformance must not only include compliance to the individual standards, for example the naming convention for active low signals, but must also address consistency of the standard across all views. In other words, signal A in the register transfer language (RTL) code must also be called signal A in the graphic data system II (GDSII) and in all of the documentation.

Automation support should be provided for internal reuse standards, and their certification should be adopted and widely used within a company. The list of explicit requirements can be quite large as can be inferred by the discussion of internal reuse practices in the previous section. Manual certification compliance to these standards can place an onerous burden on the IP developers without tooling to support the effort. There are commercially-available tools, such as lint checkers, that can be used to help this process and that can be customized for the developing organizations-specific IP standards requirements.

Internal Reuse Data Management

Designing IP for reuse serves no purpose if other users are not aware of its existence. At a minimum, a means to catalog the IP and its pertinent information must be implemented and readily visible and accessible throughout the company. This should include a unique IP name, as well as a brief description of its functionality and pertinent metadata or descriptive information about the IP. Information should also be included on how to obtain the IP itself. Several procedures are acceptable and widely used: the user may be able to directly download the data from the catalog site, a pointer may be included to link to the IP data archival location, or contact information for the IP owner may simply be provided.

IP Name
Description
Security Level
Revision
Owner Name
Owner Contact Info
Language
Hardness Level
Bus Interface
Certification Grade
Technology
Gate Count

Figure 9.4: *Metadata Example*

Users interested in utilizing reusable IP must have a mechanism to evaluate the IP contained in this catalog for blocks that meet their functional criteria and be assured that the IP package that they receive is complete and consistent. Users should be able to either browse the available IP, or search for IP based on its metadata to start their search for an IP suitable for their end application. The implementation of a standard catalog hierarchy greatly assists the end users in their search for reusable IP.

- Analog/Mixed Signal
  - Amplifiers and Filters
    - Op Amp
    - Filters
  - Tx/Rx
  - Mod/Demod
  - Data Converters
- Digital
  - Cores
    - Std MPU
    - Std DSP
  - Peripherals
    - Arithmetic, Mathematic, and Bit Manipulation
      - System Management
        - Clock Management
        - Interrupt Controller
        - Performance Monitor
        - Timers
        - Keyboard Controller
      - Receiver/Transmitter
      - Bus, Interface, and Control
- Memory
  - NVM
  - SRAM

Figure 9.5: *Catalog Hierarchy Example*

One should consider several categories of metadata that must be maintained with the IP to aid in search criteria, the first of which should be general, common information and included for all IP. This category should include the following:

- IP name
- Release/revision identification
- Quality or certification level
- IP ownership information: either organization or individual, and contact data
- Applicable technical properties: language or technology, bus interfaces, etc.

- Available deliverable views

Different types of IP, for example soft IP and hardened IP, will have additional metadata information specific to them that should be communicated. Soft IP may list the implementation language, that is Verilog, VHDL, or a proprietary format, and hardened IP must list the process technology. In addition, hardened IP may include such data as the number of metal layers and yield information. Both soft and hardened IP may include area and performance information, but for soft IP, these numbers will be preliminary since the end user has some control over the final implementation, whereas these results for hardened IP are fixed. Additional metadata specific to sub-categories of IP types may be identified and standardized upon within the company. Here, the consistency and availability of the IP type information throughout the company is of primary concern.

In managing IP, one should also consider the security and accessibility of the information. For global companies, export restrictions may limit the sharing of IP to specific locations or personnel. As a result, access control should be considered when giving rights to the actual design data.

Lastly, perhaps the most important aspect to consider for internal reuse IP management is the tracking and communication of bugs and issues associated with the IP. Users need a mechanism with which to report issues with the IP. These may be related to functionality, a specific deliverable such as documentation or a supporting script, or simply a clarification request. Ideally, once an issue is reported and a workaround has been either fixed or identified, this solution will be available to the entire user community. These bugs and resolutions need to be maintained or linked with the catalog since this information may also factor in future decisions to reuse.

9.3 Third-Party IP Quality

Third-party IP reuse has many of the same considerations as discussed in the internal reuse section. Standard quality development techniques need to be followed, and the same care needs to be taken to develop and deliver the documentation and views associated with the IP. However, while generic reuse standards can and should be followed, it is impossible for a third party vendor to rework their IP to conform to multiple companies internal reuse standards. The issue for the end use customers is to evaluate the third party IP for its applicability into their environment and applications.

The QIP metric from the Virtual Socket Interface Alliance (VSIA)[3] is used by the end-use customers to evaluate similar IP from multiple vendors through an apples-to-apples comparison. It is also used by the providers as a simple quality control checklist of implementation details and deliverables to objectively evaluate their product quality and to refine their designs and methods for successful reuse. It facilitates feedback between the users and the IP providers and fosters quality process improvement. The QIP premise, verified by experience, is that good designs are a product of good design practices from IP conception all the way through customer delivery and support.

The QIP provides a means to quickly, easily, and consistently provide the information pertinent to the successful reuse and integration of an IP from a third party. While a provided score is derived from the completed questions in the metric, this number is not an unequivocal indication of an IP's quality or its suitability for an application. The end user can gain some insight into the relative quality of the IP in that a significantly higher score is obviously better than a low score; however, the same conclusion cannot be reached for score differentials of a few points. The QIP is primarily intended to present a common language and mechanism to communicate the high order factors that pertain to quality IP between the providers and the consumers, while highlighting the areas that may need more investigation, improvement, or resources for the IP to be successfully integrated into the SoC and working silicon.

Chances are that no single IP will be perfect for every application, but if its application shortfalls are known, the end user may be able

[3] VSIA: IP Quality Pillar. http://www.vsia.org/QIPDownload/

to develop a mitigation plan. This may include scheduling additional internal resources with the needed skills to address the identified gaps in the IP or working with the vendor up front to satisfy the requirements. The most important aspect is to be able to simply and consistently determine the true state of IP early in the evaluation process.

An end-use customer can view the completed QIP for third-party IP, and in a few minutes, can have a high-level feel for the quality of the IP. By devoting an hour or less, the user can investigate the complete details that factor into the quality of the IP and determine the risk that may be associated with its use. This information provides the basis for internal mitigation plans or additional discussions with the vendor.

QIP User Usage Model

There are many components that factor into the reusability and quality of an IP that range from absolute show stoppers to "nice to haves." Business factors may be just as important as technical issues. The customer ideally would like to have confidence in the vendor and the vendor's standard procedures before engaging on a technical level. Once the customer has confidence in the vendor, he may be more interested in continuing the technical IP investigation.

Several factors should be considered when evaluating third-party IP vendors. Has the vendor instituted quality practices throughout the company? This can be corroborated by being certified to an industry quality standard such as ISO9001[4], capability maturity model integration (CMMI), or others. Another aspect of quality can be demonstrated by the standard quality assurance (QA) procedures that are followed, such as independent QA audits and a consistent methodology for QA, including integrating the IP using typical design flows and into a typical SoC application. However,

---

[4] ISO 9001:2000 – What does it mean in the supply chain? http://www.iso.org/iso/en/iso9000-14000/explore/9001supchain.html

the aspects that may be of utmost importance to the end users are the revision control, release, and support policies. This will help ensure that the correct IP views are delivered and integrated as well as set expectations for vendor support should issues be encountered.

Once the end-use customer is comfortable with specific vendors and begins the technical evaluation, he becomes concerned with discovering the items most relevant for the integration of the IP into his application. This includes the maturity of the IP, and what deliverables are provided. At this stage, it is important for the customer to discover any potential hiccup that could hinder the integration of the IP into the SoC, affect the planned tool flow, or even the system functionality. If this initial technical level of discovery is satisfactory, the user may wish to delve even deeper into the IP.

The final level of discovery into the quality of third-party IP delves into the detailed aspects of IP development which can provide insight into the block usability and maintainability. The standard development methodologies followed should indicate the care taken to develop an IP that can be considered high quality. This includes requirements, verification and project plan documentation, in addition to the coding constraints and verification methodologies followed.

The QIP metric provides this common baseline to communicate the vendor and IP information that directly affect the quality and reusability of the IP. It provides "at a glance" scoring summaries for the answered questions, and it displays the categorized quality measures with the associated sub-topics. The supporting spreadsheets break down the quality category measures into further detail, giving running totals for the associated questions, as well as color changes to immediately indicate the potential risk level of the individual answers (In *Figure 9.6*, the lack of colors is not important as this is only the summary view). A red cell indicates that the answer infers poor quality, and it is likely to be considered unacceptable by an integrator. An orange cell indicates that the answer infers that quality could be better, but is not unacceptable.

A green cell indicates that the answer indicates an acceptable standard in this attribute.

My IP

My Company

Copyright © 2003, 2004 VSIA

Summary Report

Version 2.00 Beta

Technical Support

VSI Alliance

Scoring Summaries	
Vendor Assessment	80%
IP Integration Assessment	79%
IP Development Assessment	82%
Total	79%
Imperatives not satisfied	
Rules & Guidelines not satisfied	65
Satisfied Imperatives, Rules & Guidelines	217
Unanswered	6
Total	290

Type in the name of the IP Vendor	My Company
Type in the name or part number of the IP	My IP
Select the type of IP you are evaluating	digital soft IP
Select the type of assesment you are undertaking	Vendor, Integration & Development
Select the amount of information you wish to display	By category

	Score	%
Vendor Assessment	174	82%
VENDOR: Vendor Assessment	174	82%
IP Integration Assessment	874	80%
SOFT IP: IP Ease of Reuse	320	88%
SOFT IP: IP Maturity Assessment	0	0%
SOFT IP: Documentation Quality	172	95%
SOFT IP: Ease of Integration	148	88%
SOFT IP: Design & Verification Quality	117	64%
SOFT IP: Design Quality: Design Detail	52	50%
SOFT IP: Verification Quality	65	81%
IP Development Assessment	944	79%
SOFT IP: IP Ease of Reuse	81	94%
SOFT IP: Ease of Integration	81	94%
SOFT IP: Design & Verification Quality	391	76%
SOFT IP: Design Quality: Internal IP Design Documentation	75	83%
SOFT IP: Design Quality: Design Detail	107	72%
SOFT IP: Verification Quality	209	76%

Figure 9.6: *QIP Summary*

QIP Provider Usage Model

There are a number of obvious benefits for the SoC integrators (or, end-use customers) to use QIP. Other than responding to customer demand, how do the IP providers benefit? In general, they now have an objective means in which to showcase the quality and the long-term considerations of support and maintenance of their IP. The QIP may formalize standard internal design practices that factor into the quality of an IP and provides a mechanism for communicating the information. For some companies, QIP can help identify areas of improvement, potentially resulting in a reduction of reported issues and improvement in the turn around times for making bug fixes or enhancements to the IP.

IP vendors want to sell a product, and the most successful ones realize that a collaborative relationship with their customers is beneficial to both parties. The vendor is not only interested in the success of the initial project, but would also like to see repeat business. Two-way communications and consideration of IP quality factors facilitate continuous process improvement, thus benefiting the vendor's quality process and helps address the customer's needs.

9.4 Quality Specific to IP Types

Different types of IP have different attributes that are critical to ensure quality. While silicon IP (SIP) typically start as soft, synthesizable cores, once that IP is hardened to a specific technology node, the key reuse parameters change. And, what about the different types of hardened IP? A phase-locked loop (PLL) cannot be judged with the same set of attributes as a hard digital block. While it may be synthesizable, verification IP (VIP), by its very nature, may have different key reuse considerations from a functional soft IP. Arguably, software is somewhat analogous to soft IP, but it is not constrained to a language subset as soft IP is for synthesizability, but has its own unique factors that must be considered.

Soft IP

Many IP blocks either start out as soft IP, or have a hardware description language (HDL) model. While it is relatively easy to write legal code, it requires greater attention to detail and an understanding of potential downstream implications when writing code that ensures correct functionality across different platforms and tools. For quality IP, some coding techniques should be avoided as they may cause differences in simulation, either between different simulators or between RTL and gate level simulations, or impact downstream tool usage such as synthesis, DFT, or static timing. Obviously, tool specific constructs, such as specific synthesis pragmas, should not be included.

Simulation differences may be an issue, including start-up or race conditions. Different simulators make different assumptions on initial values and may compile or interpret concurrent statements differently. As a result, one must take sufficient precautions to ensure consistent operation by initializing flops to a known state through a direct set or reset, not via an initial statement. Statement execution order can be ensured through the proper use of non-blocking assignments and complete and correct sensitivity lists.

Several issues can be encountered when synthesizing a design, from sub-optimal performance or results, to unintended functionality. Several coding techniques can result in unintended latches or tri-state gates, and the incorrect construction of state machines, such as not defining a default state, may result in unpredictable behaviour. In addition, the amount of code contained in a single synthesizable block will impact the quality of results; too much logic and the machine run time may be unacceptable, while a block that is too small may result in a larger gate level design because the tool did not have the opportunity for complete optimization.

Code considerations should be made to facilitate downstream tools such as DFT and static timing. Internally-generated clocks and resets complicate the scan insertion process, and should either be avoided or controlled during scan. Combinational feedback loops cause problems in both scan and static timing.

Reusable design techniques should also take into account other partitioning issues, in addition to logic block size. Clock generation circuitry and bus interfaces should be partitioned into their own blocks to facilitate retargeting the design for other system applications. Additional logical partitioning, including separating logic with different speed goals, state machines, and critical paths into their own blocks help facilitate quality synthesis results.

In addition, some implementations may require the usage of the soft IP in a language other than which it was originally developed. Tools are available to translate implementation languages, yet doing so may introduce unintended issues. More specifically, names differentiated by case alone may result in name space collisions and thus, regardless of case, all names must be unique. The RMM is a good reference for more information and guidance on soft IP reuse considerations.

Hard IP

In addition to the reuse considerations mentioned in the previous section, when IP is hardened and is represented in a technology specific physical layout, another set of quality considerations need to be taken into account. Hard IP can provide a much quicker time to market than soft but, by its very nature, it is targeted to a specific technology. This may eliminate some of the tool dependencies that can be seen in soft, but it introduces many more physical dependencies.

Obviously, the hard IP should have clean design rule checking (DRC) and layout versus schematic (LVS) runs, but this may not be possible with some IP. However, any known errors and exceptions must be documented and communicated to the end user. Additional documentation should also be supplied that relate to the physical implementation, including routing restrictions, IP orientation, layer and device mapping, and power sequencing. Finally, certain types of IP, such as physical layer (PHY) transceivers, should include the appropriate validation waveforms, in this case a clean eye diagram.

It may be more difficult for a customer to evaluate the suitability of a hard IP under consideration for integration. Timing and functionality are generally known, characterized, and based on the process technology, but models for interconnect and performance, or evaluation boards that demonstrate the functionality and timing are often delivered with quality IP. Either of these verification validation approaches can greatly improve the SoC implementation and verification effort.

Verification IP

VIP is used to monitor or control a specific protocol or architectural interface, and it is usually delivered as soft IP. As such, the same development considerations should be given to VIP as to soft; however, since VIP is an integral piece of the block and SoC verification effort, additional considerations need to be included to support and communicate progress and errors.

There are two primary classes of VIP, which may be used in conjunction with the other. The first are drivers and responders which are active VIP, which either converts transactions streams or stimulates an interface in response to transactions. The second are passive devices, such as monitors and checkers, which simply observe and report on the interface or functional block.

Both classes of VIP should include several levels of checks. The first is obviously the functional interface it is designed to validate which should include full interface functionality. Yet, additional checkers should be included to validate the integration, including warning of untested or invalid configurations and parameter ranges.

Robust, quality VIP has the capability to detail time, coverage, and bus trace activity. Monitors should be able to recover from bad stimuli, and drivers should be able to inject errors on the interface. Both types of VIP should be standalone and operate on only a single interface. VIP should include unit, integration, and compliance test cases. Additionally, a corner case test suite and test cases targeting manufacturing test should also be supplied. All of the supplied test cases should have the capability to be extended by the end users.

Embedded Software and Drivers

Software has long been the differentiating component in systems and has traditionally been designed by the in-house developers. However, with the advent of more complex standard IP, third party vendors are now supplying standard software to support functional IP. Again, many of the concepts used in developing quality soft IP are applicable to quality-embedded software. This includes documentation, code portability and maintainability constructs.

Embedded software may include an abstraction layer to provide an interface between the application and control programs. If so, complete descriptions must be provided for the application programming interface, including its inputs, outputs and example implementations.

Since code might be converted from C++, C, Verilog or VHDL to any of the other formats, the keywords from the superset of these languages should be considered as reserved ones. In addition, as for soft IP, a standard, portable language should be used. Data types and data operations must not be dependent on language implementation details. For example, in C/C++, bitwise operations must not be used on signed integers, and the internal representation of floating point numbers must not be accessed directly. In cases where dependencies cannot be avoided (for example, signed integer division in C), they must be well documented.

Verification of the provided software is also a key consideration. Regression testing should be run on the software to validate all requirements and provide a sufficient level of stress and corner case tests. The software verification must include messaging for progress, completion, and failure identification.

9.5 Future Quality Considerations

Higher levels of abstraction, as well as higher levels of integration, all contribute to the changing quality landscape. Such examples of this include new languages, such as SystemVerilog, new testing methodologies, such as assertion–based testing, new process technologies, and possibly most importantly, shrinking process geometries.

Design for manufacturing (DFM) and design for yield (DFY) are emerging as areas that will improve reuse and manufacturability. DFM attempts to bring pertinent process and yield information considerations earlier in the design cycle. There are several approaches that are being evaluated, but it may be inevitable that the foundries will become more tightly coupled with the front end designers.

The trend has clearly been towards higher levels of integration as block–based reuse has evolved. It is becoming more common to reuse a "platform" which includes one or more processors, peripherals, and software targeted to a specific application. This level of reuse involves validating not only the quality and functionality of the individual components, but also the quality of the interaction of the whole.

Another trend is for higher levels of abstraction of the source code, for example, designing and synthesizing logic from C/C++ as opposed to the traditional Verilog or VHDL. Some of the approaches translate the C/C++ into an HDL from which traditional design techniques are followed, but there are emerging high level synthesis tools that also work from other languages.

The definition and measurement of quality is not static, and they must be updated as tools, methodologies, and technology change. For single source-defined tool flows, the vendor may be the best source of requirements for quality implementations. However, as seen with Verilog, tool vendors may have varying specification interpretations which greatly impact the reusability of a block in other environments. To avoid these issues in the future, the standards committee must ensure that the standards themselves are robust and that their implementation is agreed upon by all of the tool vendors.

Part III:
The IP User

Chapter 10

IP Selection

Adam Traidman, President, Chip Estimate Corporation

Pallab Chatterjee, Chief Technical Officer, SiliconMap, LLC

10.1 Introduction

If intellectual property (IP) were like other electronics components, such as computer parts, one would simply visit the local IP retailer, peruse the shelves, perhaps consult an online review site, and then purchase the selected IP. Install would be easy, solely consisting of "plug-and-play" and one would be ready to go. Unfortunately, the world of IP, as it stands today, does not yet address this ideal scenario.

The popular belief is that systems-level creation is more challenging than block design. It is not that the latest microprocessor part isn't complex, nor that it is inexpensive to incorporate. Rather, it is a standard product component with expected functionality that is verified to conform to a specific documented instruction set. Although an IP component that conforms to a functional standard may be utilized, such as the universal serial bus (USB), there will inevitably be challenges surrounding the customization, verification, integration, and quality of the implementation. Therefore, in IP selection, the old adage about, "an hour spent now could save you weeks in worry later" could not be more correct.

Perhaps the most crucial decision in IP selection is not which component to choose, but instead the process of choosing the right component for a given function. This process, as described herein, spans technical and business challenges, and it requires a vision to assess which combination of IP options will ultimately lead to design success.

The key is to create a selection criteria before actually "pushing down" to the technical comparison of the actual IP.

10.2 The Process of IP Selection

The IP selection process has changed in the midst of technological progression. In the early days of the semiconductor business, the IP blocks were small, and they were only offered by integrated device manufacturers (IDMs). These initial IP blocks were low-level primitive cells and input/outputs (I/Os), and they were only implementable by the IDMs since there were no documentation or interoperability standards in place at the time.

The next IP evolution was initiated upon the rise of the merchant semiconductor foundry model. This model separated the IDMs hold on the whole flow of price, performance, support, and capability and allowed a new segment of the supply chain to emerge—the standalone IP provider.

The IP provider was focusing on supplying a higher level of integration, beyond primitive cells, to the new category of fables semiconductor suppliers that were designed to work with one or more foundry processes. At that time, there were no standard documentation or quality and validation criteria in place in the industry.

The next IP evolution, from a selection point of view, was the platform IP which was a higher level of integration with an associated software development environment. This was presented as a minimally foundry dependent IP block with documentation, support, and defined function. Some of the advanced versions even incorporated tool flow specific and optimized versions to improve interoperability. The claims for this IP stage were generally more highly engineered than the IP itself.

Current IP has to address a combined foundry process and IDM business model, technical application capabilities, verification and quality guidelines, tool and functional interoperability, development support software, shortened time to market, and derivative product development issues. As a result, the historical methods for selecting

IP are actually no longer valid. The following guidelines on the selection criteria for 21st century IP have been used to produce successful high-volume production designs and can be easily modified to address application specific issues.

10.3 IP Sourcing

The first step in the IP selection process is to identify where to acquire the IP. IP sourcing has changed throughout the years due to changes in technology and business models. The changes were both evolutionary in regards to technical requirements and IP development globalization that has taken place in the industry.

Where to Look

Originally, all non-internally created, third-party IP was found and presented by the IDM partner that was manufacturing the end product. As things progressed to the foundry model, most of the IP was directly available from the foundry. Further, as processes moved to deep sub-micron (DSM) technologies, standalone IP companies became a major supplier of blocks for systems-on-chip (SoC) designs. The issue of where to look has a couple of different answers depending on the process node in questions and the sort of IP being sought.

For large geometry processes (standard complementary metal oxide semiconductor [CMOS] greater than 0.8um channel length), most of the IP is either created internally by the user or made available by the manufacturing partner. The sourcing is limited to the cells that they wish to make available, or a process specification which allows whatever cells are needed to be created. The only choice available for usage is what is presented in this case.

On modern DSM processes, there are a couple of sourcing options based on the function of the IP. Most SoCs have three main components, which include one or more central processors, embedded memories, and specialty-function IP. The central processor is very application-driven (automotive, video, etc), and it is primarily sourced by the software development and application library available. In this case, the IP itself is not selected as much as software and firmware development, testing, and release environment is selected. As a result, these blocks are not sourced as IP in the traditional sense.

The remaining two types of IP—embedded memories and special function IP—use traditional sourcing techniques. These blocks may be available from the foundry or manufacturing partner, and they can traditionally be located on those sources. However, these IP blocks are also available in multiple forms with different design optimization aspects based on the IP provider making the block. Sourcing through the list of providers is a more robust methodology and will typically result in the IP selection that is more optimized for the specific application.

The question is now how to identify these other suppliers. As most global suppliers have product representation available online, the results for a keyword search tends to turn out correct content, but still requires a fairly large exercise to be able to identify good sources of IP. The keyword searches in a general worldwide Web target space will typically not return results based on IP functionality, specification, or design goals. *Figure 10.1* shows a typical Google search result for an IP block where, in this instance, IP for a peripheral component interconnect (PCI) express interface was searched for. The results by adding the keyword "IP" to the search string returned a still un-reviewable set of results of over two million matching Web pages.

Figure 10.1: *Google Search for IP*

A better way to source the data is though an independent listing service such as ChipEstimate.com. Their site is free to use, and it lists IP in the form of data sheets from a large number of suppliers such as foundries, IDMs, electronic design automation (EDA) suppliers, and standalone IP providers, and utilizes specification related search criteria to help sort the products. *Figure 10.2* is a partial snapshot of the representative list of IP suppliers which allows for search by specific vendor or partner. *Figure 10.3* shows the results for a sample search on product type. As in the Google example, we are once again searching for the general term of "PCI Express." In this case, the result is 95 items which is a concise reviewable list.

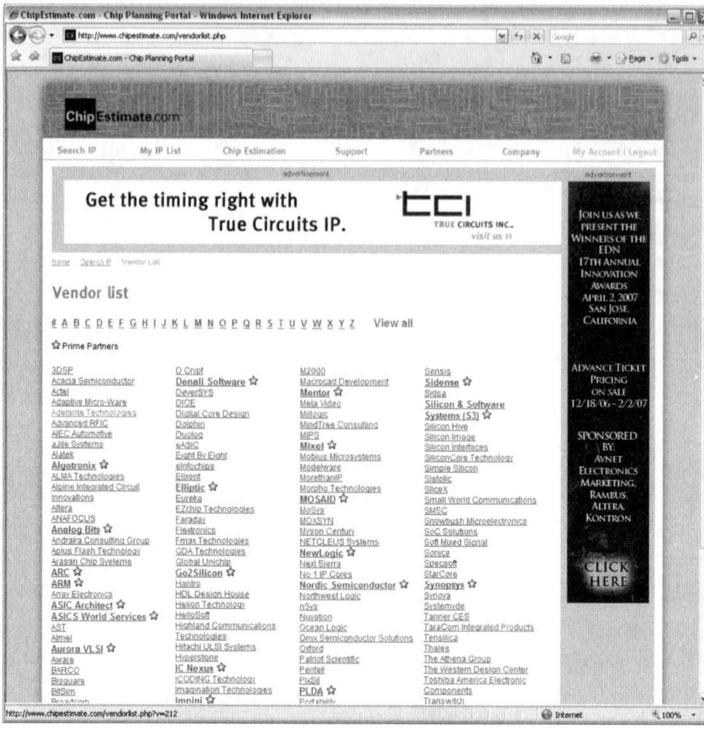

Figure 10.2: *IP Vendor List*

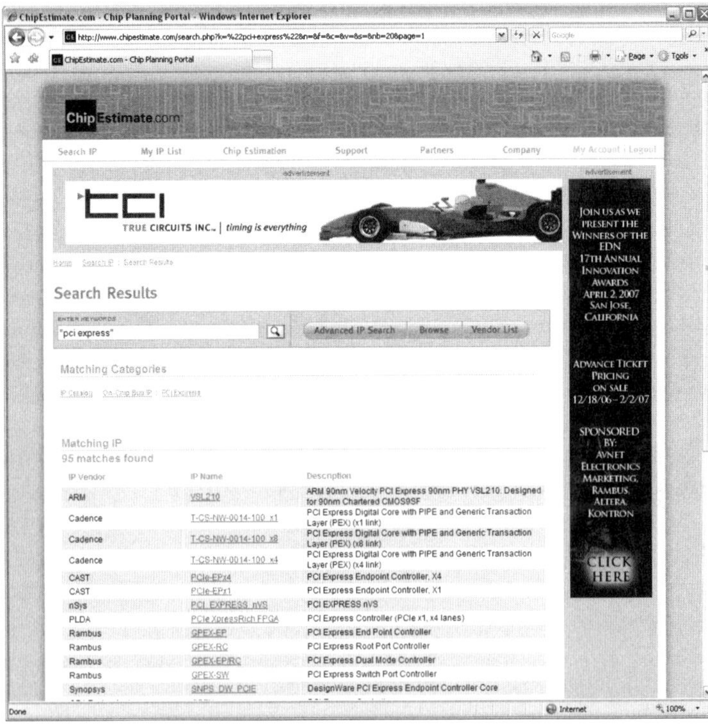

Figure 10.3: *Product List*

This list is further annotated by type of block (end point controller, transceiver, etc.) and the version of the specification (1.0, 2.0, etc.) that is available. This small list is further reducible with the advanced design search which allows for matching based on process node and target fab. This advanced search is shown in *Figure 10.4*.

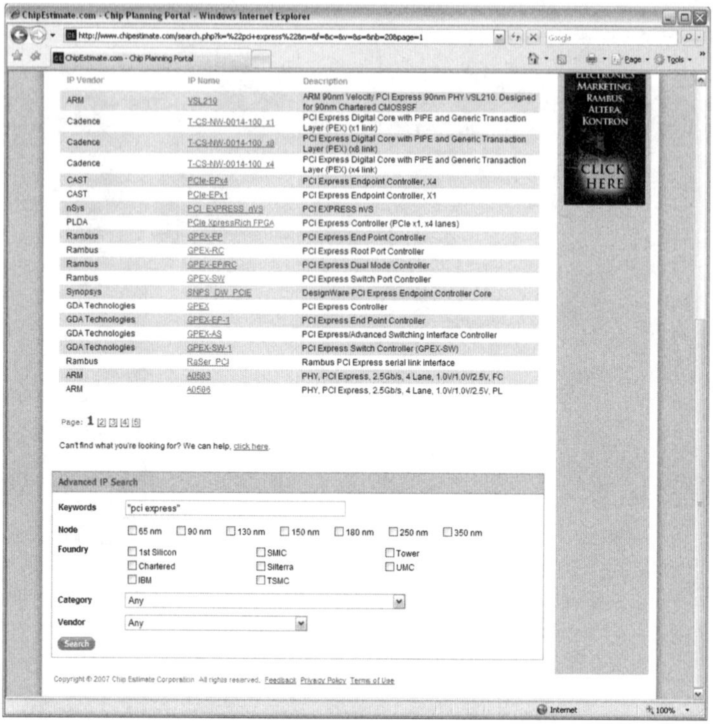

Figure 10.4: Advanced Search

The method of utilizing portal–based searches provides uniformity to the engineer by presenting IP block diversity in a format that allows for optimal comparison of their tradeoffs. The use of these portal sites for sourcing also allows the SoC designer to decide, based on IP availability, which process to use, and which manufacturing/foundry partner to select. This is the equivalent capability that was available to systems designers in the late 1970s to 1990s in the form of distributor data books for the various suppliers they represented.

Internal versus External IP

As the result of new business and technical dynamics for DSM processes, there are two different driving conditions that determine the internal creation versus external licensing of IP. The two different conditions for internal creation are the need for proprietary design data and the unavailability of the needed IP.

Historically, the drive behind the creation of internal IP has been to design a function block that provides a marketplace differentiation for the resulting product. An example would be two different companies building end-product MP3 players. While the products utilize a number of standard functions, interfaces, and memories, the only real market differentiation can be in the form of a custom IP block. As the end function for the product is standardized by application, the user interface (UI), power performance, and data reconstruction methods are the only things that can be designed differently from the other products on the market. These blocks, typically in the form of custom data path components, have long been the focus of internal custom IP creation. The blocks were created for typically non-standard aspect ratio, size, power, and performance features that required full custom logic and physical design. The need for product differentiation is still the leading motivation for architectural selection of internally created IP.

A new driver is the unavailability of IP for a project. With the reduced time to market of many new projects and the rapid release of new process technologies, there is a requirement for complete designs to be made available before foundries, EDA vendors, and other third-party IP suppliers have had a chance to create, test, and provide blocks needed for a design. In addition, with the large number of foundry and process options available today, there is a strong possibility that a particular block, while available in a licensable form, may not be tested and qualified on the particular process option that has been selected for a design. Thus, most companies on the bleeding technology edge, or those pushing process capabilities, have to resort to custom internal IP development.

Terminology regarding internal versus external IP is also blurring. With multiparty process and technology development now increasingly popular, the co-sharing of IP between companies is becoming more and more common. In this scenario, the platform IP products, with large fixed portions and smaller regions of customization, is exactly where external IP is headed.

What You Will Find

IP and other virtual components have an associated ecosystem unlike past times where data sheets were sufficient to describe all the functional and interoperability aspects of an electronic component,. This ecosystem includes multiple design views (transistor level, gate level, power, register transfer language [RTL] level, system/behavioral level), physical or manufacturing views (layout, place and route [P&R] abstracts, design for manufacturability [DFM], litho aware, signal integrity [SI], reliability), test/verification views (design for test [DFT], built-in self-test [BIST], component and systems level test, fault coverage, assertions, automatic test pattern generation [ATPG]), and software views (electronic system level [ESL], firmware/middleware development, timing and power validation, physical synthesis/APR and encryption /decryption) of the IP.

Based on the licensing and use model selected, the transferred IP is usually a tar file archive of the necessary views needed for the tools in the customer's design environment to complete and validate the SoC design. In addition to the design transfer package, there is validation data as defined by the Fabless Semiconductor Association (FSA) and the Virtual Socket Interface Alliance (VSIA) standards and live support personnel to assist with usage of the block.

10.4 Selection Criteria

The selection criteria for IP are separated into two segments—external and internal IP. This section will discuss some of the key criteria of the selection process for third party, along with externally-created and licensed IP. We assume here that any design team using external IP blocks that have significant failures in meeting the requirements for use, will resort to internally created blocks whose selection follows standard company specific design review procedures.

The following information applies to large publicly-traded and private IP supplies as well as to smaller firms, startups, and independent design consultants.

Business Models and the Engagement Process

Similar to many newly created open markets, the semiconductor IP sector was born with a grand vision and promising business plan which has gone through its fair share of gyrations over the past decade. Spawning out of the mainstream semiconductor industry, early IP pioneers sold venture capitalists on grandiose stories of royalty–based business models. This made some sense since other semiconductor and general electronic components were sold on a per-unit basis. However, as competition rose and price pressure mounted in a commoditizing electronics industry, the lucrative royalty–based model was soon only accessible to the top IP players while the rest explored new strategies.

In today's market, IP is generally licensed under a one-time, multiple, or unlimited use model or the desirable royalty–based model. Creative business negotiations have produced untold additional derivatives of these basic models, and since each IP purchase is sealed behind closed doors, many are inaccessible to the general buyer who does not have large volumes or other superior buying power.

One-time use models enable an IP integrator to utilize a piece of IP on one semiconductor chip design. This includes re-spins for manufacturing and design defects, but it generally does not include derivative designs that are considered separate chip designs. Individual licensing agreements may tie the actual design to a part number or set of part numbers which utilize the same die image. There is often a specific relation to a final tape-out. It is important to spend adequate time analyzing one-time use licenses to ensure the flexibility afforded by the licensor suits the needs of the integrator. For example, multiple instantiations of an IP generally incur additional costs. Trading flexibility for cost over a reasonable chip volume one-time use licensing models tend to bear the least financial burden.

Multi- or unlimited use licensing models come at a financial premium, but they provide maximum flexibility in utilizing an IP component several times on the same die as well as across various chip designs and their derivatives. It can be considered almost as a "package deal" where the IP integrator is buying in bulk at a higher up-front cost. These licensing deals will often have some sort of cap tied to a timeframe, number of designs, instantiations, or other limitations. IP integrators with strong buying power, combined with numerous designs requiring similar IP, will often pursue this model. This model generally requires the largest up-front financial commitment.

Royalty–based models are common only in the higher echelons of semiconductor IP. High value, standard IP (e.g., such as ARM microprocessor cores) or basic building block IP (e.g., standard cell libraries) may be licensed to various parties under royalty–based terms. Negotiations are highly variable depending upon forecast chip volumes, actual royalty pricing, and any required up-front non-recurring engineering costs.

One should always consider these options in the IP selection process. Price points will vary by IP function and licensing model, and depending upon the anticipated production lifecycle, volume, and cost of your chip coupled with the roadmap for derivative designs, weighing these options early in the design process is often critical to the long-term profitability of your chip.

Quality, Credibility, and Reputation

One of the most important factors in the technical selection of an IP block is based on their past track record. Similar to the investment industry, "past performance is not necessarily an indicator of future performance," but this is presently the best we have to work with. This history consists of various projects done on a process with a particular foundry and the number of similar projects carried out. The types of IP that are relevant for history comparison have a standard functional specification, whether or not it is a hard of soft IP block, and the number of times a block has been used in different applications. The project history has to be substantiated by the IP customers, not just the provider identifying the customers they had for the IP. In fact, solely purchasing IP does not automatically lead to the successful integration and fabrication of that IP.

A company's size is not necessarily indicative of the quality or total track record of an IP block. IP blocks are currently created the same way as complete chips once were in the early days of the semiconductor industry, that is, by a single designer. The track record of interest is that of the designer, and the willingness of the company that currently employs him, to support his products. For larger designs, the single designer concept is expanded to a team that includes circuit designers and software development environment designers. Once again, the willingness of the company currently employs the team to support their products.

One of the metrics for quantifying quality is the quality IP (QIP) metric which is an initial step to provide guidelines for IP acceptance. However, its guidelines are not passed by most of the IP today. As a result, the actual values of these metrics are not being provided on either a forthright or timely basis by most suppliers. The IP's historical yield of being used in production devices, not just test chips, and the resulting yield from the end application chips is one of the few quantitative measures of quality for an IP block.

A recent trend in creating credibility for a trusted design is the industrial partnering between multiple fabrication facilities, EDA tool vendors, test companies, and design IP firms. In this scenario, which is the predominant model for 65nm and below processes, the qualification of the IP track record is guaranteed by the fab rather than the designer. This switch is due to the concurrent development of IP, along with the stabilization and manufacturing development of the process.

The reputation for IP and its providers should be interpreted on a per product basis. There are currently no IP providers that can provide high quality, yield, and reliability IPs for the full range of processes, tools, and design applications (memory, central processing units [CPUs], standard I/Os, and mixed signal). As a result, every IP vendor will exhibit strengths and weaknesses in their reputation. This range is related to the extent to which they "reach out" from the core competency that their designers have verified in silicon, in order to meet a customer's requirements.

One of the last subjective issues on credibility for a project is the financial stability of the IP provider, their likelihood of continuing in the business, and support over the product's life. This issue of product life cycle must also consider the longevity of the process for which the IP is targeted, and if there are alternative manufacturing sources should the current foundry discontinue services on the process.

Modification for Application

In older technologies, 0.18um and larger, most IP was available in hard form (fixed physical implementation) and did not change in the context of different applications. Smaller geometry designs have additional layers of interconnect and increased sensitivity to placement in the design. The smaller geometry processes additionally have DFM and design for yield (DFY) criteria. These new criteria impose changes on the design at mask time which may impact the functional performance of the block. Such examples of these changes include metal and diffusion fill generation, wire spreading via insertion, tap insertion, and metal overlap of via changes.

These DFM and DFY issues are not uniformly instituted in a design but are related to what is adjacent, connected, and the overall placement on the chip. As a result, most hard and soft IP delivered to a customer cannot be used as it was delivered. The final design that goes out will be different from the working design unless it has been qualified to the silicon level with the fab in a partnership.

The IP vendor must provide guideline documentation, test procedures, and sufficient design views to allow these application-specific modifications to be made and must be able to verify that the end design still conforms to parametric and functional specification.

An exception to this concept is mixed signal IP. Analog blocks including data converters, transceivers, clocks, and references are generally built to be foundry clean through manufacturing. In addition, these blocks are marked as boundaries with full internal "keep out" areas for the EDA tools and shielding from higher vertical layers of interconnect.

Figure 10.5 shows a typical mixed signal or soft IP modification flow, and this also displays the post layout re-verification for modern designs. IP providers should have a structured, documented flow that can be followed by the end customer in order to safely modify a design for manufacturing if required.

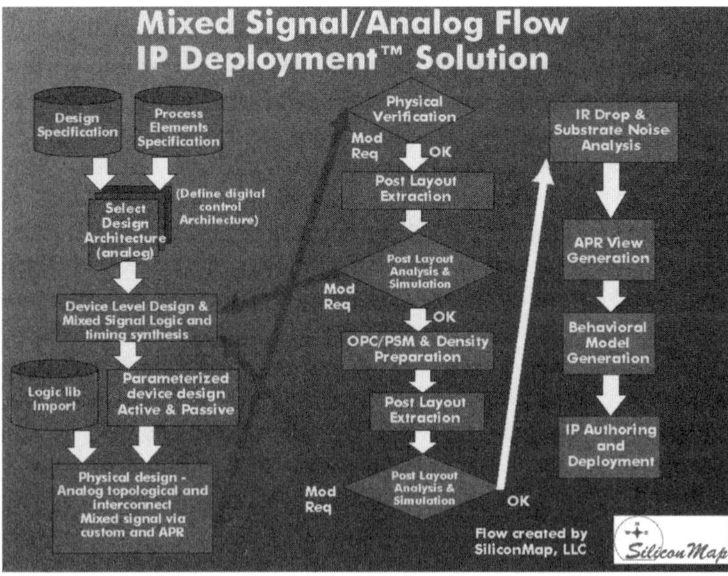

Figure 10.5: *Soft IP Flow (source: SiliconMap, LLC)*

The area of IP modification related to correction of the basic functionality and timing/power performance of the block should be part of the "applicability" criteria for selection. If the IP needs to be fixed or changed to carry out its intended purposes, then the IP is not really usable for the application and should not be considered.

Design Flow Integration

Another critical aspect of IP selection is the IP's ability to be utilized within an existing design flow. With dozens of EDA vendors providing tools which span the design cycle from ESL-type architectural tools to backend physical verification systems, IP components must ship with deliverables which enable design teams to successfully integrate them into their design infrastructure.

Depending upon the type of IP (soft, firm or hard) and the general functionality of the IP (analog, digital, or mixed signal), IP suppliers may provide views for simulation, functional verification, timing analysis, synthesis, physical estimation, physical placement or layout, physical verification and many more IP and flow specific views. Many of these views will reside in industry standard file formats such as Synopsys Liberty™ and Si2's LEF while others will be in EDA vendor specific formats such as a Mentor Graphics Calibre design rule checker (DRC) deck.

With an increased focus on semiconductor power reduction over the past several years, new standards such as the Cadence common power format (CPF) and the Accellera unified power format (UPF) are also emerging. New tool flows, advanced dynamic, and static power analysis will most likely require these types of views moving forward.

As such, design teams must consider what deliverables are available in IP selection. One should not assume that an IP supplier will provide certain views and may even find some suppliers asking for additional financial commitments to supply certain formats. Understanding these tradeoffs up-front and detailing the required views specific to your design flow is a required step in the IP selection process.

Interoperability

For most design applications, interoperability relates to compatibility with EDA tools and tool flows. For IP, the terminology refers to the idea of application compatibility with other blocks in a single SoC.

Standards–based IP utilize normal synthesizable logic interfaces to the core of the chip. There are very little interoperability issues for these blocks. Large custom logic blocks (based on synthesizable primitive and register cells) have historically had interface issues with embedded controller macros and some data and system busses. ESL and high level models, in addition to tools such as those available from Mentor, CoWare, and Carbon Design, are addressing this marketplace and successfully removing it as a design issue.

The use of these tools is actually shifting the scope of the IP blocks to be larger and to incorporate both the logic and the micro-controller core into a single placeable instance in the design. Major IP suppliers such as Mentor, ARM, IMG, and others have already moved to this trend for video, multimedia, and automotive designs. In the new arrangement, the interoperability issues have been moved up to the firmware and application control software level.

In the hardware implementation aspects of chip design, there are still two main interoperability issues. The simpler and most often discussed, is IP that does not work or does not do what it is supposed to. The major issue with all IP, even highly-qualified IP, is the proper functioning (parametric performance) in a customer application. Most IP is created as standalone blocks with a test bench being the stimulus. In the context of noise from adjacent circuits, variability in power supplies, signal skews, and cross-die variation, some tested blocks may not work. As a result, block placement in a design is critical to validate the design interoperability.

The second most common interoperability issue is power supplies. Older designs tended to have two modes (active and off), and two power grids (I/O and core). Modern designs have multiple isolated power grids (e.g., CPU, memory, data I/O, analog, system bus, display, etc) and multiple operating modes (full active, test, full active with no display, standby, sleep, power-saver, off, radio frequency [RF] only, etc.). As a result, IP blocks which are designed for a single power pair, may need to be modified for interoperability with the power mode control as well as for connectivity to the correct power grids.

Support and Warranty

As previously mentioned, the industry perspective is that support and warranties are only accessible to a select few who have some "outside the normal business cycle" access to the engineers needed. This perspective is not correct, in that "being married to the designer's sister" does not guarantee quality support nor access to the necessary data to make the IP work in the application.

The real situation is that support and warranty are subjective criteria based on the working business relationship between the SoC designer company and the IP provider company. The key to the working relationship, is the creation of a business environment (e.g., the contract) that has an ongoing benefit for the IP provider to make sure that the customer's SOC makes it into production and most importantly, yields well and hits or exceeds its volume target and price points. In this scenario, access to the needed design information, technical staff, and tool resources will be made available to the customer to complete the project. This access is provided based on the simple realization that the IP company makes more money providing what the customer needs over the product life cycle, compared to what it costs them to provide the resources.

One should avoid any scenario where the IP provider solely wants up-front money and no royalty or trailing receivable or wants on-going revenue that was not in the original discussions. This is not an IP model, it is a design service model, and the ownership and re-use rights should be based accordingly. In this scenario, there is no positive return for providing a block with any quality metric and no contractual tie to the outcome of using the block.

There are many companies listed in the IP portals that provide quality support and end applications guarantees (debug time, modifications due to process, etc.) and are professional IP suppliers. The buyer should beware of companies that both produce end products and sell the SAME IP blocks, as well as companies that only have design services references.

Documentation and Deliverables

The standard documentation that is transferred during the evaluation and engineering phases of the design consists of simple data sheets and standard (per FSA and VSIA guidelines) process design kit (PDK) transfer documents. From the deliverables side, it should consist of a design data archive that has all the design views, such as physical, logical, verification, and software, needed for one's own design flow.

10.5 IP Evaluation

IP evaluation is best performed on a purely technical and "check box"–based methodology. A sample of a check box methodology is the QIP worksheet that can be seen in *Figure 10.6*. These worksheets are part of a program that is being coordinated as information provided by the IP supplier and support soft IP. For hard IP, the FSA has a HIP program—the flow and evaluation can be found illustrated in *Figure 10.7*.

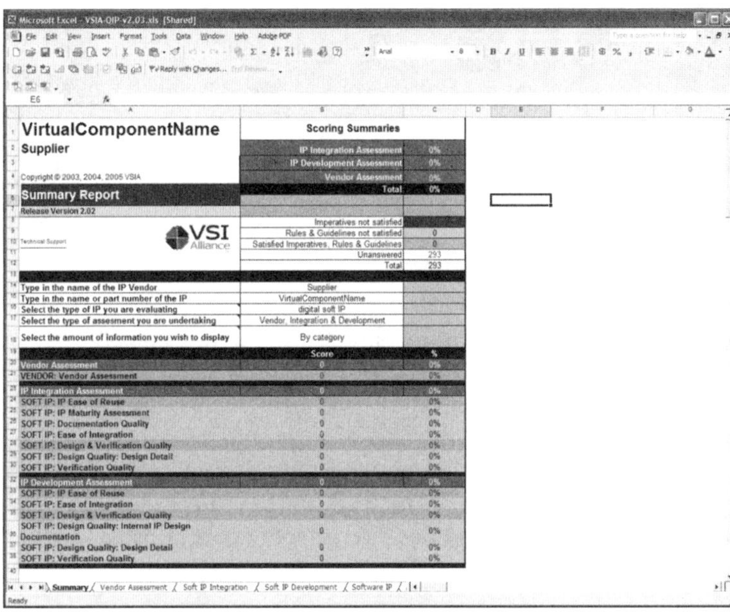

Figure 10.6: *VSIA Selection Spreadsheet*

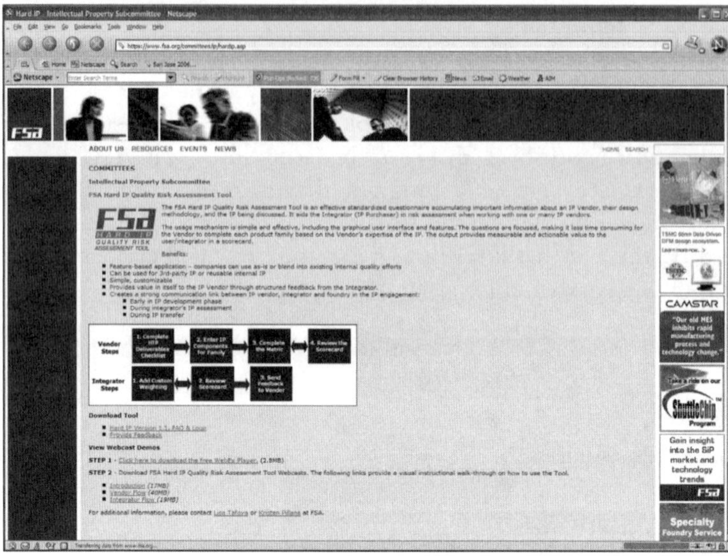

Figure 10.7: FSA Selection Website

Manufacturing availability, strategic partnerships, design interoperability, and cost all influence the final selection of IP that is used in the practical world of business. The evaluation criteria should be independent of the selection determination and either be used as the validation requirement for the suitability of an IP block for a particular design, or as the basis for the "to do list" of items to be completed or modified to provide conformance of the selected IP to the design.

Timeframe

Two time frames, business and technical, are involved in the evaluation and signoff of IP. These two schedules are not independent and require concurrent manpower allocation to complete.

The technical schedule is a single block and usually finishes under 30 calendar days. It is a straight-forward engineering effort with the process selection in place and a functional/parametric specification for the IP block. This is mostly targeted at confirming the IP provider claims for parametric conformance of the block. The secondary target is to verify that the IP works in the context of the other IP blocks and the application.

The business timeframe is two segmented blocks that are typically 15 days each in duration. The two segments are identifying the IP block guidelines for the evaluation and negotiating the cost model (payments, support costs, and royalties) for the IP once it is selected. In the event that the IP does not pass the technical portion of the evaluation, then the second 15-day period is not required.

An assumption on these time frames is that the company providing the IP has local support and business offices to the end customer, has existing business models to refer to, and the end product which will utilize the IP has realistic and predictable production volumes and price targets.

Vendor and Tools for Evaluation

The vendor evaluation is based on subjective, or soft, criteria rather than hard evaluation criteria as outlined previously. The criteria for selection of track record involves IP-provider release criteria, business model for IP licensing (including costs), financial stability of the provider, localized support for the IP, and its application in the end design, anticipated lifetime of the IP and of the IP provider, and finally support and modification debug capability.

These subjective criteria are reviewed not on a "go/no" go model, but rather on a performance scale (0 for not available to 9 for excellent) based on input from the business and marketing groups in addition to engineering. These factors are then weighted in importance based on business and time to market needs of the project. These two metrics are then multiplied together to achieve a singe aggregated score. Selection is usually based on the higher aggregated score with the assumption that the IP block is actually technically suitable for the application.

The tools that are required for the technical evaluation of both internally generated IP block and externally licensed IP are the same standard tools used for a product design review. The methodology of tool use in the evaluation process should not differ from product creation tools. The concept is that an IP block is a standalone block, like a chip, including the I/O for the block. Thus, the tools required for the evaluation require at least, but not limited to:

- Architectural what-if analysis engine (size, power, leakage, yield, etc) such as InCyte from chip estimate
- Device level simulator such as HSPICE
- Behavioral model simulator such as ELDO AMS
- Gate level/RTL level STA tool such as PrimeTime
- SI and power aware STA such as PrimeTime SI
- Mixed mode and high capacity simulation such as high-speed interface module (HSIM)
- Functional verification such as Incisive
- Fault simulation such as VCS
- DFT/ATPG tool for compliance testing of block/BIST Insertion such as TetraMax
- Physical Abstract to logic model and APR – custom script
- DRC/ERC/LVS tool and control files such as Calibre
- DFM/DFY screen tool and control files such as Calibre, Blaze MO
- Package model and power dissipation analysis

A new breed of architectural analysis tools combined with a growing ESL offering from leading EDA vendors enables higher level functional verification and physical estimation of IP components. One such tool, Chip Estimate's InCyte, is designed to be used at the earliest stages of the design flow to perform what-if analysis for a potential chip design. This includes quantifying the tradeoffs that various IP will have on a specific design. The InCyte software is connected live via the Web to the ChipEstimate.com chip planning portal enabling access to thousands of hard and soft IP components. With tools like InCyte, design teams have immediate access to IP data and can quantify how an IP component will impact their specific design in terms of size, power, leakage, yield and even packaged chip cost—all before the actual design process has begun. Early insight can be invaluable in exploring which IP component will ultimately work best for a given design.

Final Selection

The final selection of an IP block and the vendor providing it has been outlined as a systematic task that is repeated multiple times for an SoC. Guidelines on the vendor selection for large scale DSM technology designs (130nm and below) is one vendor for every seven 7 vendors for a 10M gate design.

For most designs, the number of unique IP blocks is quite varied. IP is available in both soft and hard forms and, as such, can constitute up to 95 percent of a design. The remaining 5 percent contains clock and signal distribution for the design and power control circuitry. About 15 percent of this in an IP design is internally created due to either unavailability from third parties or for the proprietary nature of a design.

Making a final decision on an IP component is a major step forward, but it is realistically the first step of a much larger design process. Once the component has been selected, the business deal finalized, and money exchanges hands, only then the arduous task of integration and verification truly begins.

Chapter 11

Block-Level and Platform-Level Reuse

Jeff Freeman, Member, Technical Staff,
Transportation and Product Standard Group, Freescale Semiconductor

11.1 Introduction

Such market pressures as time-to-market, increased complexity and integration, various feature sets, and lower development cost have driven the industry toward finding ways to "do more with less." The industry response to manage these pressures is to create products with increasing levels of reuse[1].

Reuse is not free regardless if it involves internal reuse or the use of third-party IP. It requires up-front planning, and it can actually delay first products containing the new IP that have been designed and verified through reuse techniques. However, if the correct reuse methods are implemented, the derivative and other products that reuse IP can benefit tremendously.

11.2 Architecture

For a design that will be reused in product derivatives and across product lines, an architecture roadmap is critical to ensure the design covers the market and customer requirements over time. The roadmap gives the "big picture" of where the design fits into the product portfolio as well as a beneficial security to the customers by demonstrating to them a path for growth.

Architecture begins with a top-down design approach, looking for existing IP to buy, reuse, or recycle. The intra-IP standard interfaces and external interfaces are selected during this

---

[1] IP reuse is using IP more than once as opposed to recycling where the IP is changed substantially from the original in a shorter time than it would if the IP were designed from scratch.

exploration. In an iterative process, the market and customer requirements drive the architectural specification over time. If the IP exists, the maturity of the design and verification IP will influence the overall verification plan of the IP.

For new architectures, a "white-paper" or specification is typically created to show the customers what is proposed given their requirements and the market requirements. The specification includes block diagrams of internal hardware, standard interfaces, proposed instruction sets, if the design is core–based, modeling results from running code from sources such as benchmarks (e.g. Embedded Microprocessor Benchmark Consortium [EEMBC][2]) or customer specific applications. Architects will then develop modeling equations to predict system performance on multiple design proposals so that system trade-offs can be made. These equations typically include variable terms for modeling different memory configurations and memory response times, number, and type of communications channels and bandwidth, number of processors and their executions rates, etc.

The market and customer may demand multiple product points such as:

- Low-cost (both in terms of hardware and software)
- High-performance
- Low-power
- Highly-integrated, highly-differentiated solution
- Market-specific (transportation, consumer/industrial)
- Application-specific
- Any combinations of the above

Architectural reuse needs to be considered early on in order to minimize the development and support cost and to keep up with the reducing development time. In the next sections, both block and platform reuse methodologies will be described.

[2] http://www.eembc.org

11.3 Block-Level Reuse

Architecting and designing IP for reuse requires careful research of the market requirements and feature set both for the current product and for future products. The focus cannot be just on the present product, rather it has to include the planning of how this IP might be reused later.

From a design perspective, reuse can be designed using hardware configuration pins, software programmable registers, and design parameters. These approaches can affect the performance, system complexity, power, and size of the design. This affects the verification, test, and build processes of the design, and it is overall not an easy tradeoff as it involves balancing multiple issues.

- Hardware configuration pins are flexible, but for the current product, the extra logic can introduce timing degradation (extra decode and loading on critical signals), and it can consume extra power and area.
- Programmable registers, similar to hardware configuration pins, can be used to change the IP functionality, but by software instead. For example, a bus arbiter could have programmable registers to select different arbitration schemes. The programmable registers also have a similar but greater impact.
- Design parameters, pre-processor, or hardware description language (HDL)–based, can be used to select the hardware and feature set before the product implementation. This does not have the impact above, but it does fix the selection for that product, especially since it cannot be changed by the end user. This may mean that complete verification of all possible configurations is not performed during the first design iteration.

During the IP design process, industry trends, external standards groups, and standard interfaces should be considered with future reuse of that IP. Hardware prototyping can help with early customer feedback, faster specification refinement, and more thorough verification.

Figure 11.1 shows some flexibility options that should be considered for a block.

Feature set
Memory map
Performance, power, area trade-off options
Bus interfaces sizes
Optional hardware
Standard interfaces
If core present: instruction set, performance (e.g. cache) options
Application block interface
If memory interface present: size, performance, or type options

Figure 11.1: *Block Flexibility Options*

11.4 Platform-Level Reuse

The next level of reuse extends to the platform, where the system design of the system-on-chip (SoC) can be architecture independent. A platform is a collection of reusable IP targeted to a specific application domain that may be reused in multiple design configurations without having to incur the schedule and risk to re-verify the platform for each use. The industry does not yet have a standard definition of what a platform is [2] [3] although the industry standard taxonomy [4] does provide a number of definitions that are commonly used today.

A typical platform includes one or more cores, internal or external memory interface, arbiter, peripheral IP, real-time operating system (RTOS), and test logic for embedding the platform in a SoC. The boundary includes standard bus interfaces and a debug interface.

Why use a platform? It provides faster time-to-market and inherent quality through repeated reuse, but it has limitations on choices of options or features. A traditional application-specific integrated circuit (ASIC) or full-custom approach would not have these limitations, but it would take longer to market and introduce risk to quality because IP is not being reused. Furthermore, once a product family is established based on a given platform, the software associated with that platform can be reused for derivative products and possibly new products based on that platform. The platform is thus an extension of the trend that started many years ago and eventually stopped producing custom processors for each design; instead, they standardized on a number of architectures and instruction sets. Such components as a bus, arbitration, memory, and most designs are necessary for a processor to operate and need some common peripherals including timers, interrupt controllers, and basic input/output (I/O). Since these do not provide much in the final product differentiation, having them pre-integrated and verified can significantly reduce the overall design effort. Nowadays, a number of platforms exist that also contain collections of more specialized peripherals that target a particular end market, such as a consumer or even specific product, such as a cell phone.

Using hierarchy to implement a natural divide-and-conquer approach, the platform is more complex than an IP block since it is made up of multiple independent blocks. Flexibility to choose the independent IP blocks for the product is traded for a pre-verified (used in other products) and less flexible solution. A computer motherboard is a good example where its components are expected to work together. One expects that, since the platform has worked in other products, it will most likely to work in the current product.

In order to address time-to-market, the platform must be architected to solve target problems and be flexible where appropriate. It is possible to add too much flexibility and thus create problems for verifying the product. It is natural to raise the productivity from traditional chip integration to the use of platforms in *Figure 11.2.*

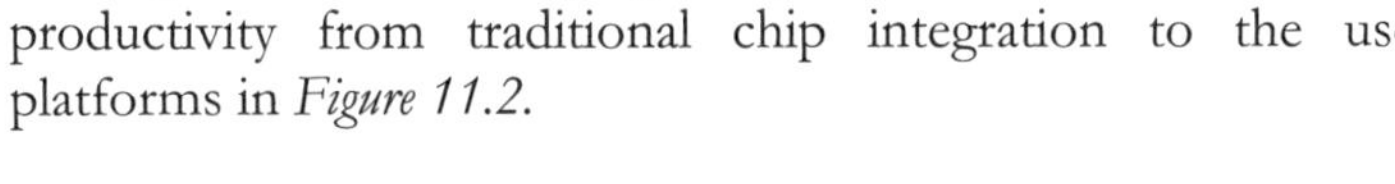

Figure 11.2: *Component Evolution*

As chips increase in size and complexity, platforms abstract the complexity using a divide-and-conquer approach. A reusable platform is more valuable that a single-use solution. Considering that customers tend to be more interested in product features and system performance than the traditional core architecture, the platform provides the needed abstraction to give them a solution, rather than an architecture.

Figure 11.3 shows flexibility options that should be considered on a platform.

Feature set
Memory map
Performance, power, area trade-off options
Bus interfaces sizes
Standard interfaces
Optional IP blocks
If memory interface present: size, performance, or type options
Inherit limited set of options from independent blocks

Figure 11.3: *Platform Flexibility Options*

The flexibility options at the platform-level are very similar to those at the block-level. However, the combinations can be excessive when combining multiple blocks to make up a platform. As a result, it is important to pick the right option subset from the viewpoint of the developer as well as the integrator.

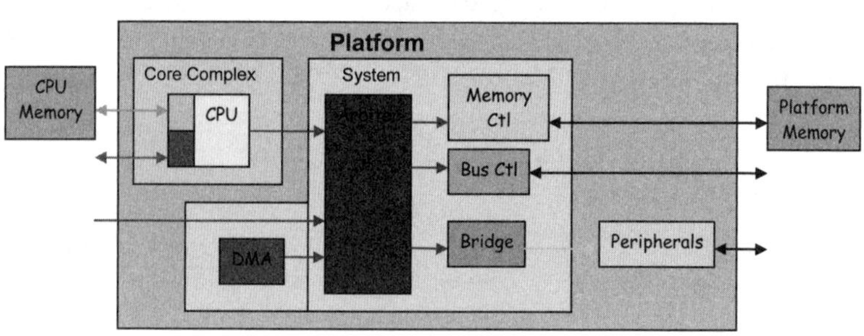

Figure 11.4: *Platform Block Diagram Example*

Much of the system performance is defined by the core and memory interface, and this is typically the driving identity for each unique platform. From an architecture point of view, the peripheral set is tied to the targeted market segment.

Limiting the platform choices manages the complexity of products and verification of those products. For example, for a given platform, the customer has three choices:

- Low-cost platform
- Baseline or reference platform
- Higher performance and feature or enhanced platform.

For soft IP that will be synthesized and reused, it is usually a wise idea to target multiple technologies to wring-out any technology dependencies that the IP may have. Of course, if the platform is obtained from a third-party vendor, one would also want to make sure that a wide range of technologies are supported, including the one that will be used. You do not want to lock yourself into a future product direction if the supplier of the platform is not showing their willingness to support the platform into the future. Here are some things to consider:

- Constraints for reuse
    - Based on clock period where possible
    - Parameters used for technology dependent values
    - Parameters to conditionally select constraints
- Register transfer level (RTL) checks
    - Coding
    - Style
    - Partitioning
    - Analysis
- Implementation checks over multiple technologies

11.5 Verification Reuse

As the block or platform is designed, there is a parallel process of verifying that the IP implementation matches the specification. The process assumes that there are both design and verification reviews; reusable design IP implies that the verification IP will need to be designed to be reused.

The verification of the flexibility options, as seen in *Figure 11.3,* can use reusable testbenches, monitors, responders, and drivers, by borrowing object oriented practices from the software world. For every design object, the associated verification objects, should also be present. Standard interfaces between design objects are objects themselves, and they are an ideal place to attach associated verification IP. The verification IP should include appropriate parameters and be intelligently partitioned so it can be vertically reused within a project and horizontally across projects. The more it is reused, the better the perceived quality of IP.

Plan

A comprehensive base verification plan is usually written for a new product line. However, for derivative products, a shorter plan can be created by referencing the base plan. For similar cross-products, the base verification plan from one product can be recycled for the new product. The plan factors-in the IP maturity, market segment (which affects the quality goals), and complexity. Many of the same reuse parameters described earlier ultimately find their way into the verification IP. Verification complexity can quickly get out-of-hand if too many parameters are allowed to change so the values and parameter choices need to be bounded.

Testbench

The testbench should follow object-oriented principles and provide multiple ways of verifying the IP because each way may introduce unique assumptions in the testbench that could lead to defects that are not caught. If the IP contains a core, then the testbench should be able to either load the core with a program or replace the core with a bus functional model, all while being capable of driving the core-interface through constrained-random transactions.

The testbench should include monitors, responders, drivers, and checkers. A resource manager in the testbench may be needed when multiple IP blocks are vying for a common resource such as memory. *Figure 11.5* shows an example of testbench components around an IP block.

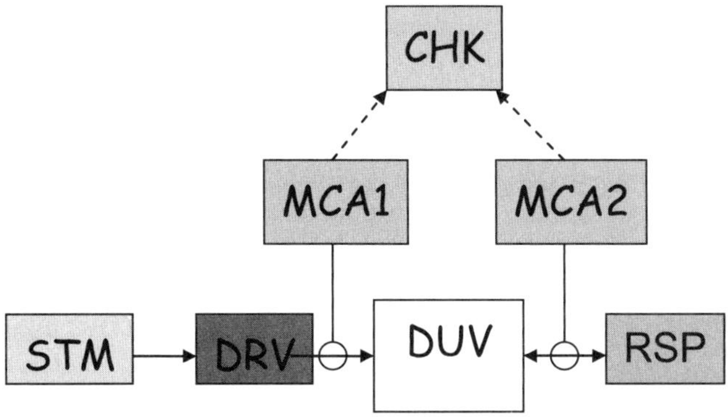

Key	
CHK	Response Checker, Functional Coverage
MCA	Monitor, Interface Coverage, Assertion Coverage
DUV	Device Under Verification
DRV	Driver
RSP	Responder
STM	Stimulus

Figure 11.5: *IP Block Testbench Components*

As verification moves up the hierarchy level, the responders and drivers are replaced with connected IP and new responders and drivers at the next level. However, the monitors and checkers can be reused at the higher hierarchy levels.

Stimulus

Unless the stimulus is coming from a core, the stimulus should be written to drive an abstract interface so that it can be potentially reused for similar products (that is, the driver would change, but the stimulus might stay the same). There are at least four types of stimulus that should be considered:

- Directed tests for functionality and performance
- Constrained-random
- Industry benchmarks
- Customer application code

The directed stimulus are traditional manually written tests covering boundary condition tests, sanity or simple start up tests, or tests that cannot be covered by an automated constrained-random-driven approach. The constrained-random tests are driven by today's high-level verification languages (HVLs) such as SystemVerilog or SystemC. The constrained-random tests are good candidates for reuse since they are high-level and can make use of parameters much like the design IP does. The industry benchmarks and customer code are reused more from a software point of view.

11.6 Documentation Reuse

Documentation reuse is just as important as design or verification IP reuse. The documentation needs to be kept in sync with all of the flexibility options available in the IP. For example, if there is an optional hardware unit on the platform, then the documentation needs to include this; otherwise, the documentation should not include it. Recalling that reuse introduces some design complexity to manage the flexibility options, the same applies for the documentation. Instead of design parameters, the writer is faced with managing conditional text or file inclusion or exclusion. It does not take long for one to realize that a relational database is needed to manage the design and verification IPs, libraries, derived deliverables, and the associated documentation.

11.7 Points of View

The IP developer uses parameters and preprocessor directives to create design and verification IP for multiple products. External interfaces connect to abstractions of the external environment and/or portions of the real environment are connected during verification. The verification of the IP block is very comprehensive.

The IP integrator is more concerned with what parameters and directives are available to tailor the IP block to their requirements. There is a fine line between easy-to-integrate and flexibility to meet the customer requirements. The verification in this case is focused on correctness of interconnect hookup and interaction testing between IP blocks.

The end customer is interested in system performance, low cost, and low power since they want a product with good debug features, available software drivers, programmability, and differentiating features to set their product apart from the competition.

11.8 Conclusion

Block-level and platform-level reuse continue to be a key component to "do more with less." The lack of widely accepted platform standards for hardware and software abstraction layers indicates that more research and standards collaboration is needed (SPIRIT and the Institute of Electrical and Electronics Engineers [IEEE]). There is a balance between ease of use and reuse as too much of one consequentially impacts the other. By selecting the appropriate degree of reuse freedom based on information from this chapter as well as personal experience, block and platform IPP reuse can yield significant productivity and quality benefits for all.

References

[1] Ramesh Chandra, "IP-Reuse and Platform Base Designs," http://www.us.design-reuse.com/articles/article6125html, August 2003.

[2] Richard Goering, "Platform–Based Design: A Choice, Not A Panacea," http://www.eetimes.com/story/OEG20020911S0061, September 2002.

[3] Claasen, Theo, "Platform Approach Can Smooth Way to IP Reuse," http://www.eetimes.com/op/showArticle.jhtml?articleID=18303945, April 2000.

[5] Brian Bailey, Grant Martin, and Thomas L. Anderson, *Taxonomies for the Design and Verification of Electronic Systems,* Springer, 2005.

[6] Gabe Moretti, "Blocks and Buses Lead the Way," http://www.synplicity.com/literature/pdf/821_EDN_platform_based_designs.pdf, August 2003.

[7] Jeff Holm, "Achieving Reuse with both Modifiable IP and Configurable IP," http://www.eetimes.com/news/latest/showArticle.jhtml?articleID=51201865, November 2004.

[8] http://www.spiritconsortium.org

Chapter 12

Integration, Test, and Debug

Neal Stollon, Principal Engineer, HDL Dynamics

12.1 Introduction

As with most things, an intellectual property (IP) block is not an island unless its size is very large. Ultimately, IP is acquired to develop a product, which mostly consists of the IP and its added value (be it one's own IP, software, among others) which must be integrated into a system and ultimately tested. One of the first technical areas that has to be addressed after making the decision of what or which IP is needed for your application is deciding how to integrate it into the system and starting to figure out the specifics of how the IP will be tested, verified, and debugged. For most savvy IP users, these are issues that should be addressed far before the final decision is made on acquiring IP. In many cases, the integration-related features of an IP block will be an important consideration when making the purchase decision and in decided among different competing IP offerings.

In most of this chapter, we will focus on programmable and fixed function blocks digital IP. Analog IP integration and test is a potentially more complex topic due to the diversity of its analog functions, the diverse integration and test methodologies, their typical availability only as process or vendor-specific macros, and their need for attention to process specific parameters such as noise analysis.

At the simplest level, IP integration into the larger systems environment is based on interfaces; there needs to be a mechanism to get information into and out of the IP block or else the IP loses its usefulness. For digital IP, this information may be data used by the IP block (control words or instructions in the case of a processor IP block), data being operated on by the IP block (data in), or data provided by the IP block to the rest of the system (data out, as well as status). For analog IP, this may be voltage levels or waveforms of varying complexity.

It is not possible in one brief chapter to give you the answers to all of IP integration and test issues. The best approach is to alert you to the issues that you may encounter. For each of the topics, there is a plethora of more detailed information available on-line and in-print that can help you to understand nuances and find solutions. One should never underestimate that, while relieving you of some of the detailed engineering design, IP–based systems require as much engineering judgement and investigation into how everything fits together and works as any other complex design.

12.2 Integration Issues

While having an IP–based system (sometimes) means that you do not need to spend time analyzing the individual blocks, you do need to make sure that things are working at the systems level. IP interfaces, even when based on standards, do still need to be modelled, simulated, analyzed, and prototyped—sometimes there are inherent problems or incompatibilities, sometimes integration problems creep in as a by-product of human error (even yours), especially when interfaces get complex. Most of the time, IP integration is an iterative process, consisting of as much (or, as shown in *Figure 12.1*, at times more) time in analysis and verification as design. At some point in the integration, the verification stages (loosely define as confirming how you think things should work correctly), starts to be replaced by testing (loosely defined as confirming how you know things should work correctly). This in turn turns into debug stages as you determine and fix what is not working correctly. The flow and transition between these stages is never precise and exact, since part of a design may need to be tested and debugged before additional stages of integrated can be started. One can determine when the integration stages of your design are done by looking at when all of your verification steps are replaced by their corresponding tests.

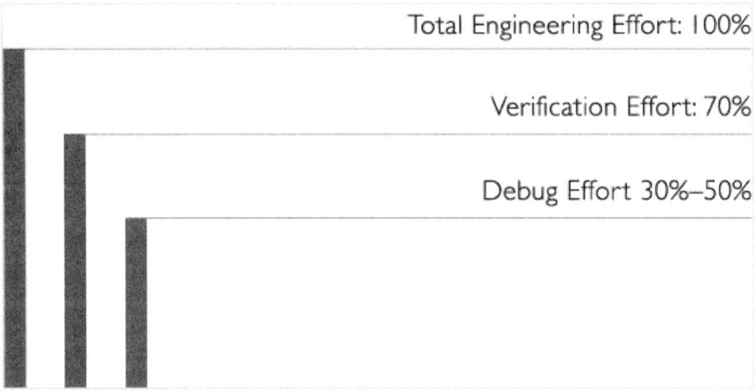

Figure 12.1: *An Estimate of the Analysis Costs of IP Design (source: Novas Software, Inc. 2005)*

There are two ways to go about this integration, test, and debug flow. This can be put together on some hardware platform and one can proceed from this point forward. Alternately, you can take advantage of a variety of modelling, simulation, and analysis tools—here, this will be referred to as a virtual prototype—to start your analysis at your desk, rather than in the laboratory. In reality, most folks work out their unique tradeoffs between virtual prototypes and hardware–based verification.

Modeling and Verification

This brings up the question of what types of models are needed for analysis, verification, and test and make sense at which stages of integration. The IP evolution has been paced by a corresponding development of model views that address corresponding size and complexity. There is a pretty wide range of model types that are provided by IP vendors, depending on a lot of factors.

SPICE models—If you are using analog IP, this is probably a default deliverable, but this is not the case for digital IP. Transistor-level models are usually based on SPICE, which is still considered the gold standard for simulation in many camps. For digital IP, SPICE makes less sense, only because it is very compute intensive and for most purposes is overkill. SPICE analysis at the integration interfaces, in many cases, does make sense, especially for high speed systems. SPICE level analysis can help to find physical integration problems, such as noise, reflections, and oscillations, but the problem in relying on this level of integration is that for integrated circuit (IC) design, there are no specifics involved such as block placement, routing, etc. until pretty late in the design. Generally, this type of analysis requires the support of the application-specific integrated circuit (ASIC) or Foundry vender, rather than a local IP vender.

Gate level models—This may be in a hardware description language (HDL) format (Verilog or VHDL) or electronic design interchange format (EDIF) netlists. This is sometimes a deliverable for digital IP, although a gate level model is highly dependent on constraints and directives that were used to create (or, synthesize) the gate level model and a primitive gate library that is often specific to a given technology. Gate level models are most often provided when the IP vender wants to limit the customer to a particular silicon target. As one common example, field programmable gate array (FPGA) vendors strive to get the customers loyalty, and they often deliver models at gate level to keep the IP specific to their technology. More general digital IP is delivered as register transfer language (RTL) models which, for digital IP, are the pre-synthesis model deliverable most commonly provided. RTL models have pretty much standardized around VHDL and Verilog, and they are written in a synthesizable subset. There is a lot of overlap and interoperability between RTL and gate level models, which allows simpler integration, as well as IP-specific testing schemes such a formal verification. In fact, formal verification approaches, such as model checking, add a lot of value in certain cases of model verification, but not so much in integration.

Models can be written at other levels, and there is wide debate about the value of system level models, which run through a range of varieties, based on timing or cycle accuracy, level of model abstraction, and other considerations. There is a wide array of opinions on system models and system integration usefulness, and some engineers are loyal to system level modelling as a key to successful design, while others feel that it does not provide sufficient value.

For analog IP, where it is often relatively simple to model the IP, system-level models are typically linear equation–based where, such as Matlab which would be a popular approach. They provide a valuable way to start looking at the integration and interactions of analog IP blocks into a system, before digging into the details.

For digital IP, system level models run a range that includes bus functional (models the interface, but not the essence of the IP), instruction set accurate (models the execution of a programmable IP, but not its specifics), transactional (models the timed or untimed data transactions of an IP, but typically not the signal interactions) to name a few. System models can be provided in VHDL, Verilog, and now increasingly, at least in some applications, in SystemC, which is a C++ class library developed specifically for digital systems level modelling.

Electronic system level (ESL) models and tools have been an area of critical, yet perennial, concern since the mid 1990s, in part to help address these issues of IP integration and verification. Although many parties would consider any pronouncement of an ESL solution as premature, SystemC seems to have taken the foreground by developing a model environment that especially for processor IP, addresses three of the major system analysis requirements.

1. It can span several levels of model view abstraction, from very abstract transaction levels to virtually synthesizable RTL.
2. It has a manageable learning curve, moderately low cost of entry level tools, and is reasonably extensible to development of new features.
3. It allows integrating software models in a hardware environment, enabling relatively fast simulation of larger systems and more efficient hardware/software co-design.

In general, interoperability of different types of system level models, from different IP vendors is not always a given, and integration at the system level can involve much effort, but when system modelling works, it can provide the capability to look at integration and systems tradeoffs and alternatives not feasible with other model views, including the ability to see the interaction of your software applications with the hardware architecture.

One should also be aware that the same kinds of analysis can be carried out with a model of lower abstraction as with a higher one. The converse is not always true. One of the main values of more abstract models is computation time and resources needed to use them in simulation. SPICE models take more time and memory in two to three orders of magnitude than their corresponding gate models. RTL models are one to two orders faster than gates. System models result in a couple more orders of magnitude speed-up but, in each case, speed is obtained by trading off some level of model resolution and detail, and when it comes to verification and analysis, it can be said that it is all about the details.

In terms of integration, one should also consider the interoperability of the pre-silicon modeling and verification environment to the in-silicon debug and testing that will be required once you have prototypes. All too often, users find problems in the hardware that did not show up in modeling. Planning to reuse the verification environment to leverage it as part your silicon testing will streamline and increase the quality of your IP testing in hardware. These in-silicon debug and analysis issues will be discussed later in the chapter.

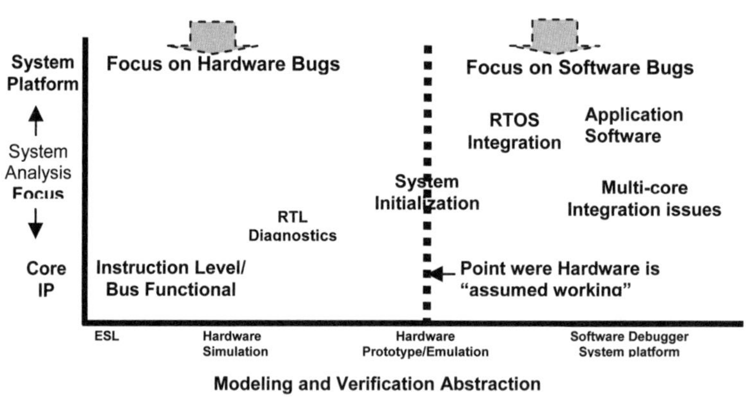

Figure 12.2: *Integration Analysis for (Processor) IP Systems*

IP Integration Deliverables

There are no fixed rules on what your IP vendor needs to provide in the IP release package, but typically they will include the following:

- Synthesizable models for soft IP—These are typically provided as RTL files in Verilog or VHDL. For synthesis purposes, it should not matter much what type of RTL file it is; for simulation purposes, it is obviously more of an issue.
- Structural models for hard IP—This may be provided in different formats, it may be a fully compiled macro cell. In some cases, it may be in a netlist format that still needs to be mapped to a given IC technology which is a not insignificant amount of additional work.
- A model generator for configurable IP—Some types of IP have configurable options and are delivered as model generator that will generate the IP blocks rather than as set of fixed files. This is the case for some processor and memory IP, along with other types. The obvious issue here is that you the user now have to learn about the tool and how to get best results with it, in addition to learning what is needed about the IP itself.
- Design scripts with constraints—For synthesis and simulation, for high performance IP, possibly physical tools, etc., a significant amount of time can be spent in setting up a design for the proper tradeoffs in gates versus performance, which is made more difficult by not having the in-depth information into critical paths, clock interactions, and related specifics. The IP vendor should have this information, and he should be able to provide it in a well-documented format.
- Documentation—It is hard to overstate the importance of good documentation in the process of integrating, testing, and debugging IP. In this case, "good" means that it needs to be both comprehensive and clear in content. Documentation should be available for every stage in the integration and test process.

- Simulation and verification test environment—This should include, at the minimum, test benches, or test verification models for exercising the RTL as well as all subsequent netlist or gate level models. The simulation test environment should provide a high (and documented) level of test coverage of the design, as well as relevant assertions that are specific to the IP (more on test coverage and assertions later in this chapter). The test environment should be robust enough to allow you to substitute or add your own application specific test vectors or stimulus. If you have specific test environment requirements, formal verification or support for a specific verification language as an example, deliverables may need to be discussed with your IP vender.

There are other items to consider which may not be part of the standard IP release packages but may be available if you ask:

- System level models—These may be critical, or not, depending on the methodology in use. There are many types of systems models, so integrating the IP vendor model into the system analysis environment may not be trivial.
- Reference designs—Both in terms of how does the IP operate in a typical application and on how does it integrate (at a hardware and software level as appropriate) at a system level. Reference designs may just be RTL, even better are those that can may be targeted to a FPGA or other specific prototyping environment.
- A slew of application notes (the more the better, it hopefully shows that some one has been thinking about and using the IP), some of which address integration, test, and debug issues.

Local versus Global System Integration

IP block integration typically takes two basic forms: (1) local integration of one block to another, using some direct wiring and (2) global IP interconnection to many other blocks, typically over some bus architecture. These decisions are typically made during architectural design, based on factors such as data bandwidth and communications between blocks. From the integration point of view, local interfaces are obviously the simplest approach, especially if the connection is direct. Bus–based integration is more of a proverbial double-edged sword, that is to say, the interface increases in complexity, since now, instead of just having two IP blocks, there is an intervening bus fabric, associated bus arbitration, and the issues that need to be considered start to increase. However, since many bus interfaces are standards–based, the IP vendor now provides its consumers the interfaces of a larger community of people who have previously used the interface.

In many cases, one can simplify, at least in the beginning stages of the integration efforts, the global system integration by treating it as a simpler locally-interconnected system. The complexity of the bus system is partially due to the fact that it is hidden by use of socket–based interfaces—as will be discussed later.

Standards–Based Integration

Leveraging standards–based technologies and interfaces facilitates several systems-on-chip (SoC) issues, including IP core portability, design reuse, leveraging common peripherals and processors, improving automated verification, and increasing customer capabilities to create products from a common design platform.

Using subsets of a standard is appropriate if it is allowed by the standard, but the compliancy of the entire system has to be guaranteed. Different blocks may only support a relevant subset of a bus interface standard, however, if different blocks are communicating and both have different subsets, this can lead to trouble or, at least, a whole lot of analysis. In the best case scenario, things will just break apart and it will then be easy to identify the problem. In a worse case scenario, however, everything will work perfectly, but performance will be much lower than the optimum that the standard supports or will not operate correctly in some obscure case that is critical to your application-operating condition.

12.3 Digital IP Integration

There are obviously a diverse range of interfaces for IP, but most on-chip digital systems communicate through some bus–based system architecture where, in fact, reference designs are important and useful. If your IP vendor cannot address how the IP is interfaced through generally relevant and industry-adopted standards, this can be considered a red flag. If, for example, the IP only interfaces to a vender proprietary bus interface, this may force you to purchase another vendor's IP in order to get maximum performance or to avoid having to develop your own standardize interfaces. If standard interfaces are not provided as part of the IP, then, by default, the responsibility will fall on you as the user to address this integration. While developing an interface gasket between different bus types is not an especially complex design activity, there are nuances in each of the several types of standardized bus architectures to choose from, and having to spend design effort integrating a common bus interfaces defeats at least part of the goal of minimizing your integration by purchasing.

Bus Level Integration

Arguably the most critical system decision that needs to be made early on in architectural development is what type(s) of bus architecture to base the system on. There are four widely discussed standards–based on-chip bus architectures, in order of widespread adoption: advanced microcontroller bus architecture (AMBA), open core protocol (OCP), Coreconnect, and Wishbone. All these buses are essentially free to use, and they have a range of IP vendors who are ready to sell compatible cores, models, and verification-related tools. There are also a variety of less standard and vendor proprietary buses, which typically have more limited support and in an age of increasing standardization, are difficult to justify using unless you have some extreme communication requirements that can not be addressed otherwise.

On-chip buses differ from board-level buses in that the latter can tri-state the interfaces, which simplify integration—basically you only enable the chip bus interface when it is needed. For on-chip bussing, tri-state logic is much more difficult and expensive to implement. As a result, at the simplest level, basic on-chip buses are multiplexer–based or are point-to-point buses. More complex bus implementations may employ embedded routers, first in first out (FIFO) buffers, arbitration, and other more networking related options to increase performance and more advanced bus features.

AMBA is a family of bus architectures which includes AMBA advanced high-speed bus (AHB), advanced peripheral bus (APB), and advanced extensible interface (AXI) which is the most recent AMBA variant; they are all managed, but not licensed by, ARM Holdings PLC. The AHB is arguably the most widely-used on-chip bus protocol, and it provides support for multi-master arbitration, multilayer and pipelining support, bursting support, etc. The AHB model assumes that signals are driven from each master and multiplexed onto a common address/data/control bus by a multiplexer controlled by the arbiter. The multiplexed output is then fanned-out to the AHB "slaves." Similarly, data from each slave is multiplexed onto the common bus and sent to the SoC masters. Simpler AHB-lite systems work similarly, but since there is only one master, arbitration signals and logic are not needed. Multi-layer AHB systems typically use point-to-point, rather than multiplexed interfaces to allow concurrent multiple master transfers. An AMBA advanced system bus (ASB) is a largely obsolete dual edge bus protocol, AMBA APB is a simpler static bus architecture for peripheral systems, and AXI allows multiple outstanding transactions and other advanced features.

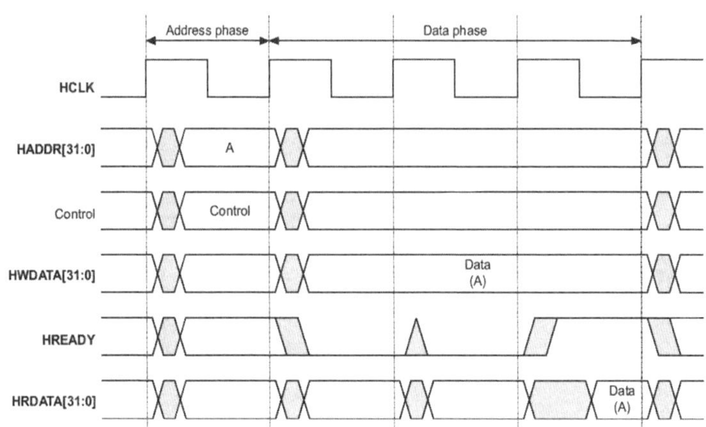

Figure 12.3: *AMBA AHB Operations*

Figure 12.3 shows some of the basic concepts of AHB operations, but does not show any arbitration operations or specifics of control signals. Operation is divided into a control or address phase where a request is made and a data phase where data is transmitted. The data phase may last an arbitrary number of cycles, until data is accepted or valid data is returned, as shown by the *hready* signal. AHB supports many other modes of operation and options which increases both performance and complexity.

OCP is a bus architecture that is defined and managed by the OCP international partnership (OCP-IP). OCP defines a range of complex multi-core and multi-channel interfaces that address pipelining, multiple outstanding transactions, threads and tags, bursting support, etc. OCP is based on a concept of socket–based interfaces that decouple the IP interfaces from the bus fabric to a large extent, the interface defines the IP to socket level interface, while keeping the actual bus fabric implementation open and user defined. This adds a lot of flexibility but by allowing a large set of optional OCP interfaces in addition to a smaller configurable set of required signals. OCP also allows the incorporation of user defined interface signals to address application specific requirements. The OCP debug signals discussed in the following section are one example of a set of side band signals that are, in fact, now in the process of being incorporated into future generations of the OCP standard.

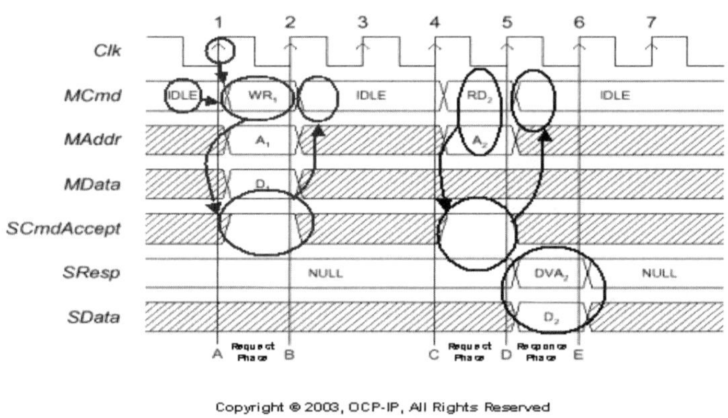

Figure 12.4: *OCP Bus Operations*

Figure 12.4 shows a simple example of OCP read and write operations. One should note that one difference between AHB and OCP is that in write operations, control, address, and data are all presented to the bus in the same clock cycle. Although not shown in *Figure 12.4*, the read response phase can last an arbitrary number of clock cycles until valid data is placed on the bus.

Coreconnect was developed by IBM and is most widely seen in systems based on IBM PowerPC cores. It is supported by Xilinx, as one on-chip internal bus architecture in part since some high-end Xilinx parts have integrated IBM PowerPC cores. Coreconnect defines a set of different buses, such as processor local bus (PLB), on-chip peripheral (OPB) and device control register (DCR) bus for different applications. Each bus component of the Coreconnect architecture is optimized to achieve specific on-chip bus architecture goals. The PLB provides a high-bandwidth, low-latency connection between bus agents that are the main producers and consumers of the bus transaction traffic. The OPB provides a flexible connection path to peripherals and memory of various bus widths and transaction timing requirements while providing minimal performance impact to the PLB bus. The DCR bus is a mechanism for removing system—such as IP initialization, configuration loads, setup of memory address resources and configuration—transaction traffic from the main system buses. Direct memory access (DMA) controller and interrupt controller cores use the DCR bus to access normal functional registers used during operation.

The Coreconnect specification is managed by IBM and to a lesser extent by the Power.org Consortium, and its implementation model differs from other on-chip buses in that IBM provides—under a fairly obligation free but formal license agreement—access to RTL and system models for different Coreconnect configurations.

Wishbone is a specification, managed by non-commercial Opencores, that was created for open-community on-chip bus architectures, and it is mostly seen in conjunction with Freeware IP blocks. Wishbone is considered a well thought-out architecture in the industry, and it shares many characteristics with OCP in having a single bus specification with a number of optional configurations. As of yet, the author knows of no production silicon having been designed using Wishbone, but it is however popular with the IP freeware and shareware communities with many freeware IP blocks being developed with Wishbone interfaces.

AMBA, AHB, and OCP are arguably the most well-used types of embedded buses, and they all share similar characteristics, can have different fundamental concepts, and can vary widely in their implementations. At a fundamental level, AHB and OCP share a similar transfer and response multi-cycle protocol (although implemented differently). AHB defines a large number of control and configuration signals defining bus size, enables, mode handshaking, etc. but does not include a native facility for extending or customizing the AHB protocol or signals. As a result, there are many AHB implementations that implement either (or both) a subset or user-defined (but not standardized) extensions to the AHB specification, resulting in many AHB variants in use. Differences in variants and extensions, especially from different vendors can impact integration and verification efforts and increase complexity for creating or reusing test benches.

OCP takes a different conceptual approach of having a relatively small set of core bus signals, but it also defines a number of optional extension sets that address special bus command modes, burst operations and multiple data tags, data alignment, multi-threading, and debug operations. OCP also includes a native capability in the specification to include user defined "sidebar" signals. The end result is as many functional variants in bus implementations as with AHB, bus with many of the options predefined in the OCP specification, potentially resulting in consistent signal set definitions (and less confusion) during integration and verification.

Even with these differences, it is feasible in many cases to develop bus bridges that allow common bus architectures to support both AHB and OCP interfaces. In the next section, we discuss some socket–based IP integration schemes that are used by bus fabric IP vendors to facilitate support of IP with different interfaces.

Socket–Based IP Integration

On-chip interconnect systems for integrating IP blocks into an SoC solution has received much attention in recent years, with a number of multi-layer, cross-bar, and network-on-chip (NoC) alternatives being developed both by IP vendors and internally by integrating companies themselves. The goal in most cases is to address within reasonable wiring and size constraints, an increasing amount of bandwidth required for complex applications, optimized communications between different block of IP, both with each other and with shared resources such as memory and peripherals. In many cases, these interconnect architectures contain sophisticated internal complexity, which have tunable parameters, to allow tradeoffs and optimization of the interconnect features for a given architecture and application. The interfaces to these interconnect systems is typically implemented at a socket level, using one of several bus interface standards (OCP, ABMA AHB, and AXI, being among the more prevalent) as a modular, bi-directional socket interface between an IP block and other interconnected blocks.

At least three commercial companies offer interconnect and bus structure automation and IP tools, with several other SoC-centric interconnect approaches being used as proprietary customized solutions by SoC silicon venders. Sonics offers the most mature commercially available solution, with its third generation server macro eXpansion (SMX) and related SMART interconnect architectures. Offering alternative approaches are Arteris, offering a NoC interconnect architecture and Silistix offering a self timed (clockless) interconnect.

Socket–based interconnect is a standards-oriented approach that focuses on adding value to the interface socket between the IP block and the bus fabric. Socket–based interconnect is an underlying principal in many OCP–based architectures, but can also be applied to other bus architectures. Since many bus architectures allow various bus options to be added and selected in order to increase the functionality of the bus interconnect, using a socket–based interface simplifies addition, removal, or accommodation of the bus interface to the IP blocks, and the development of test suites to address verification and optimization of the design.

The sockets accept initiators (masters) and targets (slaves), with the circuitry of the socket encompassing the necessary state machines, gating and multiplexing circuitry, and wiring to effect the desired data flow (quality of service [QoS], multi-threaded non-blocking communication, security features, dynamic power gating, etc.) characteristics. The socket consists of a set of agents that provide the necessary signal and protocol management to address the specific interface needs of a core to the more general resources of the interconnect fabric. Socket–based bridges may also be defined to other interconnect environments between OCP, AMBA AHB, and AXI and bridges for other arbitrary existing interconnect structures can also be developed to simplify utilization of legacy hardware. The advantage of this socket–based integration approach is that it allows a clean point of separation between the core IP function and the interfacing logic needed for the system integration.

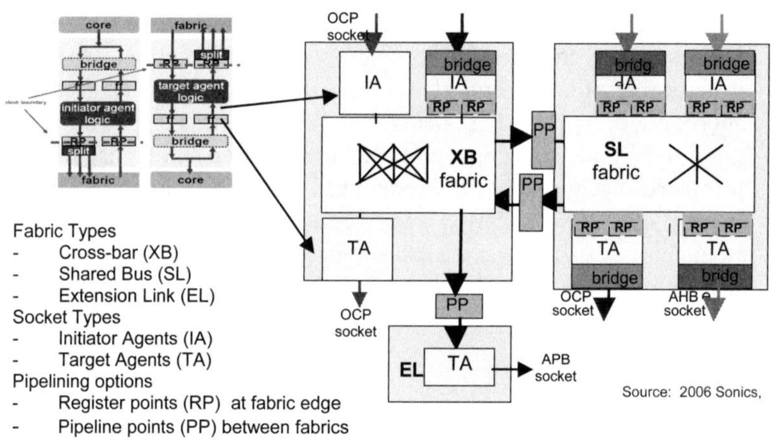

Figure 12.5: *A Socket–Based Interconnect Design Example*

If your IP provider is providing standards–based IP interfaces, some, or most, of this socket level functionality may already be implemented as part of the bus interface socket that they provide.

Integration Make or Buy—Pre-Integrated IP

An emerging trend in IP development is providing pre-integrated IP solutions, also known as IP platforms. They extend the argument for IP value from the block to the sub-system level. As discussed in the previous section, much of the complexity of hardware IP integration is related to defining requirements for and development of the interfaces between a core and a bus architecture (the fabric of which may also be IP block) connecting it to another IP. Examples of pre-integrated IP platforms vary widely and may include Mac and physical interfaces (PHY) for various high speed interfaces, integrated multi-processor configurations, processors integrated and cache controllers, DMAs, or other memory interfaces, and an increasing number of application specific platforms for specific functions (e.g., audio, video or other media processing) that integrate processor and associated relevant peripherals. The value of this pre-integration is that, for instance, while many peripherals are relatively straightforward and considered "lower value" IP, their integration results the same test and verification required as more complex "higher value" IP. Verification and test time is reduced by pre-integrating a suite of these IP blocks. This same logic can be applied to more complex IP sub-systems and several IP vendors do offer platform–based solutions in addition to cores themselves.

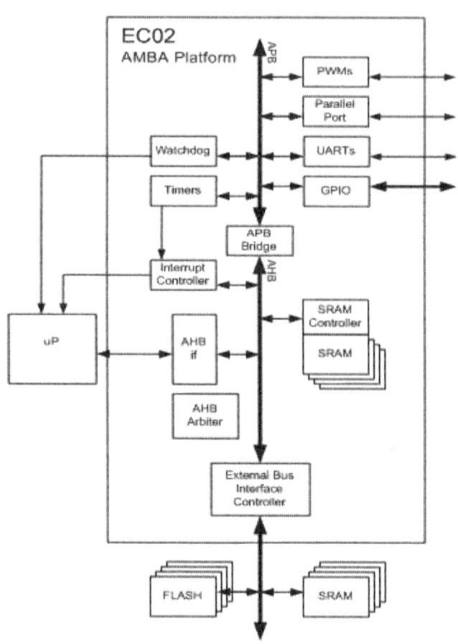

Figure 12.6: An IP Peripheral Platform

This example of an integrated AMBA peripheral platform shows the level of integration and potential verification that can be avoided by using a pre-integrated IP platform for commonly used peripherals. The value of a platform is somewhat determined by your system design. If part of the value of your design is in specific types or integration of the peripheral subsystem, then platforms are probably not the way to go. Alternately, if you are using a fairly standard set and configuration or peripherals in your design, then you may want to look if there is really value in designing, integrating, or verifying your own peripherals and whether you get more value, time to market, or quality by using a pre-configured solution.

Integration of Asynchronous IP Blocks

There are two types of asynchronous IP issues that need to be considered. The first is integration of different IP blocks running over different clock domains. There are at least two major issues that you need to consider. The first is synchronization over the clock boundaries, to ensure signals are transferred without any loss and without potential issues of metastability, aliasing, or signal glitching. There are a number of proven means to address these types of problems, such as by double buffering signals between the domains. For data transfers, there are also issues that ensure adequate bandwidth exists at both ends of an asynchronous interface to support required operations. This may involve resizing or further buffering the data buses to accommodate required data requirements. Determining if there will be bandwidth congestion over the long term or steady state operation is one area where system level models can become useful. These asynchronous interfaces are an area where you may want to focus additional simulation and analysis efforts during integration.

The second type of asynchronous IP to consider is essentially clockless systems. Many types of analog IP fall into this camp, however there is also an emerging and growing amount of clockless digital IP that are now being developed. The pros and cons of clockless systems is a complex topic, but both speed and dissipation can be improved in certain cases using clockless architectures. This does, however, open up a range new analysis and verification concerns over race conditions are other subtle quirks. Clockless IP integration is, to a certain extent, still cutting-edge technology and, as we all know, the cutting edge cuts both ways.

12.4 Analog IP Integration

Analog IP can range over a plethora of functions, but it is typically much more limited in complexity than digital IP. It is seldom programmable or configurable in a complex way. This is not to say however that integration is easy. While most experienced high speed design engineers understand the basics of analog input/outputs (I/Os), impedance matching, reflections, and the like, the more complex magic of analog systems integration still remains a relatively specialized domain. This is an area where experience counts more than anything else, and the best advice is to thoroughly perform due diligence on your analog IP provider. As a general rule, they should have some gray hair, either from years of analog design (or from the resulting stresses of prior projects thereof).

12.5 Integration of Soft IP versus Hard IP

Soft IP, or IP made available in RTL form, has better visibility into IP operations for both simulation and debug purposes. It is synthesizable and allows the customer to retarget the IP to different types of silicon. However, in this case, the customer now owns the responsibilities for all of the downstream (e.g., post-RTL) design tasks of synthesis, place and route, etc. to meet the performance and area goals of the design.

One of the key things to look at for soft IP is RTL quality, along with supporting scripts, test benches, and flow-related deliverables that will make your part of the downstream design as painless as possible. One should always expect some pain, especially if you are trying to get high speed operation with minimal area. The IP vendor may get better performance/area numbers than you can, they also can afford to spend a lot more time optimizing and tweaking the flow to get the best numbers. If in doubt, you should contact the IP provider since most of them have consulting experts at hand that will find optimizations that you may otherwise not.

One should always be sure to correlate and verify your design as you move through the flow to the IP "golden" model. This should be, but not always, the pre-synthesis RTL itself. An IP provider should be able to provide a mechanism for verifying and mapping the netlist and physical design back to the RTL. This may require some specialized EDA tools or third-party support.

Although you have access to the RTL, it is usually advisable to resist any urges to improve the IP in any way. If there is a compelling need to make any changes whatsoever to the IP, work it out in conjunction with your IP vendor. Keep in mind that, regardless of how long and hard you have worked with the IP, you are usually licensing, rather than buying IP, and the IP is still the vendor's property. Any changes made without their express approval may have legal implications and will almost certainly void any warranties for the IP operation.

Hard IP, or IP provided as a technology specific netlist or as a physically designed macro cell, eliminates some the design tasks related to getting the IP ready to go to a foundry. Since the macro has been synthesized and optimized to be "good-to-go," the work involved in basically the final drop in and place and route at the SoC stage. This work may be done by the foundry or by consultancies specializing in this aspect of the design flow.

The downside is that it is a very limited block in terms of foundry and process, since all the technology and process specific stages of the design have been added already. There is no changing your mind and moving to another foundry or process node for your next revision (without repurchasing the IP).

For analog IP purchases, hard IP is basically the only way to proceed, since most types of analog IP, support for soft IP flows, such as analog synthesis tools, barely exist.

For digital IP, hard IP is basically cheaper to purchase than soft IP due to the limitations in use. This is in many cases however to the customer benefit, since if hard IP does the job, it is arguably has much of the pre- integration and verification already completed and if it is more limited in terms of interfaces, these are interfaces that have been proven out with the foundries that the hard IP is targeted for.

When purchasing IP, one should always look for good models, such as availability of RTL, and some proven use of the IP. Most IP is verified by a foundry using some limited proof of functionality flow. This may or may not be adequate for how you will use it.

12.6 Test Issues

IP test, by its very nature, is a difficult and thankless business. Let us consider the scenario: the user has an IP block, no one on the team completely understands the innards of it; in some cases, the IP is a classic definition of a black box, and now we need it to such a point of confidence where we are ready to include it in a product that is, by definition, the most important product for the company's survival.

In talking about IP test, we need to look at it in two contexts: (1) testing at the pre-silicon stages, which as mentioned earlier, is integral to, and the point of, completion of the IP integration process, and (2) testing of the IP in silicon, the actual product. The in-silicon testing, can in turn be considered in two sub-contexts: (2a) what is being tested in the lab and (2b) what is being tested in production. You want to have a test plan that integrates all these different scenarios together and allows for an integrated test flow where each stage can leverage and be referenced back to work done previously.

Developing IP Test Plans

It is reasonable to assume that any IP of a given complexity has some design flaws, ambiguities, and errors which is the nature of most complex systems. The main concern that the IP customer needs to have is that if any of the imperfections in the IP impact its own employment. The IP customer does not even have to consider the testing aspects of the IP that do not closely relate to the application. Unless you want to do it as an academic exercise, it would not be wise to even think about testing aspects of the IP that do not relate to your application.

For complex IP, you will be taking on a potentially open-ended problem that is not your concern. It will not be appreciated by the IP vendor, nor by the management, since they are typically more concerned if the IP is working in the application just as it has been designed. It is also always useful to remember a difference between you and your IP vendor - you are shipping a product and he is shipping a model (e.g., RTL or otherwise). Models tend to be tested under ideal conditions. Silicon is reality and funny things happen to models under the stresses of reality. It is a good bet that you will find problems in a week that he has not faced in a year. This is not to say that you are good and he is bad – it is just that you are testing the IP in a different way and with a different focus than he has.

IP Test Infrastructure—The Test Toolbox

There is a wide range of testing, as opposed to verification, strategies available for IP, and an entire infrastructure of tools, service providers, and other capabilities that are at your disposal. We cover only some of the more common IP test approaches here, which the IP providers should be able to support their customers with.

Assertion–Based Test

Assertions are essentially in-process checks and tests that are added to an RTL design at the verification stage. These essentially trigger a message or other actions in the simulation or formal verification environment when a violated design constraint is encountered or when a defined illegal condition is detected. These assertions can be written and interfaced into the models in a number of ways. VHDL, for example, has a native assertion construct that allows inline assertions to be defined. There are a number of verification-related languages that support arbitrarily complex and sophisticated assertions which assess both functional and timing constraints. These include Vera and *e,* which are proprietary solutions with more standardized, but less mature, approaches being offered such as property specification language (PSL) – Institute of Electrical and Electronics Engineerings (IEEE) 1850 standard – and as part of SystemVerilog (IEEE 1800). Developing assertions required for different pieces of IP to be integrated can be time-consuming, in defining all the possible cases and in both writing and adding the assertions to your model. There are several vendors that do, in fact, provide assertions for standard functions, such as bus interfaces. IP providers can provide assertion packages as part of the RTL deliverables.

Coverage–Based Test

Coverage–based test is employed to address whether a given test bench or set of test vectors exercises all of the code and functions of the IP model. There are a number of coverage types and cases that are addressed at different complexities whether a test includes the operation of a given RTL statement or function (statement coverage), whether it fully exercises the function for all input and output conditions (condition coverage), and whether it exercises all the paths that a function is part of (path coverage). Coverage–based test is usually not inclusive of all IP operations, but it is a necessary and valuable metric to determine how well the verification has been performed. The IP provider should be able to provide coverage numbers for the IP they are selling. While structural coverage analysis is a fairly mature EDA technology, recent advances in functional and assertion coverage imply that there are new types of coverage metrics now available which are, however, not as mature. This means that one has to examine and understand exactly how these new types of coverage have been applied. Closely-related types of coverage analysis are also performed for software and should be considered for any IP that has both hardware and software components. Coverage analysis should be a definite part of integration tests to ensure that all the paths between different IP blocks are being exercised by the given tests.

Testing IP Corner Cases

If the IP turns out to be fairly mature, testing the mainstream operation of an IP block becomes very straightforward. The tricky parts come in the corner cases, unusual operating modes, or sequences for the IP or the integrated IP system that are important to select customers, but perhaps not used by a majority of them. Focusing on the relevant corner cases to determine that all modes of operation are being tested is part of the art of test and should be one of the areas of focus in your own test plans.

Directed and Pseudo-Random Testing

One of the ways to address corner cases is through directed and pseudo-random testing. While not appropriate for all types of IP, this does allow automated test of different corner cases, including ones that may not be intentionally considered, which may be a good or bad thing. In essence, directed random, or pseudo-random, tests exercise a combination or sequence of test conditions that one designates as being valid and provide constraints on the types of test combinations and sequences that need to be performed. This can include a large amount of combinations that can be automatically exercised. The operation results are then compared against a "golden model" to determine if the operation is as expected, and discrepancies are flagged. One of the advantages of verification languages is built-in features for these types of testing; one can also conduct pseudo random testing in standard HDLs, but it typically requires more work and access to specialized test function packages. There are specialized tools available that address directed and pseudo random test for different domains, for example, which exercise instruction sequences in a processor design.

Test Integration for IP Systems

The second aspect of IP involves adding the test features to both the IP and the systems to facilitate the engineering and manufacturing test of the design. We discuss two approaches to this: (1) Joint Test Action Group (JTAG)–based testing, which is the workhorse of all silicon testing and built-in self test (BIST), that provides a method of automating the verification of your system; (2) design for debug (DfD) as another method for adding analysis and verification capabilities to the IP or the integrated system to allow more in-depth testing and system optimization.

Manufacturing Test Issues

Once your design is complete, there is still the little matter of making sure that all the product it is based on is working and continues to work both through additional stages of development and in the field. IP–based product coming out of the foundry can be tested at several levels which include at the wafer, the die, and at the packaged part level. A small quantity of test vectors is typically provided as part of the foundry price; however, these tests are mostly for basic functionality and operation and will not test the specifics of the design. If you are planning any significant volume product manufacturing, it is a good idea to design in additional features to automate and improve the testing process. The extra, and usually minor, expenditure in design time and gates will usually pay for itself manifold over the overall life cycle of the design.

Test Interfaces—JTAG and Others

For most significant IP blocks, a test interface is a useful, if not mandatory, feature to allow simple(r) access to internal information and test features of the IP block. The test interface is typically via some fixed port and, in many cases, only a serial interface is needed to access on-chip test registers and built-in features. Of all interfaces, IEEE 1149.1 JTAG is the most widely used, and it has a long history at both the chip and board level as the main test interface standard. JTAG is a four or five wire serial interface consisting of a JTAG CLK (TCLK), mode select (TMS), serial input (TDI), serial output (TDO) and optional reset (TRST). Operation is controlled by a JTAG state machine function that sets the serial enable, read and write of registers defined for a JTAG test access port (TAP). One should note that a design must be developed with JTAG as a consideration, e.g., including JTAG compliant registers where required and adding and compiling any JTAG options into the design for JTAG to be useful. An IP provider can typically provide all detailed information and guidelines on including JTAG and on any specific JTAG features that are included as part of the IP. Your foundry may have specific JTAG guidelines as well, in particular as they apply to boundary scan, I/O continuity, and any other manufacturing test.

One attractive feature of JTAG is that at a system level JTAG blocks of different IP blocks can be serially connected to allow access to all of the JTAG enabled features of all the IP in a system. This allows, as an example, the ability to scan any and all JTAG compliant registers or perform JTAG–based run control on a given IP block. Since JTAG is a serial port, it typically allows access to only one JTAG function at a time, which can be a limitation for system testing of multiple IP blocks. Some of these issues can be addressed in the next section on design for debug. The JTAG standard defines a small set of required operations, but it does allow a largely unlimited number of user defined operations which many IP vendors have used to customize JTAG for specific test (e.g., programming, debug, run control, etc.) purposes. As a result, there are many effective kinds of JTAG being used, which share common characteristics in terms of port signals, programming support, and basic (1194.1 compliant) operations, but which can otherwise differ significantly.

Depending on the popularity and maturity of the extensions for a given JTAG interface, it may have probe and software support from many vendors, or it may be much more limited. As an example, EJTAG are a set of JTAG compliant processor debug extensions developed by MIPS Technologies for run control and breakpoints for all MIPS processor cores, and they are supported by many vendors. While JTAG–based interfaces tend to have the highest levels of third party support, debug blocks for other processor IP may have differing levels of support.

It is generally useful to be able to have a robust JTAG development environment, since a well-designed JTAG architecture supports flexibility to access and control registers and otherwise providing functional visibility. There are a range of test-related tools and services for predefined JTAG functions, run control as an example, which provide application programming interfaces (APIs) and higher level commands to allow access without understanding the specifics of the JTAG interfaces. There are many approaches and tools that support JTAG programming, and their usage all depends on how involved the customer wants to be in the design's JTAG development. Most probe vendors provide some programming tools and software support, although these are typically specific for the functions supported by a given probe. General JTAG functions may be described in several languages; as an example, boundary scan device language (BSDL) is the most mature and perhaps the most standardized (via IEEE 1532), while others such as core test language (CTL), which is part of IEEE 1450, provides support for additional features. Several EDA and programming tools accept these as inputs.

There are other options than JTAG for test interfaces, and other serial ports can be used for the same functionality in principal. In practice, however, other test port options have less support and can possibly "box in" the customer or limit their flexibility. As a general rule, 1149.1 JTAG compliance is the lowest risk, if not the highest performance option, in both the IP and in supporting tools.

IP Bist – Signatures

BIST is a widely-used approach to address automated testing, and it is very appropriate for IP–based systems. While most widely used in memory systems, it allows a quick and repeatable test of an entire design or some IP part thereof. The basic concept of BIST is to take regular samples of key registers during some well-defined operation of the IP block(s); this may be a test routine or other operation that provides a comprehensive amount of coverage of the design. The register values are combined, often using some linear feedback shift register (LFSR), with the result being a unique value at the end of the operation, which is called a signature. By checking the signature, a simple go/no go test is confirmed. More in-depth analysis of the signature value can indicate additional information such as which registers or portions of the design contributed to failure.

IP Test as Part of System Test

Just as the line between IP and system integration is often a blending rather than a clear transition, IP test should, in the best cases, evolve into the final system test. All of the test approaches discussed in this section are applicable both at the system and the individual IP block level. IP level test should be reusable and accessible at the system level throughout the testing process, in order to facilitate regression testing and also to simplify and add both test depth and coverage at the system level. The value of approaches, such as assertions and pseudo-random testing, increases at the system level since, arguably, the complexity of the system, its overall interfaces, and number of corner cases is increasing more than linearly to the number of IP blocks being integrated.

12.7 Design for Debug Issues

One should consider, during the design of the IP–based system, that one will soon need to test and verify the actual hardware, rather than just models. To differentiate this hardware verification from the EDA-related pre-silicon verification steps and issues, we will refer to this real-time, real hardware, in-silicon (be it prototype FPGA or final "first silicon") analysis and verification stage as in-silicon debug.

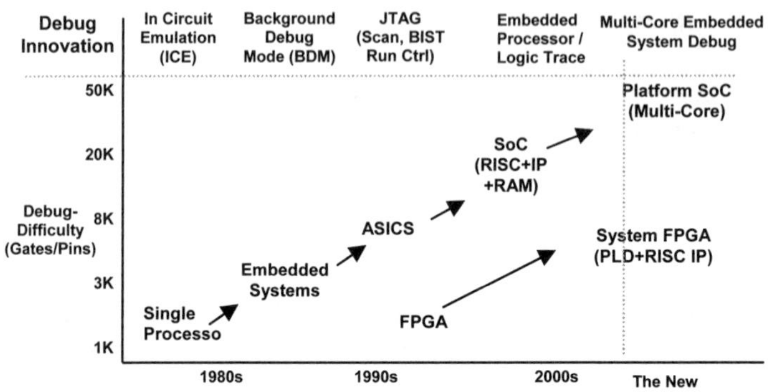

Figure 12.7: *An Evolution of Systems and Their Debug Complexity*

Figure 12.7 provides some examples of the complexities involved in in-silicon debug of leading edge silicon. There are two messages inferred in the chart. First, using a simple metric of debug difficulty, which is derived from a ratio of gates to IO pins on a device, we can see an increasing difficulty in accessing information that is embedded in a device, due to gate counts increasing much more quickly than package IO. Since an ever increasing portion of the design is largely inaccessible by direct monitoring at the IO, some on-chip support is required to make these deeply embedded signals available without significantly perturbing the operations of the system that is being analyzed. The second message is seen at the top of *Figure 12.7*. As electronic system designs have evolved into new and more complex architectures, there has also been a corresponding development and evolution of new types of on-chip debug solutions required to provide in-silicon analysis and system verification support. The increasing adoption of (at the time, new) debug tools for IP such as JTAG–based analysis, and processor and logic trace in ASIC and SoC designs, show how more complex debug solutions might come about as we move to more varied platform and multi-core IP–based solutions.

Any IP verification and integration plan should consider in-silicon debug, as well as the EDA–based verification. In particular for processor and complex bus applications, it is difficult to adequately perform the verification of applications-oriented processing operations and software data from processor sub-systems, based solely on simulation. By itself, simulation is rarely able to address all the facets and nuances of the physical IP application, such as its interaction with real life interfaces and operating environments. In addition, it is often not effective to simulate, at an RTL level, larger IP–based architectures for the extremely large numbers of cycles required to evaluate the application aspects of system operation and real world system performance. Let us consider a processor or digital signal processor (DSP) running a complex multimedia application from start to stop, to get some magnitude of the problems and simulation cycles involved. ESL models and tools provided with some IP give a workaround on this simulation performance problem, by modeling of the hardware and software more efficiently. This is carried out by using more abstracted and idealized levels of modeling, such that ESL modeling will often mask detailed problems that reside in lower levels of the detailed design. (e.g., you abstract away the details with ESL tools).

So, in addition to all other "ilities" (such as testability, reliability, manufacturability) that must be considered as part of the design effort, IP–based systems should also have a DfD strategy to address hardware–based functional verification. As more processing elements and IP functions are integrated into the silicon, the embedded complexity outstrips the capability of any standalone logic analyzer, debugger, and emulator–based diagnostic tools.

One approach to in-silicon debug during prototyping or "first silicon" verification is to add some amount of on-chip instrumentation to a verification/debug/integration platform where the IP itself is buried in the silicon and not externally visible to users on a real-time basis. On-chip instrumentation is, in many ways, a toolkit of resources and analysis philosophies that are of assistance in debug of complex systems that is generally based on embedded blocks that provides external visibility and access of the "inner workings" of the IP and external post processing or visualization tools. When properly implemented, it provides a real-time "peephole" into the operations and interfaces of key internal blocks that can not otherwise be accessed in sufficient granularity on a real time basis There are several types of on-chip instrumentation applicable to different types of IP that provide powerful general tools to embedded IP debug. Different on-chip instrumentation approaches and implementations are used for debug, say, at different aspects of most IP and system architectures; processors and buses have very different analysis requirements and, as a result, also various types of instrumentation options.

At its simplest level, instrumentation consists of one or more sets of blocks that allow collection, aggregation, and concentration or compression of selected internal data of a SoC that is desired to trace over time. The signals are exported over device pins for external post-processing or visualization. More sophisticated instrumentation blocks allow real-time data filtering, on-chip analysis of target performance, and single or multi-core triggering and breakpoint management within the target chip.

Typically more focus on debug features is seen in early (pre-silicon) debug (e.g., hardware emulation or in FPGA devices) with a lesser feature set shipped in final silicon. On-chip instrumentation can include trace capabilities and hardware event recognizers, as well as methods of trace module integration and source-level debugger integration. In making IP selections, it is often useful to compare the supporting instrumentation that is provided by a vendor against one's own debug plans, to determine if any unexpected design-ins will be required for hardware verification and in-system visibility.

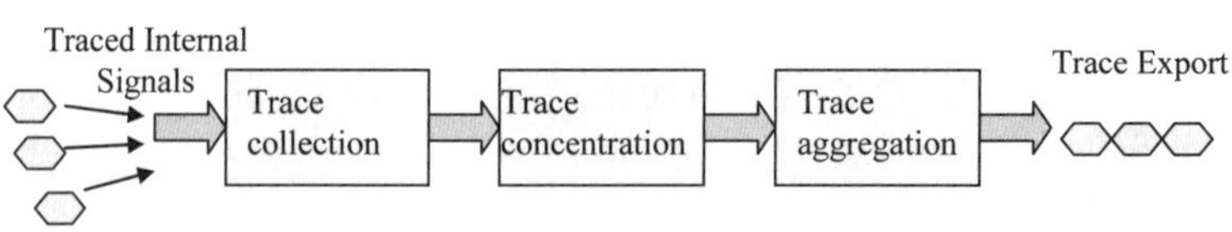

Figure 12.8: *On-Chip Instrumentation Trace Formatting and Data Export Flow*

Historically, development of each generation of IP architectures, which include 8-bit micro-controllers (MCUs), DSPs, reduced instruction set computer (RISC), application specific co-processors, among others, is accompanied by supporting debug instrumentation and tools. Most processor IP provides some form of JTAG interface for run control and basic debug (i.e., single stepping, breakpoints, etc.) functions. These are often supported by trace capabilities, usually in the form of a trace port such as the MIPS EJTAG/processor data trace (PDtrace) and ARM's embedded trace module (ETM).

These tools solely address processor IP and not system applications where the IP is used. This is where the customer's debug should be focused. Processor IP debug assumes different issues from debug of the system. Many processor IP in-circuit emulation (ICE) approaches force a non-standard operating state on the processor with different opcodes and interface features. The resulting operation of a processor in emulation mode makes it appear in a dormant state. Changing operations or disabling interfaces of an IP block for debug mode may work well for the isolated IP, but, at an integrated systems level, an interruption of the steady state performance of the system introduces (time) intrusive elements into the system operation that can complicate or invalidate the data or operations being observed. On-chip instrumentation alternatively in general provides a more non-intrusive approach to event recognition and triggering, trace, and run control.

Data tracing provides real-time information about the status and data of a system's internal signals, including examples such as, analysis of cache performance and internal memory, and data transfer operations that can not otherwise effectively be extracted from a system. In-line or post-processing of this information performance analysis of systems characteristics such as bus availability or cache hit/misses, which require a long-term steady state (measured over many cycles) system information. Additional detection of events in traced data allows the development environment to flag specific features in the trace data as it flows through the application.

IP Interfaces for On-Chip Instrumentation

Debug ports simplify controllability and visibility by providing a low overhead access to internal signals. The dominant type of debug and test control are JTAG–based ports.

JTAG allows user-defined instructions to extend the JTAG instruction set for probe or trace modes, instructions for ICE, and access internal JTAG enabled registers. While in probe mode, the processor ICE can examine and modify the internal and external state of a system's registers, memory, and I/O space. A rich infrastructure of tools environments and standardized debug schemes have been built on this foundation to provide both JTAG debug of embedded processors and other parts of an embedded system.

JTAG is the default interface for many IP blocks (it is there for test purposes, irregardless of debug features the block might support), however, JTAG is a relatively slow serial interfaces not designed for trace and other data intensive analysis. One basic instrumentation approach is to rely on on-chip memory to buffer between traced data and the JTAG export bandwidth available. This does do not need to be used carefully since buffers of modest size are easily overloaded with a large amount of trace data that is generated in cases of multiple IP blocks or internal buses, placing limits on durations of trace that can be supported.

IP Performance Analysis

Embedded performance analysis instrumentation is also used with processor IP, and it is applicable to many other IP types. The advantage of performance analyzers is that instead of following the data-intensive approach of tracing out critical information, the analyzer block can track sets of events, and provide a summary output periodically or in response to some trigger or event in the IP itself—for example, completing an operation on a video frame. The advantage of performance analysis instrumentation is that it provides quick to see and interprets long-term metrics on the system operations to identify areas for more in-depth investigation or optimization. Some common performance analysis instrumentation applications include:

- Overall I or D cache hit/miss ratios
- Cache hit/miss ratios in a specific section of code
- Instructions executed per clock

- Number of executed instructions between two points in the program, and how it varies
- Memory access utilization of on-chip or off-chip memory blocks
- Number of processor stalls versus total number of instructions executed
- Number of clocks during which processor is stalled relative to total clock cycles
- Internal bus utilization, e.g., ratio of bus busy to total clocks
- Number of branches taken relative to total number of branch instructions executed
- Mis-predicted branches relative to number of branches taken
- Number of clock cycles between location A and B in the program (A-B timing)
- Throughput of the critical system functions; i.e. how much work per unit time is being done by the processor. Examples would include network packets processed per unit time, streaming blocks of data (such as audio or video data) converted and forwarded to another stage of processing, interrupts processed, graphical data converted to pixels, etc.
- Measure loop times in critical algorithms; determine maximum duration, which may be more critical than the average duration
- Interrupt service routing (ISR) min, max, and average execution times

Processor Instrumentation Analysis

There is no "magic bullet" instrumentation approach, but rather a number of commonly required capabilities needed to provide a robust debug solution. On-chip instrumentation enables a range of widely used best practices in debugging, especially in conjunction with prototyping environments. By integrating with hardware emulation, the instrumentation provides a synchronized method of loading and debugging the code in a prototyping environment, as well as efficiently implementing run control, real-time instruction

and data trace information, real-time operating system (RTOS) support, memory subsystem, breakpoints and watch-points, etc.
A processor trace instrumentation IP block integrates with a processor core to select, collect, compress, and stream real time trace data. It consists of a JTAG–based controller and configurable trace processing blocks that allow capture of the processor bus trace, execution history, and/or other real-time information from the core. Trace blocks can be configured to allow either on-chip or off-chip trace storage, depending on trace depth requirement and can also be configured to collect profiling data for performance analysis and register/memory access via JTAG controlled registers.

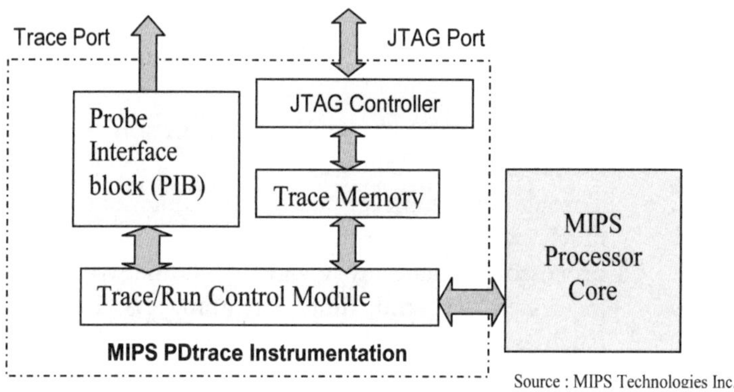

Figure 12.9: *A Processor Instrumentation Interface*

Embedded Bus Monitoring

If IP is the heart of a system, interconnect is the circulatory system and monitoring internal bus operations is the key capability to debug the IP and its interfaces. On-chip instrumentation applied to the bus signals captures bus activity that is integral to understanding and optimizing IP integration, not only in terms of finding bus errors, but also in determining critical system communication parameters that are between IP blocks—including bus latencies and tradeoffs of different arbitration schemes—which may be difficult to analyze through alternate ways.

Bus instruments allow the trace of address bus, data buses, and various control signals at the bus interfaces. We discuss an example of AMBA AHB since it is widely implemented, but the concept can be applied to any bus architecture. The instrumentation interface is configured to receive address/control and data buses data driven from the currently granted master and currently addressed slave. The bus instrumentation passively collects bus activity and transfers the collected data through a buffered JTAG or trace port to the external probe.

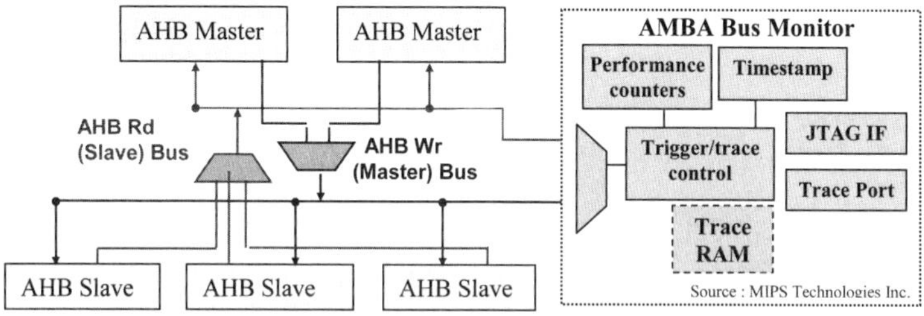

Figure 12.10: *Instrumentation Interface into an AMBA Bus*

Additional SoC signals can be included in the trace, such as interrupt requests, bus or core status, and processor control signals. The additional signals may be optionally used to compare and recognize specific IP communications outside the AHB bus and transmitted to the probe for triggering purposes.

Multiprocessor Debug

Single core approaches for debug and trace often fall short when used in conjunction with multiple cores or with processors interfacing with complex application specific IP. SoCs increasingly integrate multiple types of IP cores that may be running asynchronously or with variable or indirect communications with each other, thus making it difficult to time correlate debug information. Complicating multi-core debug issues further is the fact that IP blocks come from a variety of vendors, and usually have different compile and debug environments or levels of debug features. Such tasks as processor interfacing, inter-processor communications, run-time execution and coordination, and data presentation place a significant overhead on the debug requirements for heterogeneous and multi-core chips. In these architectures, the instrumentation block(s) must be customized to support the specific verification and debug requirements of the processors on a stand-alone basis and in a multi-processing configuration. Among the basic requirements, instrumentation blocks must be diverse enough to effectively communicate debug data with their respective cores and interface to each other to coordinate all their activities.

Instrumentation solutions for on-chip buses provide a valuable resource for visibility and debug in multi-processor designs. For systems that have multiple processors communicating over a standard bus as AMBA AHB, access to information, like which processor owns the on-chip bus, can provide valuable context as to the relative communication and stages of processor execution. With the introduction of increasingly complex bus architectures, FS2 anticipates that the future generations of multi-processor debug will rely on more extensive tracing and triggering of bus operations. This will address inter-processor communication issues in conjunction with more specific point solutions for processor specific analysis.

Looking ahead to more complex systems, the need for debug has typically changed with new types of silicon architectures and applications. As seen in *Figure 12.7*, the complexity of being able to observe internal signals in hardware has and will continue to increase in Moore's law like proportions. For complex systems with many IP components, instrumentation will need to have sufficient "embedded intelligence" to interpret information passing between cores, to be adaptive in extracting necessary information for debug and performance analysis and to support more sophisticated task aware debug for on-chip RTOS or network protocol analysis. Presenting all the diverse debug information in a coherent, understandable way is equally challenging. As in many areas of complex SoC design, new classes of instrumentation are needed to address diverse debug and analysis requirements of emerging architectures.

Standards–Based Debug

One issue in debugging IP and especially integrated IP systems is that the IP blocks may be coming from several sources and, as such, the debug features and resources will be different for each block. If you are implementing on-chip instrumentation, you may encounter different levels of support for the diverse types of IP.

Even with JTAG–based test and debug features, which typically all conform to the 1149.1 JTAG standard, the specifics of how each JTAG test access port (TAP) interacts with probes and software is typically inconsistent and proprietary from block to block and even limited interoperability between instrumentation, probes, and tools from different venders is not a given. Standardization of the debug environment is being widely pursued in many areas related to SoC and IP development. Several industry working groups are addressing different aspects and levels of instrumentation interfaces and tools interoperability problem.

Nexus 5001 Forum is the most mature of the working groups and has focused on a standardized set of instrumentation interfaces and capabilities based on the IEEE 5001 specification. These include a consistent set of auxiliary pin functions, message–based transfer protocols and standard development features to facilitate debug implementations. The standard itself is open and processor-independent, but the implementations are user-specific, within a context of Nexus (JTAG and AUX parallel port) interfaces, messaging resources (TCODEs), default registers, and support for different classes of Nexus implementations.

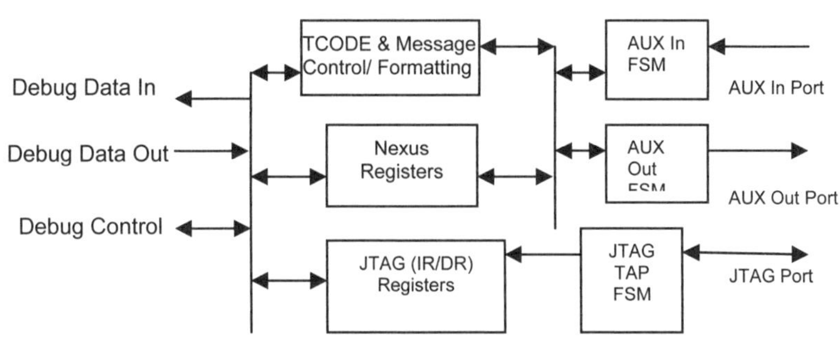

Figure 12.11: *Nexus IP Debug Interface*

The OCP-IP Debug Working Group has developed a draft set of debug-related interfaces that standardize on a common "debug socket" to allow consistent signaling for IP debug control, trace enable, and triggering interfaces. The debug signal set consists of a small number of default signals that define IP debug mode and enables and handshakes with a debugger or debug controller, along with a larger set of optional signals that are added to support particular tasks, including trace, triggering, timestamping and performance analysis. The debug socket, similar to its data counterpart, reduces the IP risks by defining and supporting debug interfaces.

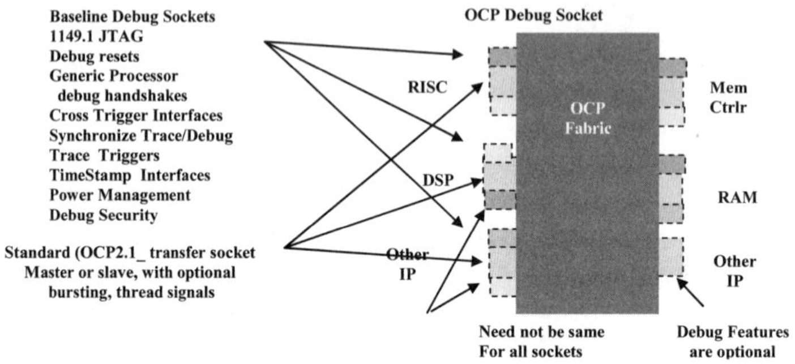

Figure 12.12: *OCP-IP Debug Sockets*

The SPIRIT Debug Working Group addresses standard formats for passing information to and between (EDA and debug) tools. The information is passed in an extensible markup language (XML) format and allows the tools to understand basic information about the IP, including IP type and configuration, IO signals, and available registers. For debug purposes, these definitions are extended to include JTAG commands and registers that are specific to an IP debug block.

There are other debug efforts as well that are worth understanding [1]. But, are debugging standards a hard requirement for integration and debug of IP? As of 2006, most of these standards (with the exception of the ubiquitous JTAG TAP) have seen limited adoption by IP vendors. As important as standards are in principle, a more important consideration is confidence that the IP provider has committed long term in-house expertise and partners that provide verification and debug support as needed.

12.8 Other Related Issues

Almost every IP–based product needs to address such issues as integration, test, and debug. There are other peripheral issues that also should be addressed, preferably early on in the design process that can have a significant impact on integration and test of the IP. Many IP vendors have some "eco-system" of complementary third-partners that provide EDA tools, software products, complementary IP, design services, and other infrastructures. These can have a significant impact on your integration and test, both in terms of simplifying your work, and as necessary, providing additional capabilities and resources that are compatible and familiar with your IP.

IP EDA Support

IP products should have a synergistic relationship with EDA tools. Most IP releases should be compatible with a range of EDA tools; however, it is not unusual for certain IP to have better support from one vender versus another.

One should consider the fact that the cost of EDA tools can easily outstrip the costs of the IP itself when determining what EDA capabilities and features are supported for different IP types. EDA tools mostly address the pre-silicon verification stages; however, they are increasingly realizing, especially among larger EDA companies that there is a real need to address the whole design cycle, including verification for hardware prototypes and for first silicon back from the foundry. This is being addressed by different EDA companies, along with other initiatives, in various ways.

One of the more notable initiatives is the SPIRIT Consortium, which drives EDA standardization, debug tools features, and XML–based descriptions that encapsulates IP and integrated platform design components, connectivity structure, and meta-data in IP requirements. This also includes the protocol information necessary to integrate a piece of IP both in a system and with a variety of EDA and debug tools.

Verification IP

Verification IP is another category of available IP that exists for most major standard interfaces, such as on-chip and off-chip interfaces. Comprehensive test, verification suites, and test benches automate and provide the application-specific information for testing standard IP functions.

Verification IP may be implemented in standard RTL languages, such as Verilog and VHDL, but it is more often implemented through verification specific language such as 'e' or VERA that include standard constructs for verification functions, such as assertions and pseudo random testing. SystemVerilog has been developed as a superset language to address both functional synthesizable IP modeling and its associated verification interfaces and environment. SystemVerilog was ratified as an IEEE standard in 2003 and, as a result, extensive support is still immature. However, it is often the preferred means of support from the IP provider, since it allows more flexibility in interfacing IP test features with a range of EDA environments. Since IP vendors may not support all verification IP types and languages, you may need to work with third parties to get the right verification IP and format to be compatible with your design flow.

A difficult aspect of verification IP is that it is closely tied to the implementation (in your chip) IP and will be your basic interface for analyzing the IP which needs to be both understandable and comprehensive. Ideally, it should be well-proven with your needed IP which has been run through several flows and resulted in tested and debugged silicon. Verification IP should be able to provide detailed coverage and metrics, to insure it is exercising all the necessary operations. It should also offer either a software (SW) support environment, or close interfaces into a standard tools environment.

There are currently mixed opinions on the fact that many IP vendors offer companion verification IP for blocks they support. There are mixed opinions on this. Just as the ideal verification environment should have its own engineers who differ from the design engineers (to avoid falling into the trap of too many a priori assumptions during verification), verification IP should be developed independently from the implementation IP in order not to test what is expected rather than what is needed. On the other hand, the company that develops your IP is probably the most familiar with the specifics and details of the operation and which functions require more verification support. Several companies, focusing on verification IP, have sprung up in recent years, and some IP verification tools have been developed either in-house or with third parties as a range of verification IP solutions. The 'e' verification tools in particular, have a number of partners developing *e*-verification components (eVCs) for a wide range of IP functions.

Software IP

If you are using some type of processor IP, then, supporting software IP for your application may be complementary consideration. There are two types of software considerations to consider:

- Applications—In some cases, it can make more sense to purchase software than develop your own. There are many software companies that offer a range of applications that have been pre-designed and optimized for different processor IP. The amount of familiarity a software IP provider has with the processor IP can have a direct impact on the quality, performance, and features of the final product. If your application is making use of standardized functions, in particular, where developing it yourself is not adding any specific value, there may be some make versus buy tradeoffs that are appropriate.
- Tools—There are a variety of tools that are required to make effective use of processor architectures. These include basic compiler, linker, assembler tools as well as a range of RTOS, other OSs, hardware abstraction layers, graphical–based design tools, software analysis tools, other middleware and hardware/software co-design tools, to name just a part of the infrastructure needed. These can range from fairly simple GNU–based tools to an almost arbitrary level of complexity and capability. Most processor IP vendors offer some internal software development tool capabilities, and many provide more extensive offerings

through more specialized SW partners. Development effort areas are migrating to extensible, standards–based platforms like Eclipse as a framework for more comprehensive interoperability and cross-platform, and multi-operating system support.

From both the applications and tools point of view, the software issue should be considered as a part of both integration and testing. Some software solutions may offer better support or performance for specific design configurations (a given processor and memory controller system as an example) which may result in tradeoffs that will impact other areas of the integration, test, and debug process and possibly your final system operation.

12.9 Conclusion

As mentioned in the beginning of this section, this is more of a primer on integration, test, and debug issues that one should consider, rather than a chapter of answers. I have intentionally tried to keep from referencing commercial tools, as well as books and papers on the IP integration, test, and debug topics. This is more to force you, the user, to find your own best solutions and to avoid offending the huge amount of high-quality work available ranging from application notes to scholarly papers since it is not possible to include all in any reference. The following three Web sites will prove fruitful as a starting point for any additional searches:

1. John Cooley East Coast Synopsys Users Group (ESNUG) Website, http://www.deepchip.com/, is more EDA–focused, but there is a lot of peripheral, but relevant, information to IP–based design and many links to more IP centric sites, including several IP freeware repositories. This is full of options, answers, and leads to the most specific and technical questions on IP integration.
2. Janick Bergeron's http://verificationguild.com/ is more verification-focused, but it contains a lot of peripheral information relevant to all aspects of IP–based verification, along with links to other IP-related sites.
3. http://www.design-reuse.com/ provides the most inclusive listing of IP and directly-related information.

Future IP Integration Test

The issue of integrating more and more complex IP blocks is central to the evolving nature of the ASIC, embedded systems, and general SoC business on which a large portion of the electronics business is based. One could affirm that we will see increasing trends for more complex IP, including larger platforms and subsystems and hopefully this will be followed by more standards–based IP and infrastructure for support. Since the curve of innovation tends to rise faster than the trend to standardization, we expect that there will continue to be non-standard interfaces, functions, and features in all areas.

Analog and clockless IP, in general, remains a growth area where the demands of very large system development may migrate to improve both speed and power in the overall system functions. After a decade or more of focusing on all digital systems, there is a realization that analog systems–based approaches may provide better and simpler solutions in some applications. Similarly, clockless IP has started to emerge as options in both the processor and bus interface domains, with some success. These are areas where older rules for integration, test, and debug may not fully apply and where new design approaches to support the IP should start emerging.

There is a strong need for more and better system level tools and approaches for IP design and integration. It is not clear how fast or strong the adoption of existing system level tools like SystemC will proceed or whether they will be superseded by other simpler or more robust approaches, which might as an example, encompass analog as well as digital systems.

The test and debug areas will undoubtedly continue to expand as the need to more effectively test and debug the increasingly complex IP and IP systems that will emerge. The design for debug area has, as an example, seen a wave of half-dozen standardization–based activities in the last year or so, driven by customer needs for better, cheaper, and more scalable solutions. Other areas are likewise undergoing a similar growth to support the needs to more complex and diverse IP solutions.

Part IV:
Case Studies

Chapter 13

Back-End Methodology and Techniques for a Multiprotocol Mixed Signal IP Design

Ken Umino, Consultant, Synopsys Professional Services

Jason Upton, Synopsys Professional Services

Ross Segelken, Design Engineer, nVidia

John Stonick, Application Engineer, Synopsys Professional Services

13.1 Introduction

Recently, high-speed serial links (e.g., serializer/deserializer [SERDES]) have become popular, such as peripheral component interconnect (PCI)-Express (2.5Gbps) and serial advanced technology attachment (SATA) 2 (1.5-3Gbps) in the personal computer market. Furthermore, 10 gigabit attachment unit interface (XAUI) (3.125Gps) has gained popularity as an application-specific integrated-circuit (ASIC) interface. This chapter covers the physical layer (PHY) and multi-protocol intellectual property (IP) that handles the aforementioned protocols. The analog physical design, along with the digital physical design, have both proved to be challenging. Select digital physical design and timing closure techniques are covered in this chapter. These techniques give insight into challenges faced by a PHY IP provider to develop a high-performance and configurable PHY.

To conserve time and engineering resources, the challenge for the design team was to develop a single design that could support multiple input/output (I/O) protocols, and at the same time, work over various processes and wide process corners. Sub-blocks are implemented since the PHY is offered in multiple lane configurations (1, 2, 4, 8, 12, and 16). These sub-blocks are designed so that a lane configuration is created by placement abutment. There are four sub-blocks, two digital, and two analog. Having only four sub-blocks allowed the design team to focus on relatively small sub-blocks; however, these sub-blocks had to be flexible enough to cover all lane configurations. For the digital side of the physical design team, the wide timing corners and multiple lane configurations posed physical and timing challenges for the sub-block designs.

13.2 Basic Architecture Description

Figure 13.1 shows a block diagram of a one lane (x1) PHY. On the left is the PAD interface, while on the right is the ASIC interface.

The top block and the transceiver lane have both analog and digital circuits. To transmit data, the transceiver lane serializes the parallel data from tx_data[*] into a single bit stream, and sends this data off chip on a differential pair, Txp and Txm. To receive data, the transceiver lane deserializes the single serial bit stream on the differential pair Rxm and Rxp into parallel data on rx_data[*].10 bits represent 1 coded byte for all the supported protocols. Tx_data[*] and rx_data[*] can be configured to be 20 bits wide at a slower clock frequency or 10 bits wide at a faster clock frequency for the ASIC interface in *Figure 13.1*. The receive path uses a slicer block which converts the high-bandwidth analog data to a serial sequence of binary data. The clock/data recovery (CDR) uses a digital phase-locked loop (PLL) to control an analog PLL which generates the clocks to feed to the slicers. This is also the source of the recovered clock which becomes the digital word rate clock.

The bottom (control) block handles the common ASIC interface and is the master control for the PHY. It includes a multiplying PLL (MPLL) with optional spread spectrum generation that generates the required reference clock to be buffered throughout the transceivers. The Joint Test Action Group (JTAG) block in the clock and control block the device in systems to be tested, and it is connected to ASIC pins (or other JTAG controllers on the ASIC).

The PHY has implemented a number of test features, as a specific example, the transmit channel can be looped back to the receive channel. Bit-error rate testing (BERT) generates and matches a pseudo-random pattern for error testing. A high-resolution internal eye measurement is available on each receive signal.

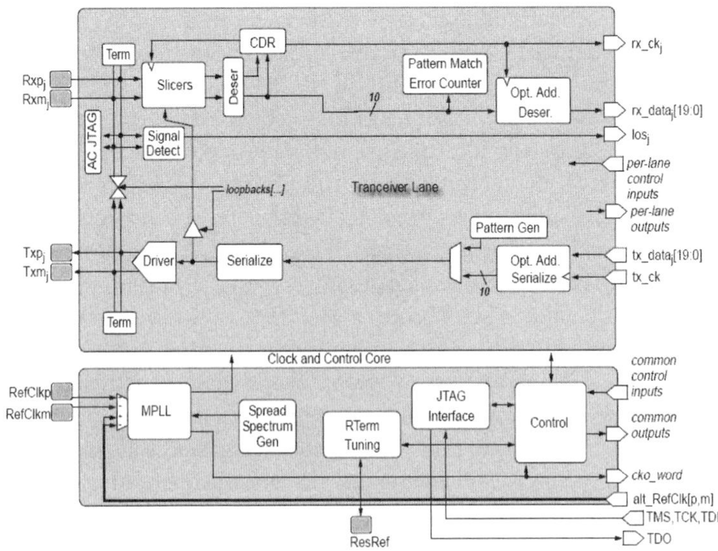

Figure 13.1: *Diagram of a PHY Lane and PHY Support for a SERDES Block*

13.3 Sub-Block and Lane Configuration Description

The PHY logically consists of one or more transceiver (or lane) blocks, along with the clock and control block. In the implementation, these two logical blocks are each broken down into digital and analog block shown here:

- Transceiver block
    - Digital transceiver lane block: up3_topd
    - Analog transceiver lane block: up3_topa
- Clock and control block
    - Digital clock and control block: sup_topd
    - Analog clock and control block: sup_topa

The physical implementations of the digital block for 130nm and 90nm are different, however, they are based on the same register transfer language (RTL) and have the same pin locations. The flows are also nearly identical, with the exception of the following: different standard cell libraries, layout versus schematic (LVS)/design rules checking (DRC) runsets, and timing constraints. The timing constraints are similar, but differ in electrical design rules like transition times, maximum capacitance, and clock skew target. In addition, the timing constraints on some of the pins are adjusted to optimize synthesis. The slowest 130nm node is about 15 percent slower than the slowest 90nm node which accounts for some differences between the 130nm and 90nm timing constraints. The 130nm and 90nm sub-blocks share the same pins and physical width dimensions, but they have different heights. The 130nm and 90nm analog blocks have exactly the same pin locations, along with the same width and height dimensions. The 130nm blocks and 90nm blocks are identical in size, unlike the digital blocks, since the analog blocks share much of the same physical designs and are custom designed at the transistor level. Further, much of the analog block is designed with the same I/O transistor which does not change significantly between 130nm and 90nm technologies. On the other hand, the digital 130nm blocks and 90nm blocks are designed at the standard-cell level, and today's ASIC flows cannot simultaneously work on different standard-cell libraries in different technology nodes. Having just four sub-blocks that can be assembled to create all lane configurations conserves on the amount of blocks that need to be designed and maintained.

Additionally, from a timing perspective, the PHY supports wide ranging technology nodes, as well as the worst, nominal, and best process-voltage-temperature (PVT) corners. *Table 13.1* shows the 130nm nodes and PVT corners supported by the 130nm design. *Table 13.2* shows the 90nm nodes and PVT corners supported by the 90nm design.

Process	Best case voltage/temp (V/C)	Nominal case voltage/temp (V/C)	Worst case voltage/temp (V/C)
TSMC130G	1.32V/ 0 C	1.2V / 25 C	1.08V/ 125 C
TSMC130LV	1.1V/ 0 C	1.0V/ 25 C	0.9V/ 125 C
TSMC130LVOD	1.26V/ 0 C	1.2V/ 25 C	1.08V/ 125C

Table 13.1: *130nm Nodes and PVT Corners Supported*

Process	Best case voltage/temp (V/C)	Nominal case voltage/temp (V/C)	Worst case voltage/temp (V/C)
TSMC90G	1.1V / -40 C	1.0V/ 25 C	0.9V / 125 C
TSMC90GT	1.26V / -40 C	1.2V/ 25 C	1.08V 125 C

Table 13.2: *90nm Nodes and PVT Corners Supported*

The following show the figures that show the dimensions of a one lane (X1) configuration. *Figure 13.2* is the 130nm version, and *Figure 13.3* is the 90nm version. The analog blocks, as stated previously, are the same size for both 130nm and 90nm. The 130nm and 90nm digital blocks are both 400um wide to be aligned with the analog blocks. The height of the 130nm is 556um while the 90nm is 430um in height. The height difference is because the 90nm digital block is significantly shorter than the 130nm block. The 4:1 aspect ratio of the 90nm digital blocks made the 90nm block more difficult to route DRC clean. The final width is padded on each side by a 12.5um wide end-cap block.

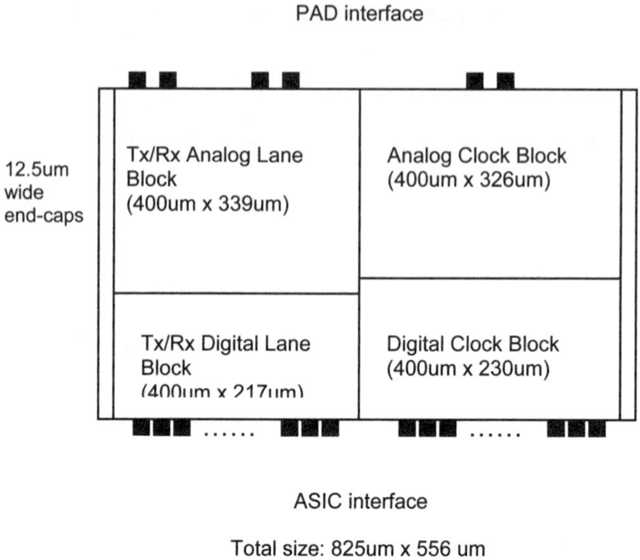

Figure 13.2: *Physical Dimensions of a 130nm X1 Lane Configuration*

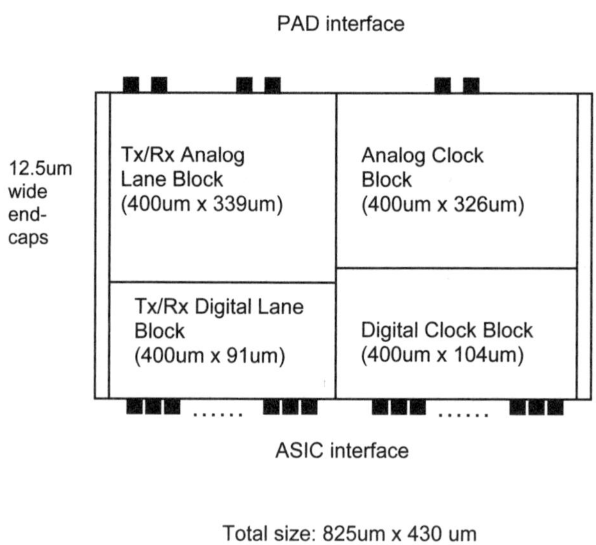

Figure 13.3: *Physical Dimensions of a 90nm X1 Lane Configuration*

From these sub-blocks, the other lane configurations can be easily created. *Figure 13.4* shows all available lane configurations. One challenge was to get all configurations to pass all timing runs. All lane configurations have horizontal feed through pre-routes that span across the transceiver (Tx/RX) lane blocks. These feed through routes run on parallel tracks for long distances, and are susceptible to crosstalk. As a result, care was taken to place and size these preroutes, and both setup and hold timing was difficult to meet because of the wide difference in lane sizes. Generally, the X1 lane configuration had most of the hold violations. The largest lane configuration, X16, had most of the setup violations as well as some hold violations.

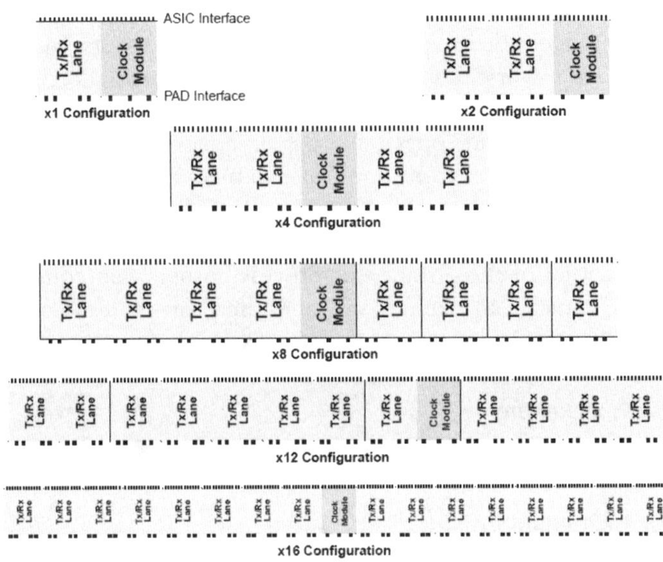

Figure 13.4: *All Available Lane Configurations*

13.4 Library Preparation Flow

To make sure that the timing is closed across all nodes and all PVT corners, the slowest library and slowest parasitic is used for set-up timings, and the fastest library and fastest parasitic is used for hold timings in the synthesis, place, and route tools.

For 130nm, the two extreme PVT corners are:

- Maximum (slowest) PVT: TSMC130G/ VDD=1.08V/ Temp=125 C
- Minimum (fastest) PVT: TSMC130LVOD/ VDD=1.26V/ Temp=0 C

For 130nm, the two extreme parasitic types are:

- Maximum (slowest) parasitic: maximum capacitance
- Minimum (fastest) parasitic: minimum capacitance

During library preparation, all maximum/minimum libraries and parasitics are identified for the placement and route tools. By using these absolute worst conditions for the implementation tools, timing could be met for all intermediate conditions. This makes the tools work harder and longer, but it minimizes the chances that manual ECOs will have to be done later to meet timing.

All corners are considered for final static timing runs. In this flow, there are a total of three transistor process corners that consist of maximum, typical, and minimum, and there are a total of five parasitic corners:

- C maximum
- C minimum
- RC maximum
- RC minimum
- RC typical parasitic

Taken all together, there are three transistor corners (TSMC130G, TSMC130LV, TSMC130LVOD), three voltage temperature corners (worst, typical, and best case), and five parasitic corners (C worst, C best, RC worst, RC best, and typical). The typical voltage temperature corners and typical parasitic corners are not used for static timing analysis (STA). The total amount of STA runs works out to three transistors corners X 2 voltage temperature corners X 4 parasitic corners X 6 lane configurations or 144 total STA runs.

Additionally, STA was run with and without SI, on-chip variation (OCV), and IR effects which added numerous additional runs and reports to review. Running this many different STA corners ensures that there are no outlier timing paths that escape detection. Notice that there are C-based parasitic and RC based parasitic files. The C is for the worst and best case capacitance while RC means the worst and best case R with adjustments to C towards worst or best, but keeping the process planar. Based on experience, it has been found that the C-based extraction provides the worst and best cases over RC for internal timing paths. This is because capacitance dominates for short wires. However, for large configurations, inter-block timing paths were often worse with the RC worst parasitic files since the R dominates for long wires, which affects the propagation delay of signals from the support block to the furthest transceiver and back.

13.5 Constraints Strategy

At the lane configuration level, there is one master constraint file for all lane configurations. In addition, the digital transceiver block and digital clock block each have a master constraint file. Having one constraint file with common constraints to handle all cases (e.g., for synthesis and STA) versus managing multiple constraint files reduces the amount of error and maintenance effort. The master constraints define all clocks, along with their latencies, transitions, uncertainties, and periods. Furthermore, I/O delays, false paths, and multi-cycle paths at the port level are defined. The master constraint file for a lane configuration supports all possible lane configurations (X1, X2, X4, X8, X12, and X16) by extracting the lane size embedded in the lane configuration name. Each master constraint file also sources a lower level constraint file that is described in the following.

A unique feature with this PHY flow is the usage of a script that extracts constraints from RTL comments. Due to PHY design and timing requirements, there are considerable timing exceptions involving internal registers. Embedding the constraints in the RTL makes it easier for a designer to see the connection between the constraint and the referenced RTL. In addition, the constraints are more compact since it is based on reference rather than instantiations. The following shows a snippet of embedded constraints in RTL (Verilog) code. The constraint extraction script keys on the comment "// %%SYNTH:" and processes the constraints after this. The designer does not need to know the instance path defined by the variable $inst.

```
// %%SYNTH:
  //  set_false_path -to [get_pins $inst/hold/lat_bit_*__slice/cn]
  //  set_false_path -to [get_pins $inst/hold/lat_bit_*__slice/sn]
  //  set_false_path -through $inst/hold/lat_bit_*__slice/q -to
[get_cells $inst/q_reg*]
  //  set_max_time_borrow 0 [get_cells
$inst/hold/lat_bit_*__slice/*]
  //  set_multicycle_path -setup 2 -to [get_pins
$inst/hold/lat_bit_*__slice/d]
  //  set_multicycle_path -hold 2 -to [get_pins
$inst/hold/lat_bit_*__slice/d] -end
  //
```

The constraint extraction script, which is extracted from the RTL, generates the Synopsys constraints as shown in the following. One should notice that variable $inst is based on the reference name so the designer does not have to worry about the constraint hierarchy.

```
foreach_in_collection instc [filter_collection [get_cells -hier] -
regexp ref_name=~cr_inline_reg.*] {
  set inst [get_object_name $instc]
   set_false_path -to [get_pins $inst/hold/lat_bit_*__slice/cn]
   set_false_path -to [get_pins $inst/hold/lat_bit_*__slice/sn]
   set_false_path -through $inst/hold/lat_bit_*__slice/q -to
[get_cells $inst/q_reg*]
   set_max_time_borrow 0 [get_cells
$inst/hold/lat_bit_*__slice/*]
   set_multicycle_path -setup 2 -to [get_pins
$inst/hold/lat_bit_*__slice/d]
   set_multicycle_path -hold 2 -to [get_pins
$inst/hold/lat_bit_*__slice/d] -end
}
```

Next, the extracted constraints are merged and modified to create a multiple lane version of the constraints. *Figure 13.5* shows the timing constraint flow. First, the lower level digital blocks, up3_topd and sup_topd, have their constraints extracted from RTL. When the master constraints of up3_topd and sup_topd are applied to their respective designs, the extracted constraints are also applied. For the multiple lane configurations, a script merges the extracted constraints, and adds the appropriate prefix to the constraint paths to create a new "extracted" constraint version for the multiple lane configuration level. When the multiple lane configuration (up3_x_top) master constraint is applied, this new "extracted" constraint is also applied.

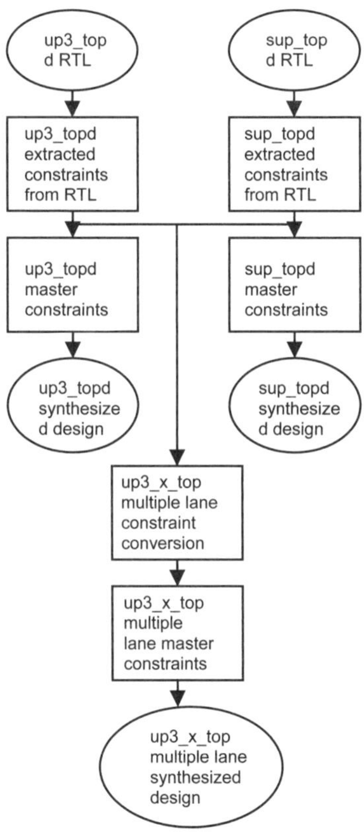

Figure 13.5: *Timing Constraint Flow*

An important aspect of the master constraints is its ability to adjust the content and parameter values of the constraints based on which tool and step the flow is at. One aspect is the handling of implementation constraints such as maximum transition, maximum capacitance, maximum fanout, don't touches, and critical range. Critical range is a constraint that allows Synopsys tools to improve timing paths beyond the worst negative slack of a path group. Some implementation and analysis tools will not understand these constraints. As a solution, when these non-compatible tools are detected, a procedure is used to filter out these unusable constraints, and then a second Tiger procedure reads in these filtered constraints.

Decreasing clock periods during sub-block creation is an important strategy to meet timing by creating extra margin. For example, during synthesis and placement, a clock period of 2.72ns (367.6MHz) is used for the transmit and receive clocks. During clock tree synthesis (CTS) and route, the clock period is increased to 2.8ns (357MHz). Finally, for final sign-off STA, the sign-off periods of 2.88ns (347MHz) is used. In silicon, the real period will be longer depending on the standard used and will range from 3.2ns to 4.0ns (between 312.5MHz and 250MHz).

Another important feature is the adjustment of clock uncertainties. For example, in the synthesis and place and route tools, the transmit and receive clocks have setup uncertainty of 200ps and a hold uncertainty of 100ps. However, in sign-off STA runs, OCV modeling is used. At this point, setup uncertainty is reduced to 50ps to account for jitter and hold uncertainty is reduced to 10ps for extra margin. For the PHY, OCV derates the clock tree by 25 percent of its delay. However, in most cases the clock skew is almost never 25 percent of the total clock latency because much of the clock tree is common between timing start and end points.

13.6 Physical Design Strategy

Long Feed through Pre-Routes

With the X16 lane configuration, the horizontal feed through pre-routes is as long as 8 x 400um or 3200um.

Figure 13.6 shows half of an X16 lane configuration where there are eight lane blocks on the left side of the clock block. The inset drawing shows a close-up of some of the metal five horizontal feed-throughs. Crosstalk can become a serious problem on routes this long, and transition times degrade over these long routes. Careful planning is required in order to avoid serious problems.

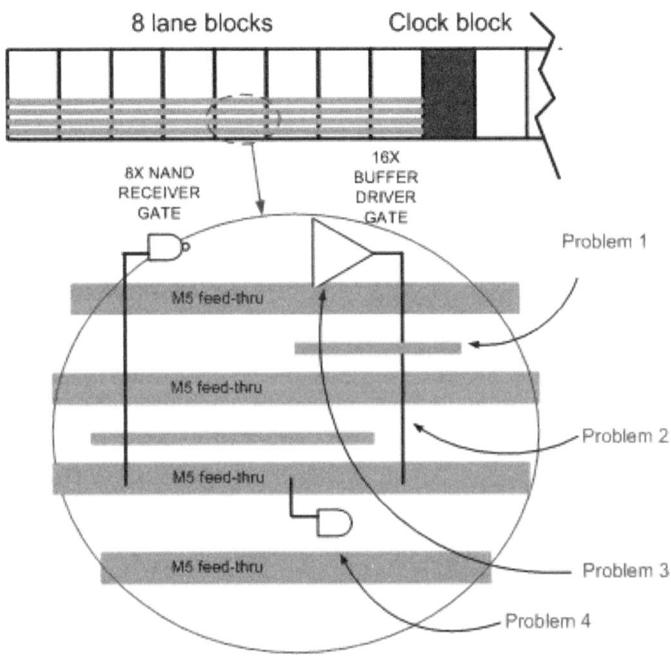

Figure 13.6: *Inset View of Feed Throughs with Problems*

Figure 13.6 illustrates four main problems:

- Metal 5 feed throughs are spaced too far apart, causing the router to put in signal routes in between.
- The driver and receiver gates are placed far away worsening transition times.
- The driver is over-sized, adding a large pin cap per each lane. Also, an over-sized receiver at the far-end from a driver will seriously degrade transition times.

- There are multiple receiver gates unnecessarily adding gate capacitance.

Figure 13.7 shows a feed through design that addresses the previously stated problems:

- Solution 1: Feed throughs are spaced over twice the minimum spacing, but not far enough that signal routes get in between.
- Solution 2: The placement tool has a bound that forces the driver and receiver to be placed directly underneath the feed through. The bound is shown by the dashed box.
- Solution 3: The driver and receiver gates are down-sized enough to minimize pin cap, and still get good transition performance. The receiver gate is reduced from an 8X NAND to a 4X AND. This is better than twice the reduction because an AND gate is actually two gates, where the first gate is smaller than the second gate.
- Solution 4: A single receiver gate is instantiated, and fixed in size and placement.

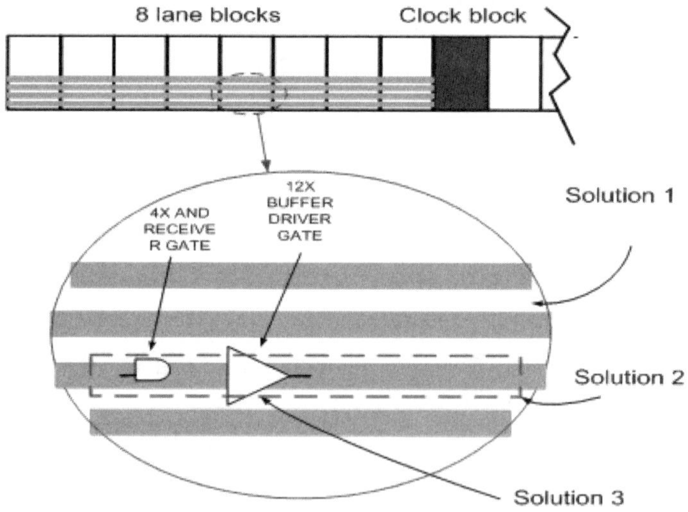

Figure 13.7: *Solutions to Feed through Problems*

Via Programming to Re-Use Physical Designs

To allow a single design for multiple lanes to be reused, it was necessary to re-configure portions of the design at a higher level. *Figure 13.8* illustrates an example of an X2 lane configuration. The clock block drives a signal through a feed to the middle lane block. The middle lane block takes the signal, and drives an output signal to the left lane block with an AND gate. The problem in this figure is that the left lane block, upon being mirrored, drives the same feed through to the right which would cause a short. Via programming is a technique where the dropping of vias in the correct places would effectively reverse the direction of the AND gate in this example.

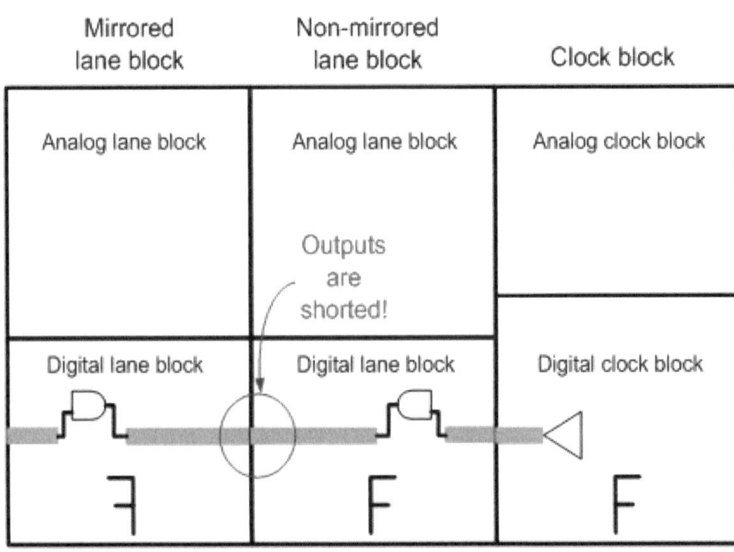

Figure 13.8: *Problematic Non-Mirrored and Mirrored Lane Blocks in an X2 Configuration*

Figure 13.9 shows a horizontal route that is reversible in direction where an input signal is driven on metal 5 bs_io2, goes through the via, drives the "horse-shoe" pin, bs_in. Next, bs_in drives a AND gate, and the output of the AND gate drives pin, bs_out. Last, the signal continues through the via and out on bs_io1. This is a right to left signal flow.

The key to reversing direction are the "horse-shoe" shaped pins, bs_in and bs_out. The signal flow direction is reversed by swapping where the vias are placed. Despite the small size of the PHY digital design, it was surprisingly challenging to carry this out partly due to extreme timing corners that must be closed for all lane configurations. In addition, supporting multiple lane configurations added challenges in terms of feed through design and via programming. Proper feed through design minimizes signal integrity problems and optimizes transition times, and via programming allowed for re-use of physical designs.

Figure 13.10 shows a mirrored lane where the vias are on the opposite sides of the horse shoe. As a result, right-to-left signal flow is brought about, and the AND gate is re-wired and drives in the opposite direction.

In the past, the design team had created a script that would calculate the coordinates of where wires and vias are dropped based on physical knowledge of all lane configurations. However, this scripting was cumbersome and required changes if the pre-routes and "horse-shoe" pins moved. However, there is a superior method.

First, the "horse-shoe" pins are made to be actual pins at the lane level. Next, a script automatically generates RTL for the lane configurations. This script has knowledge if a lane is mirrored or non-mirrored, and connects the pins of bs_in and bs_out to the correct pins of bs_io1 and bs_io2. After floorplanning is completed, the router places vias at the intersection of the "horse-shoe" pins and feed throughs. No knowledge of coordinates for the vias is needed by the script.

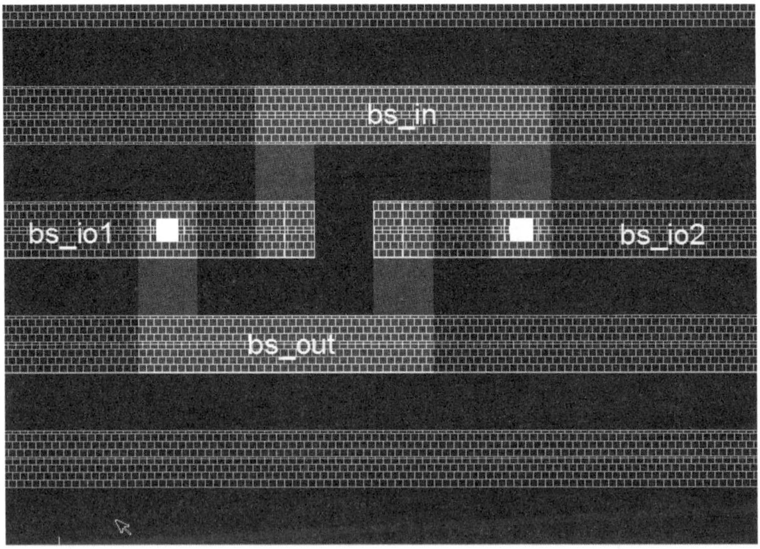

Figure 13.9: *Right to Left Signal Flow in Non-Mirrored Lane*

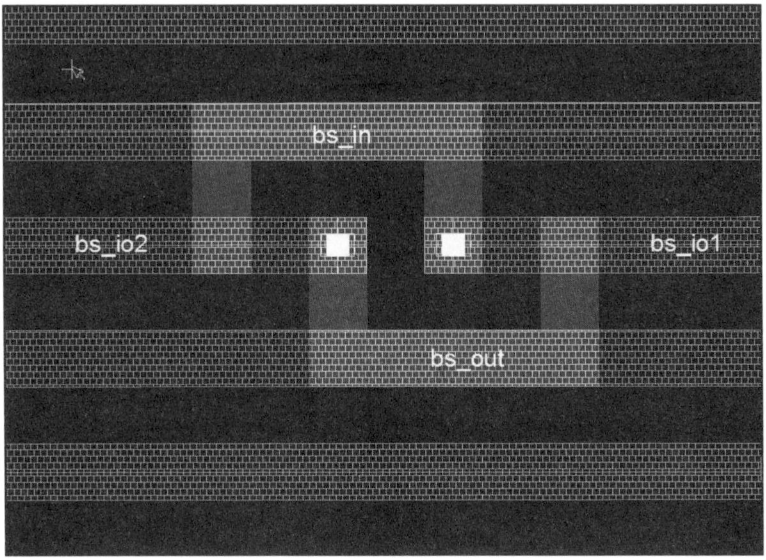

Figure 13.10: *Right to Left Signal Flow with a Mirrored Lane. Vias Have Been Moved.*

In another view, the via programming problem is reduced to relatively easy RTL changes. This same via programming technique is used to program the sub-blocks to operate with a particular protocol. Thus, via programming is effective in re-using physical designs and in putting a design in a desired functional mode.

13.7 Final Remarks

Although the digital physical design of the PHY is small, it was surprisingly challenging. Sub-blocks were created to support the myriad of lane configurations, which could be abutted to support all lane configurations. Having the sub-blocks reduced the amount of designs that had to be developed, but complicated the sub-block designs since these sub-blocks had to support a wide range of lane configuration sizes (from 2 to 16 lanes). There were other challenges as well.

Extreme timing corners had to be closed for all lane configurations. Metal pre-routes (or feed-throughs) spanning across the various lane configuration posed signal integrity and timing transition challenges. Via programming was an effective method to deal with lane blocks that were mirrored relative to other lanes.

Chapter 14

eVC for Functional Simulation of Voltage/Temperature Variation and Noise on an I/O Bus

Joseph D. Mendenhall, IBM

Robert B. Likovich Jr., IBM

Chad E. Winemiller, IBM

14.1 Introduction

Functional simulation has many inherent limitations, particularly at the boundaries of a design under test (DUT) containing a high-speed input/output (I/O) interface. As I/O transfer rates become higher, the problems that are most commonly encountered are relative bit skew and noise associated with transmission across the bus. In order to verify a design that would compensate for such asynchronous effects, we devised an e-verification component (eVC) that can drive clocks and data simulating these effects. This paper will describe our eVC, which details its structure and parameters, the mechanisms for resolving parameters down to the lowest-level bus functional models (BFMs), and a test phase completion management scheme.

14.2 Problem Description

This paper discusses a method of performing functional verification on a high-speed, self-aligning, elastic I/O design using an eVC [1] developed through Verisity's Specman.

There are typically certain path variances in I/O buses, caused by deviations in wire length and impedance, which can cause certain bits to transmit more quickly than others across the media connecting two chips or components. This delay is sometimes negligible, for example, for slow transmission frequencies or short paths where all bits can be easily sampled within a wide data window. Other times, particularly when transmission frequencies become very high, these disparities can cause very narrow data sampling windows, thus making it difficult or impossible to simultaneously sample every bit on the bus [2]. This is the general motivation behind developing elastic buses that can compensate for such skew. Some designs are capable of aligning groups of bits (each transmitted with their own clock) to each other, even if the groups have multiple periods of delay difference. Instead of relying on strict physical design rules for chips and systems employing high-speed I/O buses, the elastic interface designs handle alignment problems by utilizing training sequences at power-on and recurrent dynamic adjustments during functional operation [3].

There are other negative effects that become difficult to manage at higher frequencies. Random noise sources can cause jitter to occur on clock and data signals, which narrows sampling windows [2]. Distortion in the clock duty cycle and reference voltage can also reduce the sampling accuracy of the bus at high speeds. These signal quality issues are traditionally dealt with through analog circuit techniques such as noise suppression and filtering. Modern elastic interface logic can handle certain levels of noise and distortion without relying strictly on circuit design to compensate for them [3].

The elastic interface design we verified allows an external receiver communicating with a similar elastic interface transmitter to:

- Deskew (align) data bits within a clock-data group to account for path variance within a group
- Deskew (align) clock-data groups within a bus to account for variance from group to group
- Compensate for noise via selecting appropriate-sized sampling windows
- Adjust for non-ideal duty cycles and reference voltage (Vref) distortion, and

- Make dynamic recalibrations to keep the bus aligned and sampling correctly under situations of voltage and temperature variation.

An environment capable of driving non-ideal clocks and data onto the bus in a randomized, yet controllable, fashion is required to verify such an interface and its associated class of communications designs. The clock and data effects that we modeled in our verification environment are skew, drift, jitter, synchronized jitter, duty cycle distortion, Vref distortion, and delay element drift.

To provide effective control to model these clock and data parameters, there are a large number of switches that need to be set for each bit and clock on the bus. The bus is divided into clock groups where all the data bits in a single group are captured by the clock in that group. As a result, we needed a way to constrain the bus functional model (BFM) parameters at the bus, group, and bit levels corresponding to the different hierarchies in the design.

Both the number of clocks and data bits in each group, along with the number of groups, are very large in select bus configurations. Constraining all the parameters for each bit and clock on such a bus presented considerable efficiency issues. A manual constraint methodology would be very time consuming, requiring many lines of code to enumerate even simple scenarios. Such excessive constraint code would be highly error prone, difficult to decipher, and very difficult to change quickly if necessary. Therefore, a global and efficient mechanism was required in order to constrain the bus parameters yet still allow individual clock and data bit settings to override the global settings.

In this paper, we will describe the overall BFM structure, and associated resolution methods for the global management of the many bus parameters. In addition, we will outline some of the unique features of our eVC which enable test writers to customize the condition for a test phase termination either based on a number of cycles or specific functional events, where a test phase is defined as a section of the testcase. This serves as an efficient alternative to predefined phase-end conditions, which thus provides the test writer a variety of methods to terminate a test phase. We will also discuss some of the advantages and efficiencies gained from the development of our eVC.

14.3 Selection of Verification Language

We chose Verisity's Specman to develop our verification environment for a variety of reasons. The VHDL–based elastic I/O design had several portions of the design that modeled both delay chains and the effect of analog circuitry on the shape of clock and data I/Os (duty-cycle and Vref correction, respectively). In addition, our BFM needed to finely control the delays on the clock and data I/Os to enable skew, drift, jitter, and bus distortion. As a result, we needed an event-driven simulator coupled with a verification language that would allow us to introduce delay in the simulator resolution and not solely clock cycles to model and control signal delays. Specman was a good solution because it works with many popular event-driven simulators and provides the fine-grain delay capabilities we needed.

In addition, Specman provides a reuse methodology as well as coding guidelines that create a portable verification component [4]. It also provides an efficient mechanism to encrypt the model while still allowing a public interface for users to wire it to their unique environment and extend it if any changes are required. This was extremely important to us since designs on the chip periphery, such as elastic I/O, can be potentially used in diverse environments.

Specman also provides functional coverage capability built into the tool. Functional coverage was the primary mechanism we used to decide when we finished verifying a particular function of the design, and it was an important factor in selecting a verification platform.

Finally, Specman provides time-consuming methods (TCMs) where multiple simultaneous threads can exist at any one time in the system. The ability to start separate threads of the same TCM in parallel allowed us to easily pipeline the future transitions of a single bit on the bus as it is affected by the noise and distortion capabilities of our eVC. This was an important feature used to model the various noise and distortion characteristics of high-speed elastic I/O buses.

14.4 Elastic Interface eVC Capabilities

The eVC we developed has many driver capabilities which can be utilized by a verification engineer attempting to simulate a high-speed, self-aligning, elastic I/O bus design.

Reactive and Proactive Driver Modes

For flexibility, the eVC can be configured to implement a "reactive" or "proactive" driver. In reactive mode, the eVC accepts clocks and data from another HDL model source, which is typically a transmitter complement to the receiver DUT. It has the potential to perform certain manipulations upon the clocks and data (such as error injection or delay insertion) to simulate data corruption across the bus and a limited set of noise and distortion characteristics. In this mode, the eVC can be visualized as an intermediary distortion model existing between driver and receiver logic. *Figure 14.1* illustrates a reactive eVC.

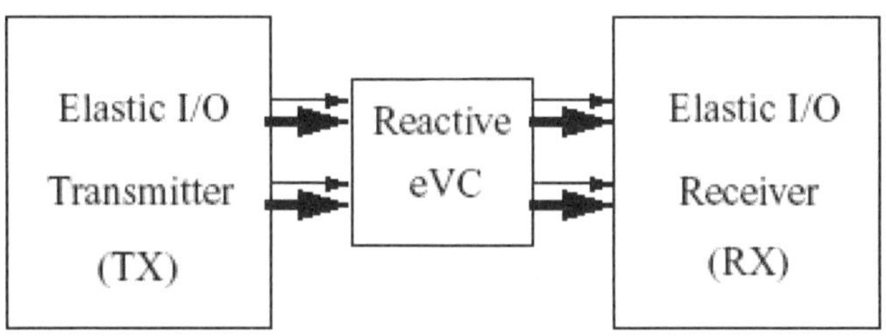

Figure 14.1: *Reactive Mode eVC*

In proactive mode, the eVC functions as a standalone driver which generates its own clocks and data patterns. These clocks and data are transmitted to the inputs of the receiver bus model after which the driver applies any error injection or delay-related distortions. A proactive driver can create any type of noise or distortion defined by the eVC due to its ability to apply "negative delay" to transmitted bits. Initially, the driver uses a reference point of zero delay (which looks like a perfect, non-distorted clock operating at a set frequency). It then adds a delay buffer amount ("n" number of half-periods, called "bit times") which it uses as a new "zero delay" reference point. The buffer amount is chosen by the user and is commonly a function of the bus elasticity. All clock and data transitions then occur "n" number of bit-times following generation. Due to this positive delay buffer, the proactive driver can add positive or negative delay to the reference point to simulate acceleration, deceleration, and jitter on either side of the reference point for each clock and data bit. Thus, proactive drivers are more versatile than reactive drivers, but they are generally slower to simulate due to the generation aspects. *Figure 14.2* illustrates a proactive eVC.

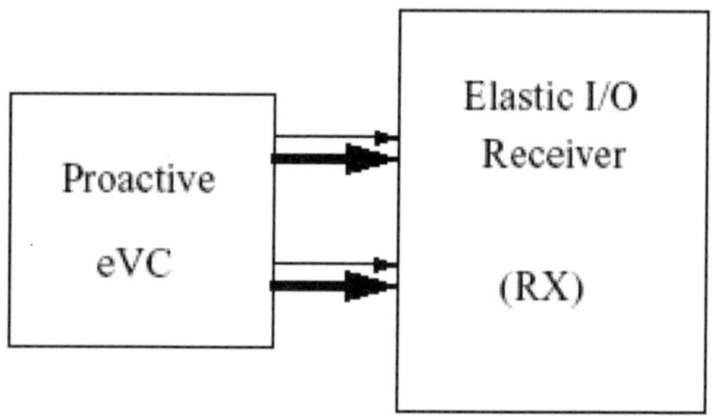

Figure 14.2: *Proactive Mode eVC*

Skew Modeling

The most basic characteristic modeled by the elastic interface is skew [2] which, in this eVC, is defined as the static delay difference of one bit relative to another bit. It models the effects that path length and impedance discrepancies between different signals would have on the device.

Figure 14.3 illustrates the result of skew between two bits. Bit 1 has a longer path delay across the transmission media with respect to bit 0, and its data arrives later at the inputs of the receiver than bit 0's data. Skew can also be implemented between a clock and data bit(s) or from one clock to another if multiple clocks exist within the system.

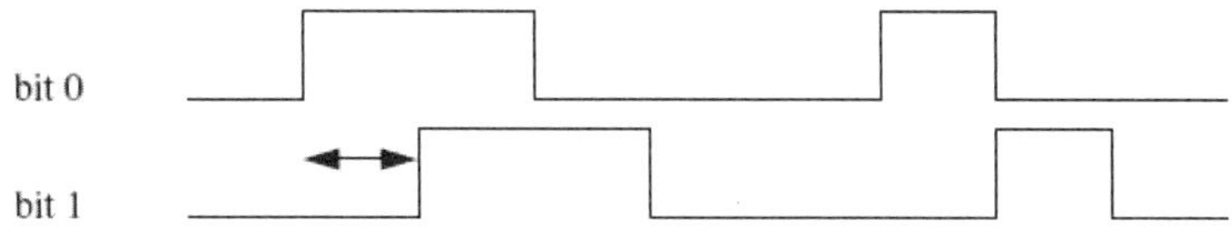

Figure 14.3: *Bit 1 Skewed Positively with Respect to Bit 0*

Drift Modeling

Drift is another characteristic of I/O buses that our eVC can model, and it can be thought of as the first derivative of skew with respect to time. As a result of thermal or voltage fluctuations over time (either across the bus itself or within its internal delay elements), the relative skew between clocks and data bits can change. If a single bit is observed, it would appear that another bit speeds up or slows down relative to its own transmission time.

Figure 14.4 is an example of what drift would look like in a system where one bit (bit 0) is drifting faster relative to the other (bit 1). Bit 0's skew is changing with respect to bit 1 as a function of time such that it begins to arrive earlier at the receiver inputs. In order for buses to compensate for this effect, the design must find a way to recalibrate itself, either statically (by halting the bus and realigning) or dynamically (on-the-fly, using functional data). Drift can occur over time at various rates and in either direction on clocks and data. The delay buffer (used in proactive mode for the eVC) allows the skew values to drift in either direction (only as far as the delay buffer in the negative direction and indefinitely far in the positive direction). In reactive mode eVCs, negative drift can only be modeled for bits which have a preset skew amount.

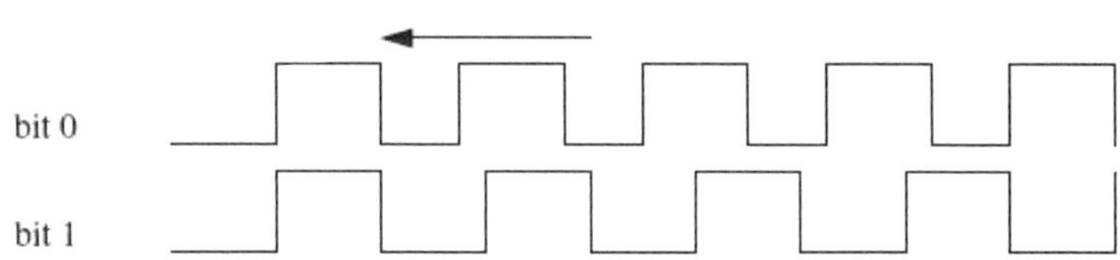

Figure 14.4: *Bit 0 Drifting Faster with Respect to Bit 1*

Jitter Modeling

A common form of noise observed on most buses is jitter [2]. Jitter is a generally un-biased, randomized form of noise resulting from transmission media imperfections and various sources of radiation which cause the transmitted clock or data bit to arrive slightly earlier or later than it would in a perfectly isolated, noiseless setting. This is observed as slight variations in the edges of received data to the left and right of the ideal transition point. *Figure 14.5* depicts an example of jitter, which is applied to bit 0, in this case, with bit 1 being the ideal reference waveform. This noise can become a difficult problem for buses operating at high frequencies, particularly if there is also per-bit skew present on the bus. Higher amounts of jitter effectively close the data window, making sampling more unreliable. A well-designed I/O bus should be able to define the boundaries of its sampling window for each bit and ensure that all the windows are aligned to guarantee accurate sampling across all the bits on the bus. The eVC can simulate jitter in both proactive and reactive modes. It should be noted that in reactive mode, the jitter may be positively biased if the clock or bit is not delayed via skew to the absolute maximum jitter amount.

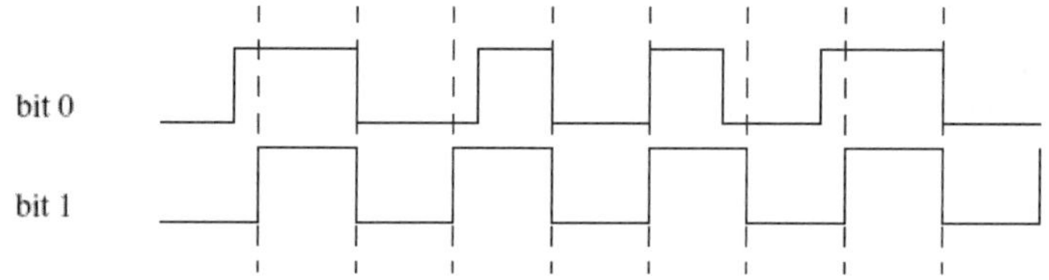

Figure 14.5: *Bit 0 with Jitter*

Jitter can also be created by the transmitter itself. In certain cases, this form of jitter creates a "common mode" jitter which causes all bits within a clock group of the bus to jitter by the same amounts at the same time as the launching clock. This has been modeled as source synchronous jitter (or just "synchronized" jitter) and functions much like jitter with the exception that the clock and all the data bits within the clock group jitter synchronously by the same amounts.

Figure 14.6 illustrates a clock (clk 0) and a data bit within that clock group (bit 0) which has synchronous jitter. Bit 1 (in a different clock group) has no jitter applied.

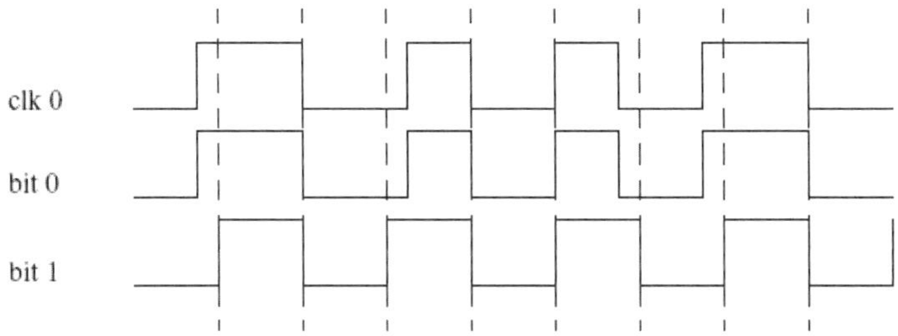

Figure 14.6: *Clk 0 and Bit 0 with Synchronized Jitter*

Most of the capabilities described until now are available in both proactive and reactive mode simulations. However, two forms of distortion—clock duty cycle and reference voltage distortion—are only possible in proactive mode.

Clock Duty Cycle Distortion Modeling

Clock duty cycles are ideally 50-50, meaning the time the clock level is high is exactly equal to the amount of time the clock level is low. For a balanced 400ps period clock, the signal would be high for 200ps, low for 200ps. In some instances, the duty cycle for a clock can become distorted such that the clock is more active at one level than another. For example, as a rather extreme case, the same 400ps clock could be high for 300ps and low for 100ps, making the duty cycle ratio 75-25, as depicted in *Figure 14.7* for clock 0. Some I/O designs are capable of correcting this form of distortion via adjusting certain analog controls and observing the result on the sampling window (a clock with the most ideal duty cycle will result in the largest sampling window). The proactive clock drivers in this eVC are capable of modeling non-ideal clocks by varying the duty cycle and even allowing the duty cycle to shift gradually while the clocks are running. Because the correction algorithms built into the logic entail analog controls, some behavioral modeling within the DUT may be necessary to make full use of this aspect of the driver.

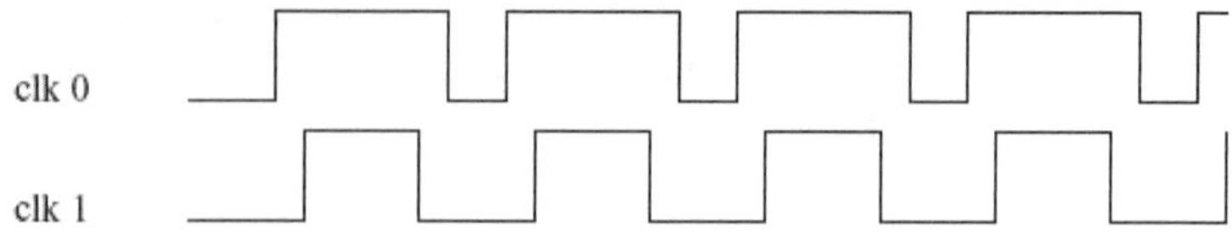

Figure 14.7: *Clk 0 with Duty Cycle Distortion (75-25)*

Reference Voltage Distortion Modeling

Just as clocks can be non-ideal, data bits can have an analogous form of distortion based on the Vref. At each bit's receiver logic, the Vref may need to be adjusted up or down to obtain the ideal scenario where the perceived switching points (during rise or fall) are equalized. Like the clock duty cycle adjustment, this can be handled by some I/O designs by varying analog controls to obtain the most ideal sampling window. The proactive data drivers in this eVC can model this type of distortion, by varying the effective rise and fall times and the switching thresholds per bit. Although functional simulation has no concept of rise or fall time, certain calculations are performed based on the per-bit parameters which create an effective switch point for each data transition at the receiver inputs. The net effects of this manipulation in the simulation environment are seen as shifts in the data edges, which can make high pulses longer or shorter relative to low pulses. Let us refer to *Figure 14.8* for an example of Vref distortion – in this case bit 0 has a longer rise time than bit 1. Again, some behavioral modeling may be necessary for the DUT to be able to compensate for the analog changes made in response to this distortion.

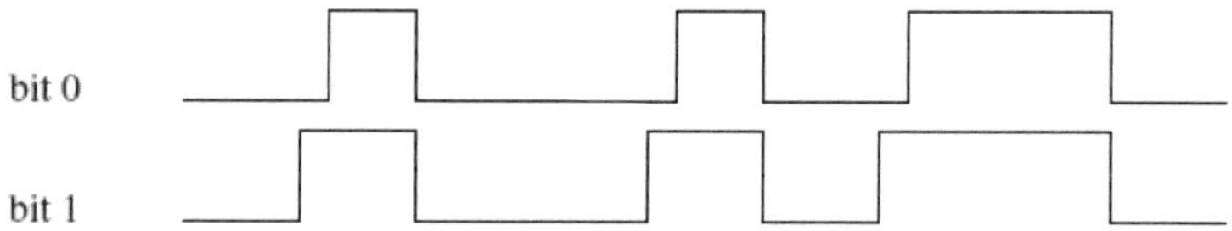

Figure 14.8: *Bit 0 with Vref Distortion (Long Rise Time)*

14.5 eVC Parameters

The parameters used to control the noise and distortion characteristics of each bit throughout a simulation are generated and stored in a single struct. This bus parameter struct contains a list for each modeled parameter that, in general, maintains the parameter values for each bit across the bus. Each bit's information is stored in its own index. This struct serves as a "sequence item" in Specman such that it can be generated during a test and driven to the bus BFM by the sequence driver. Each clock and data bit BFM (unique driver instances contained within the bus BFM) is notified upon the arrival of a new sequence item. The individual BFMs have their own fields corresponding to the fields of the bus parameter struct, and they are each responsible for updating their fields when a new bus sequence item arrives. Once the fields are updated, the clock or data bit BFM's behavior will change accordingly. *Figure 14.9* is a high-level diagram of how parameter information flows from the testcase to the BFM's in the eVC.

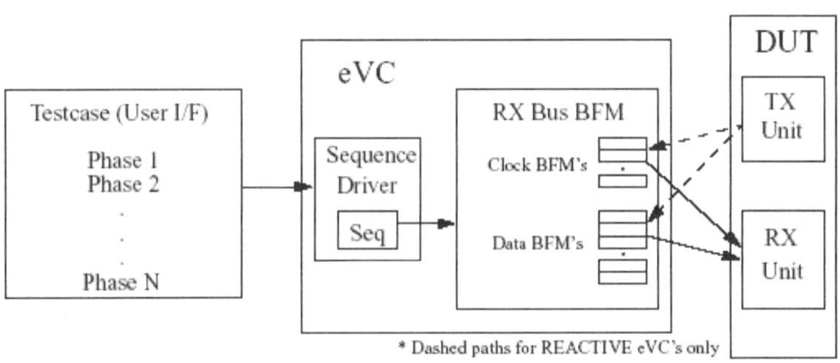

Figure 14.9: *Driver Portion of eVC*

There are several base parameters contained within the bus parameter structure that corresponds to the characteristics previously mentioned. These parameters, along with the actual struct field names, are listed here:

- Skew ("skew_amount")
- Drift ("drift_")
- Jitter ("jitter_")

- Synchronized jitter ("sync_jitter_")
- Duty cycle distortion ("duty_shift_")
- Reference voltage distortion ("Vref_distort_")

Several of these parameters have common sub-parameters that allow the base parameter to change values (or drift) as simulation time progresses. The following sub-parameters, which will be discussed in more detail, are:

- Enable—Whether drifting of the parameter is enabled or not
- Limit—The drift boundaries for the parameter
- Rate—The step size where the drift updates can occur
- Rate reverse enable—This allows drift to reverse if limit is reached
- Start time—When to enable the drift after the phase starts
- Run time—How long to allow the drift to occur
- Evaluation time—Number of cycles between drift updates
- Percent—The amount of time an update will allow a drift to occur

Skew

The first basic parameter is "skew_amount," which represents the relative amount of static delay for a clock or data bit from transmitter to receiver. Positive skew can be generated by both proactive and reactive drivers. Proactive drivers can also generate negative skew by using the delay buffer.

Draft

The next set of parameters specifies the drift behavior, which is the gradual change of skew with respect to time. The "drift_enable" parameter tells the BFM whether to allow drift to occur and "drift_limit" indicates how far in time from the reference point (zero skew) that the signal can drift. "Drift_limit" can be defined as a positive (decelerating) or negative (accelerating) integer value. A "drift_rate_reverse_enable" Boolean parameter is used to indicate whether the drift direction can change once the drift limit is attained, thereby allowing a gradual swing from a positive drift limit to a negative limit and back as time progresses.

"Drift_start_time" and "drift_time" define when to start drifting and when the drift should terminate, with respect to the arrival of the new sequence.

Several other parameters—"drift_rate", "drift_eval_time", and "drift_percent"—are combined to dictate how quickly and how randomly the drifting is to occur. "Drift_rate" specifies the step size, or change in skew, which is made with each update. "Drift_eval_time" defines how many cycles should elapse between potential drift updates. "Drift_percent" indicates what percent of the time, during an evaluation, the drift will actually occur. Using these switches, drift can be controlled to occur in a slow, smooth function, in a fast, erratic manner, or somewhere in between these two extremes.

Jitter

Jitter and synchronized jitter only have two parameters—an enable and a limit. Jitter and synchronized jitter are essentially the same, except that simple jitter gets applied to clock and data bits independently, whereas synchronized jitter causes the clock and all data bits associated with that clock to jitter the same amount at the same time. "Jitter_limit" and "sync_jitter_limit" basically provide absolute bounds for the jitter injections during simulation. For example, specifying a jitter limit of 5 (for a simulator resolution of 1 ps) will allow a bit to jitter from -5ps to +5ps from the reference point for each edge transition.

Clock Duty Cycle Distortion

For clock duty cycle distortion, there are several parameters used to define the behavior. "Duty_high_amount" can be used independently to set a fixed duty cycle on a clock. If the reference period is 400ps and "duty_high_amount" is set to 40 percent, the clock will have a 160ps high time and a 240ps low time corresponding to a 40-60 duty cycle ratio. In order to allow the duty cycle itself to change over time, "duty_shift_enable" has to be true. "Duty_high_long" and "duty_high_short" define the boundaries for the high time. "Duty_shift_rate_reverse_enable," "duty_shift_rate," "duty_shift_time," "duty_shift_eval_time," and "duty_shift_percent" all function in a similar manner as the drift parameters defined above, except that they operate on the "duty_high_amount" as a percentage value rather than as an actual time value.

Reference Voltage Distortion

Analogous to the duty cycle distortion parameters for clocks, data bits have Vref distortion parameters that control the simulated rise/fall times and thresholds for data signals. For a fixed Vref distortion, "Vref_rise_time," "Vref_fall_time," "Vref_rise_thresh," and "Vref_fall_thresh" are the only inputs necessary. For simplicity, let us consider the data rising edges. "Vref_rise_time" defines how long (in terms of simulator resolution) it takes the data signal to transition from a low value (usually 0) to a high value (such as +Vdd). The "Vref_rise_thresh" is used to indicate a percentage of that range (somewhere between 0 and +Vdd) that the receiver would recognize a switching point from low to high.

For non-distorted simulations, the rise time would be 0 and the rise threshold would be 50 percent. If either value is set to its default, the rise transition would occur at the reference time regardless of the other value. If both values are modified, however, the rise of the data signal can occur before or after the reference time according to the following equation used by the BFM:

$$\left(Vref\_rise\_time * \left(\frac{Vref\_rise\_thresh}{100}\right)\right) - \left(\frac{Vref\_rise\_time}{2}\right)$$

For example, if the rise time is defined to be 50ps and the rise threshold is 40 percent, the data signal would have rising transitions that are -5ps from (e.g., 5ps before) the reference edge. For the falling edge switch points, the equation is slightly different:

$$\left(\frac{Vref\_fall\_time}{2}\right) - \left(Vref\_fall\_time * \left(\frac{Vref\_fall\_thresh}{100}\right)\right)$$

If the same criteria is used for the falling edges (50ps fall time and 40 percent fall threshold), the fall transitions for the same signal would occur at +5ps from the reference edges. The rise and fall transitions, combined, would effectively make the high pulse width for a signal larger by 10ps and the fall time shorter by 10ps.

Figure 14.10 shows an example of these effects on a signal's shape. Here, the top waveform illustrates the real-world switching characteristics of a signal while the middle waveform illustrates the idealized waveform used for simulation based on the default 50 percent switching threshold. The bottom waveform shows the effect on the signal's shape when a 40 percent switching threshold is used for both rising and falling edges instead of the default 50 percent switching threshold. As mentioned before, the rising transition in simulation seems to occur 5ps earlier than the default (e.g., reference) case while the falling transition seems to occur 5ps later than the default case. In other words, in this example, every 10 percent change in threshold will cause a $50ps * 10/100 = 5ps$ change in rise/fall time from the reference edge. The negative or positive influence of the change results from whether the new threshold intersects a point on the non-ideal waveform that is earlier or later, respectively, than the middle switching point.

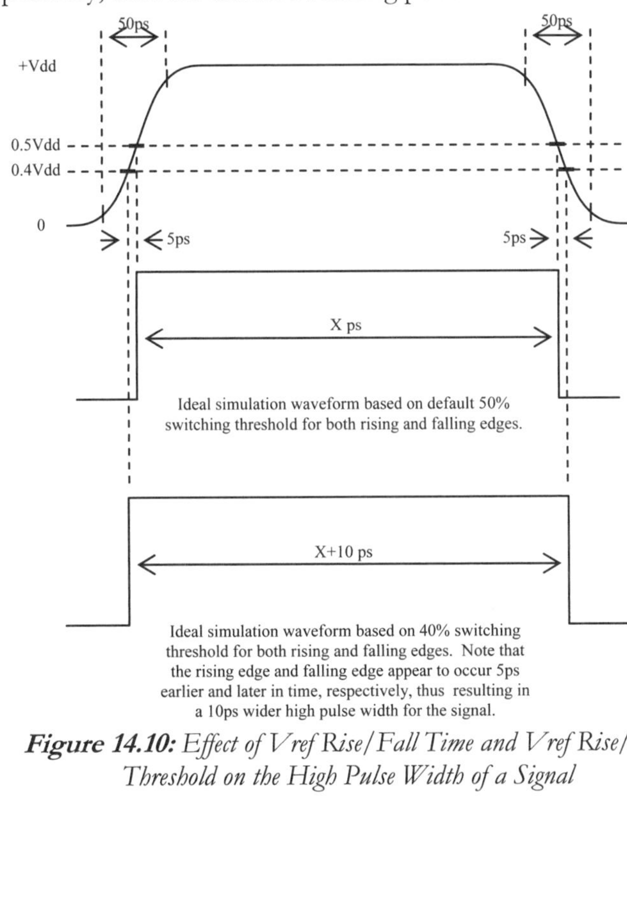

Figure 14.10: *Effect of Vref Rise/Fall Time and Vref Rise/Fall Threshold on the High Pulse Width of a Signal*

Like duty cycle distortion shifting, Vref distortion can be allowed to drift as well by setting "Vref_distort_enable" to true. "Vref_[rise/fall]_long" and "Vref_[rise/fall]_short" define the rise/fall time boundaries and "Vref_[rise/fall]_thresh_high" and "Vref_[rise/fall]_thresh_low" define the rise/fall threshold boundaries. The rates of change for both rise/fall times and rise/fall thresholds are computed using similar parameters as defined for drift and duty cycle shift.

These parameters all work in concert (e.g., they are additive) to model the ways in which changes in voltage and temperature and various noise sources might affect the timing and clarity of clocks and data signals being transmitted across a bus.

Delay Element Timing Characteristics

Another potential problem with high-speed I/O buses exists internal to the design within the delay elements which allow bits to be deskewed. These delay elements all have some inherent specified delay values which can be set by various analog controls. Thermal variations within the chip (which can be non-uniform) can also affect skew by changing the timing characteristics of the delay elements. This source of distortion was modeled by creating global signals, set by the control BFM, as inputs to the base delay elements that set their timing characteristics. In a similar fashion to the bus parameter structs mentioned earlier, the control sequence item (another struct) contains fields which are constrained in the testcases, and the control BFM acts based on the fields in these structs when received from the control sequence driver. The control sequence item contains a command field which can be set to "override_delay_values" and other fields that contain the new delay values. There are also drift fields similar to the bus drift fields defined above which allow the base delay values to gradually change over time to simulate thermal variations.

14.6 Resolution of eVC Parameters

Although there are already a large number of fields in the bus parameter struct, the scenario can become more complicated, requiring even more fields. The reason for the complication is that some buses are defined hierarchically such that there may be more than one clock-data group per bus. As I/O frequencies increase, it becomes necessary to limit the number of bits transmitted per clock. In order to achieve larger bus widths, more than one clock may be distributed across the bus, each clock being associated with a certain number of data bits. Alignment of these clock groups then becomes another issue for the design to handle. *Figure 14.11* illustrates a bus with multiple clock groups.

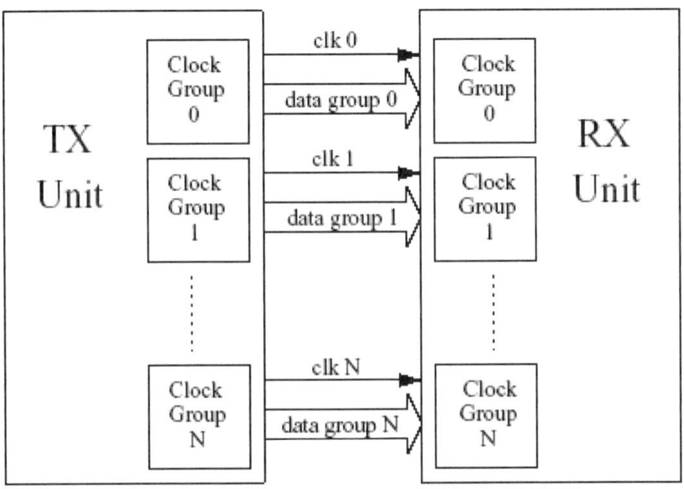

Figure 14.11: *Bus Hierarchy*

In order to make our eVC as flexible as possible for the verifier, parameters need to be definable at each level of hierarchy, that is, across the entire bus, for specific groups, and for individual bits. Several factors are necessary to allow a large number of parameters to be concisely, if not effectively, programmable for several hierarchical layers. The first is a means to allow the parameters to remain the same from one test phase to the next without requiring explicit redefinition by the test writer. This was accomplished by defaulting all integer parameters to MAX_INT such that if they are not overwritten, the bus BFM will not change their values. For Booleans, a special Boolean type was created containing the enumerations "NO," "YES," and "UNCHANGED." "NO" and "YES" translate to "FALSE" and "TRUE," respectively.

"UNCHANGED" is a keyword used to allow the previous settings to be retained into the new test phase.

The second extension was made to the bus parameter struct. A "scope" field was added to the bus parameter struct indicating whether the struct is single-bit (for individual data bits or clocks) or multi-bit (for groups and the entire bus). Then, for each integer parameter of multi-bit scope, the struct was subtyped to include a min and max. For example, for "drift_rate", the bus scope struct contains a "drift_rate_min" and "drift_rate_max." This allows the test writer to define a range for the "drift_rate" parameter across the entire bus. For a three-bit bus, the test writer could define a bus sequence constraining:

```
.data_bus_parms.drift_rate_min == -3
.data_bus_parms.drift_rate_max == +2
```

The three bits would then be constrained (randomly) within that range (e.g. bit[0].drift_rate = +1, bit[1].drift_rate = -2, and bit[2].drift_rate = 0). This mechanism will obviously save time in constraining very large buses.

A superstruct was created which contains a single data bus parameter struct, <num_groups> number of parameter structs for group settings, and <num_bits> number of parameter structs for data bit settings. Clock structs were also created for the bus and group levels within the superstruct. Upon generation of the superstruct, methods added to the bus BFM resolve each parameter down to the bit levels, which ultimately get passed on to the individual clock and data bit BFMs. There is a resolution method for integer types and another for Boolean types. The methods observe each value for a given parameter (bus, group, and bit). If a bus value is defined, but group and bit are MAX_INT (or, UNCHANGED for Booleans), it will set the bit-level parameter to the bus value. If the range parameters (<parm>_min and <parm>_max) are used for the parameter at the bus or group level, a value will be randomly generated for the base parameter between those two values. In this way, the test writer can constrain the entire bus to have a certain range of values, such as a skew between 5ps and 10ps, in a single line. Then, if a certain group or bit is of particular interest, the test writer can apply specific settings (such as skew = 20ps) to that group or bit. Basically, the lower hierarchy parameter values always override the higher level values (e.g., bit overrides group and bus, group overrides bus). If no value is set at any level for a parameter, the value from the previous sequence item will be selected and applied to the bit BFMs.

Parameters are efficiently resolved by using the struct extensions, type, and the new bus BFM methods, by allowing the test case files to be compact, efficient, and easily understood.

For example, let us assume here that the current parameter being resolved is skew_amount. The bus is configured in the following manner:

- Group 0 (5 bits - indices 0..4)
- Group 1 (5 bits - indices 5..9)
- Group 2 (5 bits - indices 10..14)
- Group 3 (10 bits - indices 15..24)

The test file might contain something that looks like the following code snippet:

```
do bus_sequence keeping {
        .data_bus_parms.skew_amount_min[0] == 7;
        .data_bus_parms.skew_amount_max[0] == 8;
        for each (item) in .data_grp_parms.skew_amount_min {
                index == 0 => item == 4;
                index == [1..2] => item == 5;
        };
        for each (item) in .data_grp_parms.skew_amount_max {
                index == 0 => item == 4;
                index == [1..2] => item == 8;
        };
        for each (item) in .data_grp_parms.skew_amount {
                index == 1 => item == 10;
        };
        for each (item) in .data_bit_parms.skew_amount {
                index == 0 => item == 3;
                index in [5..6] => item in [20..30];
```

```
        };
};
```

The first two lines of the sequence constrain the entire bus to have a skew amount between 7 and 8. At this point, all bits in all groups will have a skew amount of either 7 or 8. The next two "for each" blocks constrain the group parameter for group 0 bits to be between 4 and 4 (which will always be 4) and groups 1-2 to have skew amounts between 5 and 8. At this point, all bits in group 0 (bits 0..4) will have a skew amount of exactly 4. All bits in groups 1 and 2 (bits 5..14) will have skew amounts of 5, 6, 7, or 8. Group 3 (which does not have a group constraint) will remain at the bus constraint value range of [7..8]. The next "for each" block constrains the skew amount for group 1 to be exactly 10. Since this is the fixed value, it will override the previous min/max values. Thus, every bit in group 1 (bits 5..9) will have a skew value of 10. The final "for each" block constrains the individual bits. Bit 0 is constrained to have a skew amount of 3, and bits 5 and 6 are constrained to have a skew amount between 20 and 30.

The final skew amount values will look as follows:

- Group 0
  - bit 0 == 3
  - bits 1..4 == 4
- Group 1
  - bits 5..6 == [20..30]
  - bit 7..9 == 10
- Group 2
  - bits 10..14 == [5..8]
- Group 3
  - bits 15..24 == [7..8]

If the bus parameters' min/max values had not been designated, then group 3 would have the value MAX_INT assigned for each level, and would therefore assume the previous skew amount values for each of its constituent bits.

14.7 Phase Completion Mechanism

One challenge faced during the development of our eVC was in providing an interface for the test writer to control the conditions under which a test phase would end. "Test phases" are basically logical segments of a test which are defined by a certain mode, data pattern, and parameters which do not change until the subsequent phase. In certain cases, the test writer needs the ability to allow phases of the test to continue until either a certain number of cycles have elapsed or a combination of signal states has occurred (or a combination of both).

Specman sequences are the building blocks for phases. In our eVC, they are the mechanism used by the test writer to define the phases. Most sequences are written such that they consume a predetermined amount of time or wait for certain signal states before emitting an event which ends the processing of that particular sequence and allows the test to continue on to the next sequence. This approach, however, does not work well in all cases due to the need to terminate sequences in a variety of ways that ultimately need to be controlled by the test writer. The test writer has no visibility to the environment code, so his or her only interface to the environment is through the sequence fields. Therefore, we created fields that would be accessible in the sequences which would allow the test writer to end the phases appropriately for a particular test. Let us refer back to *Figure 14.9* for an illustration of how the test phases are interpreted within the eVC as sequences.

The phase_done_condition field was created and added to all sequences such that the test writer could control the end conditions for each phase. The phase done condition is a user defined type which can be set to COUNT, STATE, STATE_AND, STATE_OR, or any combination of these values. Other fields (phase_done_count, phase_done_signames, and phase_done_sigvals) are used in conjunction with the phase_done_condition to fully define the test phase completion criteria. Each BFM gets the next sequence item (from the test), executes its intended stimulus, and then calls an environment method to wait for the phase done condition(s). Once the phase done conditions are satisfied, it then emits an event indicating sequence completion.

The phase done conditions are organized in a list, and the conditions are evaluated iteratively. For example, let us assume the following fields were set in a test sequence:

```
.phase_done_condition == {COUNT; STATE_OR; COUNT}
.phase_done_count == 20;
.phase_done_signames == {"~/:/top_level/test_passed",
    "~/:/top_level/test_failed"};
.phase_done_sigvals == {1, 1};
```

Once this sequence is started, the wait_phase_done TCM would first wait for 20 toggles of the local clock before continuing. It would then begin evaluating test_passed and test_failed. If either signal went high, the TCM would then wait another 20 local clock toggles before exiting. The BFM would then emit a signal indicating that the sequence is completed, allowing the test to move on to the next phase.

Through this methodology, the test writer has precise control over the test phases and more flexibility in generating meaningful testcases.

14.8 Conclusion

This environment provided us with an efficient, if not extensible, platform where we could quickly and cumulatively create a multitude of noise and distortion scenarios which could be used to thoroughly verify many configurations of high-speed, self-aligning, elastic I/O bus designs. Some of the numerous benefits of using this eVC are as follows:

- We have the ability to combine any number of the distortion characteristics within a single testcase. In the early stages of development, we were interested primarily in introducing only skew into our simulations and flushing through all the design bugs associated with skew alone. As time went on, we began combining the effects of skew, jitter, drift, etc., until we had very complex and powerful simulations. With our environment, we could easily target one type of distortion or combine several types for complex corner case testing.

- We have migrated various analog effects into the functional simulation domain, by allowing logic devices, to compensate for these effects, to be more thoroughly tested at an earlier stage in the development cycle.
- Within one environment, we can drive the RX DUT directly with a full range of distortion options from our behavioral driver (proactive mode), or we can hook the TX DUT to the RX DUT through our behavioral driver and still apply several distortion characteristics (reactive mode). In this way, we can ensure that the TX logic and the RX logic work together, as well as thoroughly stimulate the more complicated RX logic via the behavioral driver only.
- We can easily apply our BFM as a stimulus to a variety of bus sizes and configurations. In dealing with our specific design implementation, there were several different bus/group/bit combinations, and our generic driver solution was configurable to handle all combinations.
- Our resolution of parameters concept enabled us to dramatically reduce the amount of time for testcase creation. It allowed us to easily target certain bits or groups in a testcase for specific scenarios, while allowing the rest of the bus to run with random (or ranged) values. Random values were applied to the entire bus where specific bits or groups did not need to be targeted. This technique gave us synergy and flexibility among all the bits in a group, and all the groups in a bus and can easily be extended to designs with more layers of hierarchy.
- Using our phase completion concept, test phases are more general, allowing a test writer some flexibility in how to end or combine sequences. We did not have to replicate a lot of sequence code to account for various phase end scenarios.

In conclusion, we have created a BFM capable of generating the many noise and distortion characteristics now so commonly seen in high speed I/O applications. When doing functional simulation on I/O interfaces in the past, we have not been as interested in seeing the distortion effects described earlier. Speeds were slow enough that any skew could be contained with relatively careful board and chip design, making elastic interfaces not yet necessary. Now, with higher speeds and the need for elastic interfaces, there is also a need to have verification models capable of creating the many types of distortion characteristics evident in such systems.

References

[1] e Verification Components, 25 e Verification Components http://www.verisity.com/products/printer/evc.html

[2] Definitions for skew and jitter http://www.synqnet.com/pdf/jitter_skew.pdf

[3] Description of elastic bus design concept http://www.hotchips.org/archives/hc11/2_Mon/hc99.s2.2.Ferriaolo.pdf

[4] Specman Elite 4.3 User Guide – Verisity https://www.verificationvault.com/docs/index.php

Chapter 15

Toward Automation of IP (Reuse) Quality Measurement

Michael Rohleder, Principal Staff Engineer,
Freescale New Product Development Center

Ahmed Dabbagh, Design-for-Reuse Team Leader,
Freescale New Product Development Center

Joachim Fader, Principal Staff Engineer,
Freescale New Product Development Center

15.1 Introduction

Keeping pace with faster development cycles and productivity gains, needed to develop state-of-the-art devices, requires the semiconductor industry to maximize intellectual property (IP) reuse. One critical factor is confidence in the reuse quality of IP. Reuse standards are metrics to define and measure reuse quality with comprehensive rules and guidelines. Adherence to such standards is as important for reusability of a design as design for test (DFT) is for the testability of semiconductor devices. The process of validating IP against corresponding requirements, otherwise known as certification, is both lengthy and tedious, and it requires automation to gain acceptance within the design community. However, related work in the past two years has shown that automation alone is not sufficient, and an appropriate classification and definition of the related rules is further needed. Another area that has not received much attention is the structure of testbenches and the resulting requirements for the testbench architecture. It is our experience that appropriately defined and structured testbench components are equally important for the reuse quality of verification components than the ubiquitous base classes and coding guidelines for the code generation. This chapter presents related work and results.

The need to greatly increase its level of IP reuse became obvious to Freescale Semiconductor (formerly Motorola's semiconductor products sector) as system-on-chip (SoC) development costs continued to exponentially escalate during the 1990s. Initial attempts of IP reuse yielded less than satisfactory results partly since it was difficult to reuse IP that was not initially designed with this objective in mind. However, it is essential to plan for reuse as early as possible. Design-for-reuse (DFR) has been a long-time concern for software development [1], now it becomes a reality for hardware design. Unfortunately, this is more complicated than providing engineers with some tools and a manual.[1] A set of best practices was compiled, which was later formalized into an internal document called the semiconductor reuse standard (SRS). This document sought to institutionalize design construct and methodologies which would enable a more effective reuse of design IP. Three major iterations were required before a satisfactory, if not accurate and repeatable, means to validate compliance to the standard could be developed. In order to be successful, the validation process must be tightly integrated into the design flow and applied early-on and often to the work in progress. Achieving this goal required automation of the certification process, which is a central problem described in this chapter.

In the early years, most effort was spent on the infrastructure required for IP reuse. After defining metrics to measure IP reuse quality and to automate the related processes, work here concentrates on its translation into IP quality. Recent and ongoing work breaks this into manageable chunks, and it prioritizes the related effort. The code and collateral required for verification IP (VIP) is often more complex than the IP to verified, which made it a second focus of our current work. First, the results identified some properties, which are unique for VIP and essential for enabling its reuse, which cover the user-centric, second part of this chapter.

---

[1] Qualis (acquired by Synopsys) gives a good overview about the factors involved [8].

15.2 Common Design Environment

The set of tools and collateral information used for hardware design at Freescale Semiconductor is known as the common design environment (CDE). Its purpose is to support several aspects of the SoC creation process which includes the following elements:

- The design system, called stingray, seamlessly integrates the essential tools for hardware design. It enforces consistent design data management practices through a common directory structure and file naming conventions, and it improves design productivity by automating routine tasks such as project setup, tool management and design flow scripting. Third-party tools and licenses that can be shared among globally distributed design teams are handled by a central system, known as the comprehensive software asset management (CSAM) server [4].
- The SRS is a comprehensive set of documents defining mandatory and recommended deliverables needed for the effective reuse of design IP. It addresses hard, soft, and analog blocks as well as standard cell libraries and SoCs with a set of rules and guidelines intended to assure quality, ease of use, and repeatability in implementation. A subset of this document is publicly available and downloadable [6]. In addition, the functional verification chapter of the SRS was donated to the Virtual Socket Interface Alliance (VSIA), and served as the basis for their *Functional Verification for Virtual Component Development and Integration Specification.* Other related activities included alignment of IP reuse standards with Philips Semiconductor and ST Microelectronics as an extension of their joint venture foundry partnership.
- The IP repository [11] is suitable for storing and delivering a wide variety of IP within Freescale, including hard and soft design IP, software (e.g., Stingray), and documentation (e.g., SRS and templates). It consists of a central relational database equipped with robust search capabilities to locate IP stored in distributed vaults. The overarching goal of the system architects was to provide users with a one-stop shopping experience while asserting Freescale proprietary

information protection policies and applicable US export control laws.

15.3 Certification

Certification is a structured process to assess the compliance of IP deliverables to applicable rules and guidelines and to document the observations in certification checklists (one for each applicable SRS standard). Once all deliverables are assessed, a reuse grade is calculated to indicate the overall reuse quality of the IP for integrators. *Figure 15.1* depicts the scheme used to calculate the reuse grade from the degree of rule and guideline compliance. The final step is to summarize the results of each certification checklist, and an overall grading result into a block certification report that provides an overview of the determined SRS compliance levels.

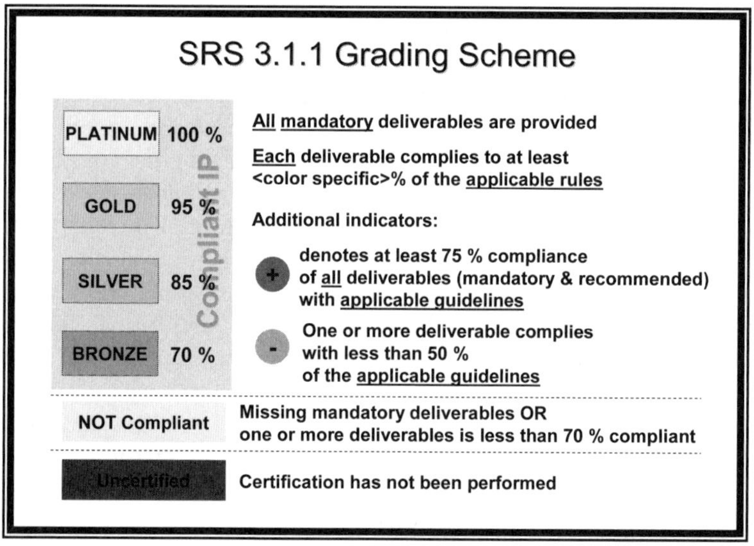

Figure 15.1: *SRS Grading Scheme*

The most recent version of the SRS consists of sixteen chapters, that cover a wide range of semiconductor design practices including hardware description language (HDL) coding (e.g., Verilog, VHDL), DFT, functional verification, and full-chip simulation. It consists of roughly 800 pages and defines about

1,660 rules and guidelines for nearly 80 deliverable types. These deliverables are unique and shared for each type of IP that the SRS covers: virtual components (VCs) of soft, hard, analog or library IP, embedded software, full chip simulation models, or a VC of a complete SoC.

Only until recently, certification has been entirely a manual process. Templates were provided for users to create the certification checklists for each deliverable. Since these templates had to cover many possible IP delivery scenarios, they often contained reuse requirements that were not applicable for a particular scenario. Given the number of the SRS rules and guidelines, along with the number of deliverables to certify, manual certification using the templates is a tedious and error prone undertaking. Most designers did not embrace this manual process because they considered it as an inefficient and excessive overhead. The feedback from one design manager having just completed certification of a large block provided a clear warning that the process was flawed: "*Going through the certification process for this device took an effort of over 380 hours of our time plus other engineers' time. To go through it again for its successor would still be a monumental task in terms of paperwork and conferring with our engineering team on our answers. Even if it only took half as long it's still way too much time to spend on answering questions.*"

With this testimony, the process became an obvious hindrance to quantifying IP reuse quality and explained why certification was bypassed in most cases. The certification effort clearly needed to be reduced drastically in order to overcome the overall resistance to its performance.

15.4 Initial Analysis

IP certification against the SRS occurred rarely, if ever, in Freescale, and the process was considered by many engineers as quality assurance work and an unproductive activity. The complexity of the manual certification process and application late in the design process brought about growing opposition. This was rarely the last step before the design data was released to manufacturing, or even later. The net result of this work habit is that non-conformities were discovered too late to be fed back into the development process, thus hindering the designers from improving the quality of the original design. It was unlikely that opportunities for improvement identified by the certification results would be incorporated into the final design iteration before release. Subsequent implementation of the IP in another product often meant that the integrator had to accept design deliverables of sub-standard quality or be forced to modify the IP before it could be reused.[2]

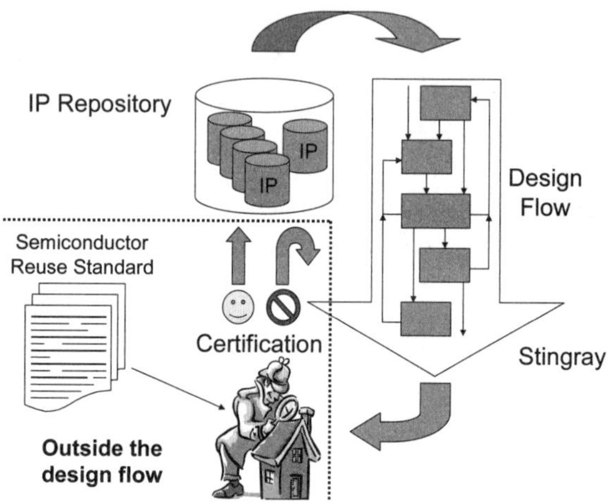

Figure 15.2: *Manual Certification and the Development Process*

Figure 15.2 depicts the previous interactions between the certification process, Stingray, and the IP repository. The certification of a design against the SRS was performed primarily outside of the design flow.

The fundamental solution to this problem is to make certification an integral part of the design flow. This requires a lightweight and simple validation process that can be performed earlier in the design process, preferably on partially complete data, while supporting repeated iterations to show the improvement in compliance each time it is checked. Any automation in this area benefits the certification process, although it is not a strict

[2] Using IP that has to be modified is not regarded as true reuse as viewed by the SRS.

requirement. With complex, if not leading-edge, SoC projects in the semiconductor business typically requiring multi-disciplinary teams, the certification process must also be scalable; for instance, gathering and combining the certification results from multiple designers and/or organizations working on different deliverables of a project needed to be quick and simple. None of these objectives can be achieved with a manual certification process. Another problem identified in the SRS certification pilots is that the standards are originally written without certification in mind.[3] Rules and guidelines were phrased in a way that makes the appropriate validation steps unintuitive or not applicable for various reasons. However, all of these points have to be considered during the certification process, and this must be documented in the certification report even if the stated requirement does not apply for a particular deliverable. Both points add undesired effort and complexity to the certification process and it quickly becomes a sizeable problem when the numbers of rules and guidelines is large.

15.5 Related Work

The team investigated what similar certification tools and methods already existed and often found results relating to the quality of the final code and its associated testbench [9], rather than the deliverable creation process. Others relied on the certification method defined by OpenMORE [7], used by many IP providers as the means to certify design IP quality. A more recent attempt to solve the certification problem was developed by the VSIA with their quality IP (QIP) metric released in 2004 [12]. Both efforts will be compared against the SRS work in this section.

OpenMORE consists of about 200 guidelines largely derived from the reuse methodology manual [3]. These questions are formulated on a relatively high-level when compared to the certification method used by the SRS. In other words, an OpenMORE question often relates to multiple aspects of an IP that are described by

[3] This is not a problem that is unique for the SRS; many other standards are similar when it comes to certification. But this issue becomes more obvious in case of a large and comprehensive standard, due to the sheer amount of rules and guidelines to be validated during the certification.

separate rules and guidelines in the SRS. The response to an OpenMORE question must therefore provide a summary judgment of all those aspects with a consolidated answer. Another difference is that the assessment method of OpenMORE relies primarily on visual inspection [10] and entering the findings into an excel spreadsheet that contains all OpenMORE questions.

VSIA's QIP [12] is a derivative work based upon the principles of OpenMORE certification after it was donated to VSIA in 2002. It focuses on ease of reuse, design, and verification quality, IP maturity, and vendor assessment. The number and granularity of the related questions is similar to OpenMORE with an IP quality score calculated automatically after answering the individual questions. Both approaches utilize a similar audit style to assess compliance to widely accepted criteria for reuse, but neither considers a company's specific requirements or practices. This is a reasonable approach to make the results relevant to IP which is intended to be reused by different companies; however, this fails to provide any type of aid with respect to company-specific needs. Yet, all companies have different requirements for their integration work, a fact that commercial solutions which the QuickUse IP qualification system from Mentor Graphics [5] attack by involving consulting services to customize their solution.

15.6 Simplifying Certification

One major difference between the SRS and other certification methods (like OpenMORE) is that the SRS has significantly more rules and guidelines due to the smaller granularity of the individual requirements. This resulted from the desire to make each requirement less ambiguous in order to leave less room for debate and to produce a more repeatable certification process. The downside of this approach is that there are significantly more rules and guidelines and more questions to answer during the certification. An undesired consequence resulting from the wide range of IP that require certification is that many rules and guidelines are not applicable to all classes of deliverables. The diversity of IP types is compounded by differences in design styles which allow multiple and legitimate implementations, that is, each one with its own, unique reuse requirements. Determining the correct set of rules and guidelines applicable to a given deliverable under these conditions becomes a major part of the certification effort, while still leaving excessive ambiguity in the certification process. The quandary boils down to a balance between a standard that has fewer rules, where every rule requires a deeper analysis and more elaborate answers, and a standard that has a greater number of specific rules, where each one must be individually evaluated for applicability.

The latter approach gives the implementer a greater insight into how to achieve a more reusable design, which was a primary objective of the SRS when its development began. An additional benefit is the certification process becomes more conducive to automation. It is much easier to generate a script or application when the set of requirements it has to validate is formulated with a very specific and narrow scope. Having less specific rules would require those rules to be interpreted by the script or tool developer in order to enable their verification by a machine. It is very conceivable that the resulting assumptions would require the detail captured by the rules and guidelines in the SRS, although there are times when even more granularity might be needed. The SRS developers believed it was preferable to provide its users with those details including justification for each rule and guideline within the standard rather than obscure these details within the code of an anonymous software tool.

15.7 Rule Pruning

This initial analysis showed that a lightweight, scalable certification process distributed over a whole team which is highly desirable. It became clear that the existing manual certification process was not a viable paradigm for the success of the SRS, and it would allow more excuses not to use the standard in an area that needed change towards a more reuse-friendly IP development process. Further analysis revealed the need to make identification of the required deliverables easier, together with the proper association of applicable rules and guidelines. The process of eliminating all unnecessary or non-applicable rules and guidelines from the certification checklist became known as "pruning."

The achievable savings are significant and easy to estimate. Let us consider how much work can be saved if two minutes are required to read a guideline, identify whether it is applicable, and document it – when only 30 percent of 1,660 possible rules and guidelines apply to a certain deliverable? Let us now multiply this figure by the number of deliverables per IP block, and again by the number of IP blocks to be certified each year, and you get some impressive numbers about the potential effort savings.

How Pruning Works

A different set of deliverables is required for different IP types. For example, device drivers may be needed for an IP block, but not for a processor core. Another example is design IP containing memory blocks, which may require some special form of memory built-in self-test (BIST), which is not needed for a block without any memories. Also, the chosen implementation language (e.g., Verilog/VHDL, C/C++, or SystemC) and style (e.g., synchronous vs. asynchronous) often determines the need for some specific deliverables, along with a set of language and style specific rules and guidelines.

Rules and guidelines may be specific to only one type of IP, such as a processor core, a memory block, or a driver/monitor used for verification purposes, in order to accommodate these differences. Furthermore, there are language-specific rules and guidelines in the SRS specific to design and scripting languages such as Verilog, VHDL, C/C++, Tcl/Tk. In the forthcoming versions of the SRS, these differences might even extend to multiple Verilog versions (-1995, -2001, and SystemVerilog) or language extensions like SystemC.

Providing a small amount of classification information about the type and implementation of a deliverable permits a certain number of rules and guidelines to be removed. By identifying those rules which apply only to specific types of IP or certain implementation languages, it becomes possible to screen out, or prune, these out of the user's view. Further refining this approach towards a more granular classification schema permits a more precise selection of applicable requirements for a given deliverable.

There are additional benefits to the rule pruning process that are less obvious, yet easy, to quantify. One precondition to automate the selection is that every guideline must be clearly assigned to one or more IP deliverables. Following this practice is beneficial, since it enforces a better organization and structure of the overall certification process. Another benefit is that the specification of the corresponding deliverables along with conditions for applying a guideline can aid in its validation; i.e. the need to define this information often leads one to rephrase the guideline in a more complete and unambiguous manner. The question now becomes what is needed to identify the pruning criteria.

15.8 Metadata

Metadata (or, data about data) is used within the CDE to identify data needed to describe important aspects of the IP being developed. The SRS defines it in the form of properties, where each property can receive a set of predefined values, a user defined value, or both. Some properties can receive multiple values; good examples are properties describing the available interface types (e.g. RapidIO, Coldfire, IP Bus) or those identifying the implementation language used (e.g., Verilog, VHDL, C/C++). Within the CDE, metadata was already used for multiple purposes such as describing important aspects of a block within Stingray or the selection criteria in the web based search and query capabilities of the IP repository.

It was an obvious choice to use existing metadata and additional properties as criteria to reduce the number of guidelines that need to be validated during certification. The metadata already available is inherently correct and does not need to be entered by the user. Since it was desirable to minimize the number of additional properties to be entered, we defined the following process to identify related metadata and to select the required deliverables and applicable guidelines:

- Automatically retrieve the available metadata properties from the CDE
- Identify a small set of basic properties that defines the major aspects of a VC and which can be used to prune away other properties

- Request appropriate values for properties missing in the data extracted from the CDE from the user and validate the provided answers
- Eliminate properties that are not applicable, dependent on the values retrieved ("property pruning")
- Select which of the deliverables of the VC block must be delivered and certified based on the selected properties (or "deliverable pruning")
- Calculate the set of properties needed for "pruning" from the pruning rules defined for the rules and guidelines for these deliverables
- Remove pruning rules that are already falsified by basic properties from the selection (or "rule pruning")

This process implements a three-staged sieve controlled by a minimum set of basic metadata properties, where the result of each sieving stage is used to control the sieve of the next stage. The first stage selects the metadata needed (related metadata) for selecting the required set of deliverables (required deliverables). This selection then identifies the set of rules and guidelines to be applied to those deliverables (applicable guidelines).

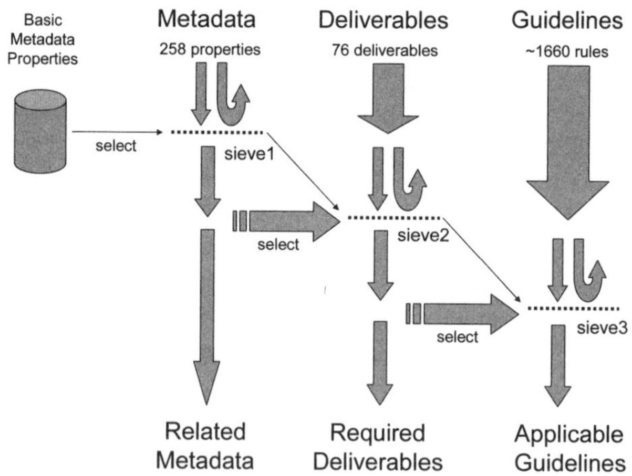

Figure 15.3: *The Pruning Process*

Figure 15.3 visualizes this process, and the result is a minimum set of metadata properties, deliverables and rules, and guidelines that can be used to control many of the automation steps introduced in the following section. It is worth noting that only values have to be provided by the user if they can not be retrieved from the CDE. Once provided, all information is preserved and does not need to be entered again.

15.9 Golden Rules

Another related activity was a careful review of the rules and guidelines stated in the SRS against some "golden rules". Rules and guidelines that did not adhere to these requirements had to be deleted, reworded, broken–up, or otherwise modified to assert an easier verification. These rules are illustrated in non-SRS specific terms in the following:

1. Deliverables must only be defined in the "deliverables" chapter (e.g., no rules that say "provide" or "deliver").
2. Every deliverable must have at least one rule or guideline.
3. Every rule and guideline must be associated with a deliverable.
4. Documentation requirements must be stated in the "Documentation" chapter
   (i.e. no rules such as "document xyz" outside of the "Documentation" chapter).
5. Rules and guidelines must contain only attribute requirements for the deliverable ("must be 7" long") and not indicate a process ("run abc before").
6. Metadata properties must be defined only in the metadata standard.
7. Pruning equations may only use properties defined in the metadata standard.

The goal of these rules is to establish a clear definition of the deliverables to be delivered (rule 1); and their relationship to rules and guidelines stated within the standard document (rules 2 and 3). Since documentation of a deliverable is considered a deliverable, it is important to clearly distinguish both items (rule 4). Development process specific rules are forbidden, because it is required to measure the quality of the result (the deliverable), not the process itself (rule 5). The remaining rules are intended to assert the correct formulation of pruning rules (rule 6 and 7).

The resulting modifications were a significant improvement, since they removed a lot of ambiguities that were hindering an efficient certification process. An issue that remained was the sheer amount of rules and guidelines to be visited as well as the related paperwork to answer the related certification questions. Even worse, the related activities are tedious and not well suited for human beings, which made them an obvious target for further improvement possibilities.

15.10 Automation

To further support the creation of re-usable designs, Freescale has integrated, or customized, third-party applications or developed helper applications within the CDE, when there was no commercial solution available. For this discussion, we will only cover aspects related to design-for-reuse[4]:

- MotoSpy is a Motorola/Freescale specific customization of Atrenta's SpyGlass RTL code checking application [2]. SRS coding guidelines for Verilog and VHDL have been coded into MotoSpy to automatically check RTL code for compliance. MotoSpy is fully integrated into Stingray/the CDE.

MotoSpy is an excellent example for the automation of the certification effort related to a subset of the SRS rules and guidelines. There are some other areas where similar automation of

[4] This discussion is not intended as an advertisement of the described tools and helper applications; Freescale is not in the business of developing tools for this purpose. Its sole purpose is to describe some first steps on automating the overall certification process by some "real world examples". It is our experience that a more abstract discussion of those terms and the involved work is far less descriptive, and it is harder to explain the achieved benefits.

the certification work is possible. Unfortunately, not all of them are attractive enough for third-party vendors to provide a tool that can be customized for this purpose. Furthermore, the development of an internal tool for such tasks suffers a similar problem; management must be convinced that the automation of the related certification efforts is a task worth attacking. Yet, this is impossible when certification rarely happens.[5]

- The size of the SRS slightly hinders the efficient identification of all deliverables, rules, or guidelines applicable to a specific design project. mySRS is a tool designed to take over this task at the earliest stage possible. Controlled by essential selection criteria and metadata properties, it identifies the deliverables required by the SRS and the corresponding guidelines. All information that is not needed for this type of IP is automatically removed through the automatic pruning processes. This alleviates the need to browse through hundreds of pages to find and learn what is required by the SRS for a particular design. This is a far more efficient way to tell a design team what it has to deliver and in which format this has to be done than requesting the team to read the whole SRS document. It also provides links to the full SRS for more details.
- AutoCert is a mirror of the mySRS facility applied towards certification. It selects the required deliverables and requests the corresponding compliance information only for applicable guidelines. This reduces the effort for collecting compliance information significantly and permits subsequent integration of automatic compliance checking applications (like MotoSpy) into the certification process. The AutoCert tool is designed to be used iteratively as designers complete their design deliverables.

[5] Obviously this is a typical catch-22 situation. Until more companies are looking into the automation of their certification efforts, the only chance for improving this will be to proof that certification does happen more often. This was anyway our initial goal …

Both applications described above exploit the pruning process and related capabilities built into the new SRS version to identify the applicable deliverables, rules and guidelines.

- Certification reports are deliverables of the certification process. The availability of the corresponding information within the AutoCert tool permits the automatic creation of these documents, in particular, the block certification report and the associated checklists for each deliverable. The inclusion of statistical information in these documents as well as the calculation of the corresponding reuse grade is done automatically. The automated and online generation of this information further permits its use as a viable metric during the design process. This enables the collection of certification information on-the-fly, by identifying problem areas while IP is being developed. It also helps to further define improvement targets for subsequent projects. An excerpt of an example block certification report[6] is shown in *Figure 15.4*.

SRS Block Certification Report												
Product Name	PilotProject											
Product Version	1											
VC Type	FCS, Soft											
SRS Delivery Compliance	93%											
Deliv. Code	**Deliverable Name**	**Deliv. Type**	**Required**		**Total**	**Met**	**Not Met**	**N/A**	**Unanswered**	**Pruned**	**% Certified**	**% Compliant**
G1	VC Packet	Released	Mandatory	**Rule-Rel.:**	3	3	0	0	0	0	100.0%	**100%**
G2	Metadata	Released	Mandatory	**Rule-Rel.:**	1	1	0	0	0	0	100.0%	**100%**
D1	Product Brief	Released	Mandatory	**Rule-Rel.:**	25	14	1	4	0	6	100.0%	**93%**
				Guideline-Rel.:	3	0	0	0	0	3	100.0%	**100%**
L1	Synthesizable RTL Source Code	Released	Mandatory	**Rule-Rel.:**	348	58	3	7	4	276	94.4%	**95%**
				Rule-Prop.:	103	0	0	0	0	103	100.0%	**100%**
				Guideline-Rel.:	108	50	2	2	4	50	93.1%	**96%**
				Guideline-Prop.:	52	0	0	0	0	52	100.0%	**100%**
L2	Synthesis Scripts	Released	Mandatory	**Rule-Rel.:**	2	0	0	0	0	2	100.0%	**100%**
				Rule-Draft:	13	2	0	0	0	11	100.0%	**100%**
				Guideline-Draft:	11	8	1	2	0	0	100.0%	**89%**
L3	Synthesis Constraints	Released	Mandatory	**Rule-Draft:**	15	3	0	1	0	11	100.0%	**100%**
				Guideline-Draft:	11	9	2	0	0	0	100.0%	**82%**
S1	FCS Model		Mandatory	**Rule-Rel.:**	21	9	0	1	9	2	52.6%	**100%**
				Guideline-Rel.:	16	3	0	1	2	10	66.7%	**100%**
S2	Build Environment		Mandatory	**Rule-Rel.:**	17	9	0	0	0	8	100.0%	**100%**
				Guideline-Rel.:	2	1	0	0	0	1	100.0%	**100%**

Figure 15.4: *Block Certification Report Example (Excerpt)*

Documentation templates for all documents defined by the SRS can be downloaded from the IP repository. The SRS predefines

[6] The block certification report summarizes all certification results for a block, detailed results are provided by the individual checklists.

the structure and corresponding content of these documents, in order to provide complete and consistent documentation as part of a design project. The existence of a concise set of templates should result in a more consistent layout and reduces the need for engineers to solve technical writer problems.

AutoDoc is a facility that goes even further, especially as it helps to identify the required documents, fetches the related templates, and populates them with information already available. Some of the related information can be retrieved from original design data by features of the Stingray design system. Other data originates from metadata that has been previously entered into the IP repository. An example for this facility is the automatic generation of the IP brief, which is a short document that seeks to give a short overview about the available IP, its features, modes, parameterization options and other technical details. The creation and subsequent upload[7] of this document is fully automated; the user only provides a graphic showing an overview of the block structure and enters some metadata that may not be available from other sources. AutoDoc selects the required metadata and identifies the appropriate format (both dependent on the VC type) before creating the document.

Although AutoDoc does not directly benefit from inserting the pruning facilities or the modifications to the certification process, it is an indirect result of those changes. AutoDoc would not be possible without all the information stored within the metadata that has been made available as part of these enhancements.

Most of the facilities mentioned above have been integrated in an umbrella tool called VCWizard, which reuses features of the IP repository and is linked tightly to the Stingray design system. *Figure 15.5* shows the various parts of VCWizard and how they are interacting with Stingray and the IP repository.

---

[7] This refers to the upload of this document into the IP repository to publish *the intent* to create the corresponding component (before creation) or to publish its availability (after creation).

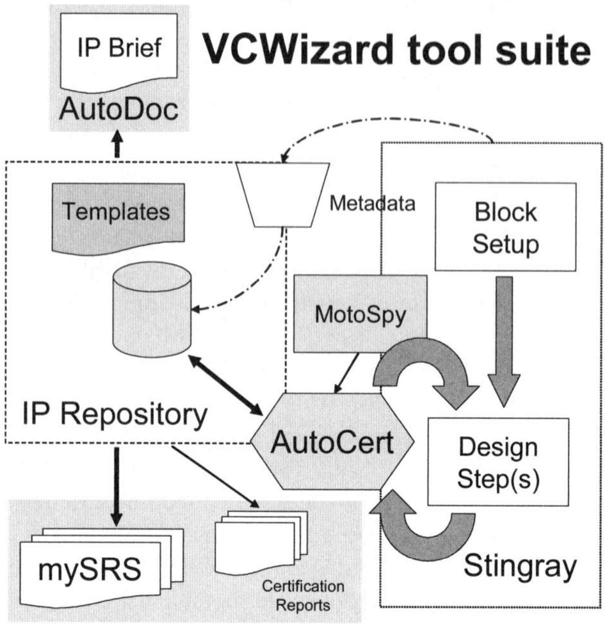

Figure 15.5: *Members of the VCWizard Tool Suite*

It also visualizes the extraction of metadata from the Stingray and the reuse of the IP repository as a central database for all data gathered by the certification process.

15.11 Integration and Data Integrity

Freescale's reuse strategy has been evolving since the early 1990s. The CDE integrates all related applications into an environment supporting a certain set of clearly defined interactions. The only exception has been the Semiconductor Reuse Standard; obviously it is much harder to integrate a document about methodology aspects into such an environment than a set of tools. The VCWizard tool tied this important piece into the CDE; resulting in an earlier and better coverage of reuse aspects and enabling further automation of certification work.

One concern has been the integrity of the SRS related information. It is virtually impossible to manually re-enter all associated data in case of a new release of the SRS, not only because of the sheer amount of data. Any update must be performed in a timely fashion to avoid users working with old data. The solution to this problem is as simple: all data is retrieved directly from the SRS documents. A set of filters has been created that extracts all required data from the original documents and writes them into a set of exchange files, which can be subsequently imported into the VCWizard. The VCWizard uses this "electronic SRS" to control all SRS related activities. Using these facilities takes less than a day to perform a full upgrade to a new version of the SRS. An important prerequisite for this methodology is that all input forms, checks and any other activities of the VCWizard are fully controlled by data and does not involve any code changes within the VCWizard. Any need to perform software modifications would result in much longer upgrade cycles.

Figure 15.6 shows how the VCWizard incorporates the certification process into the CDE. It further visualizes the capability to perform certification tasks after every design step performed within Stingray. The VCWizard shares not only many features with the IP repository, but it also becomes part of the overall design flow.

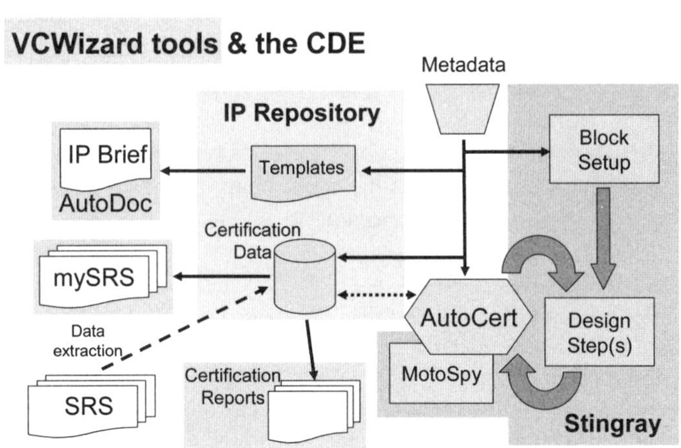

Figure 15.6: *Integration of VCWizard into the CDE*

15.12 Effort for Implementation versus Update

With an existing infrastructure that provides the most basic capabilities, the implementation effort for a solution like the VCWizard is rather low. In our case, it was possible to reuse some of the functionality implemented within the IP repository, and the Oracle database [16] that is used to implement major portions of the IP repository.[8] The powerful database manipulation language, Oracle, provides with SQL a benefit for the manipulation of the involved data, especially for the pruning effort. Further, a Web–based data entry and manipulation like the one provided by the IP repository enables the distribution of the certification effort over multiple teams and locations.[9] The development effort for this new infrastructure has been split into five main themes:

- *Definition of the work flow, the related software architecture to support this flow, and the related interfaces and exchange files.* This includes the "selling" of this work to senior management and the involved organizations. The related effort took four to six man weeks.
- *Implementation of the VCWizard, which consists mainly of some Web–based data entry forms, an upload facility, and some report page generators to display the certification data.* Most Web pages are generated by SQL scripts using data from the underlying Oracle database. This effort undertook three to four man months of an experienced Oracle developer.
- *Instrumentation of all 16 chapters of the SRS with pruning rules and related metadata.* This work has been performed by the champions responsible for a chapter and required about 2 work days effort per chapter.

---

[8] Due to the nature of the certification effort there is some need for a database to store the metadata, and information about the deliverables, rules and guidelines, as well as the provided answers. It is beneficial to reuse the involved infrastructure, especially since this enables some tighter integration of the reuse capabilities into the overall development process.

[9] Other existing infrastructures (e.g., in other companies or organizations) might suggest a different architecture or implementation, but the requirements to be fulfilled as well as the selected solutions are very likely rather similar.

- *Creation of scripts to extract all required data from the SRS chapters and to generate the missing cross-reference information.* The data files resulting from this step are uploaded into the VCWizard tool, thus enabling the support of a new SRS version within hours after its release. Creating these scripts required two to three man months of effort.
- *Verifying the data exchange files created by these scripts and performing some online verification of the resulting entry forms, help pages and the underlying pruning algorithms.* This work has been shared between the Stingray and IP repository teams, the VCWizard developers and SRS chapter champions to maximize the test coverage and bring different point of views into the verification work. At the end, there were nearly 20 people involved for about one week in verifying this new capability.

In total, this initial creation effort sums up to 318 man days spent by a world-wide team of nearly two dozen people, having different fields of expertise and a wider range of responsibilities than working on a reuse infrastructure. *Figure 15.7* shows the distribution of this effort: about one-third had to be dedicated to data extraction from the SRS (instrumentation and interface scripts), one third went into the tool creation, and another third had to be spent for the verification of the result. This initial verification was rather exhaustive, since major portions of the tool suite are completely data driven, and we had to make sure the correct data has been extracted and it does properly control the overall process.

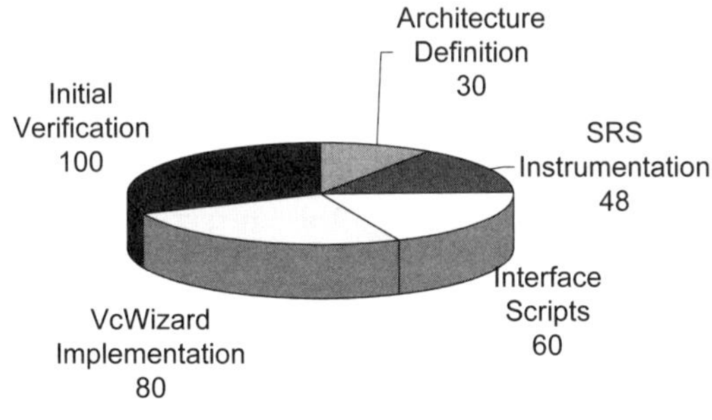

Figure 15.7: *Initial Creation Effort*

It was mandatory that only minor parts of this work need to be redone for a new SRS version; this requirement was built into the architecture by extensive scripting and mapping information. The SRS instrumentation, the VCWizard, and the scripts to extract the related data are in place. Therefore only minor updates (if any) and bug fixes to the corresponding information and code are required. The only work that needs to be repeated is the verification effort, but here only a very limited amount of work is required, if VCWizard or the input data (structure) has not been changed substantially. *Figure 15.8* shows the effort we have experienced when updating the SRS from 3.1 to 3.1.1; which accumulates to only about 6 percent of the initial effort.

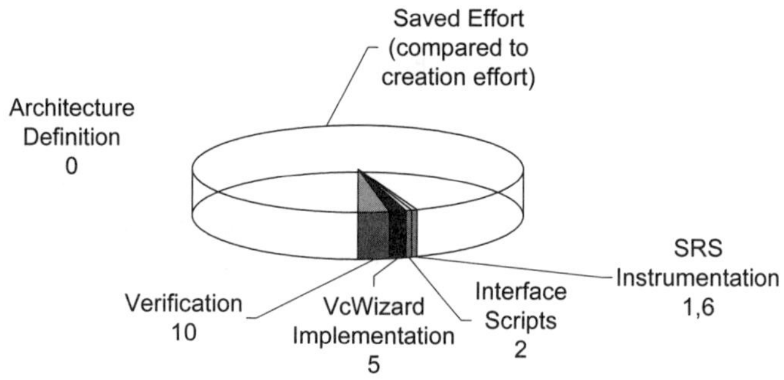

Figure 15.8: *Update Effort (Relative to the Creation Effort)*

Figure 15.9 shows the same effort to update the VCWizard distributed over the noted work areas. This distribution might be a little atypical, since we had to perform a larger amount of bug fixes for problems that escaped the initial verification phase. Besides those required fixes on the VCWizard application, the interface scripts and the related control data from the SRS, the bulk of work went again into the verification of the resulting web forms; mostly because we had to verify the effects of the performed modifications.

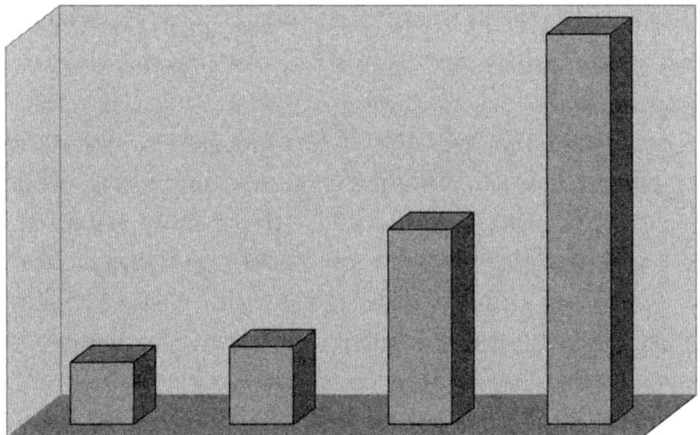

	SRS Instrumentation	Interface Scripts	VcWizard Implementation	Verification
Update Effort	1,6	2	5	10

Figure 15.9: *Update Effort (Work Distribution)*

15.13 Some First Results

There is not much real data available about certification work before the existence of VCWizard, besides the number of pages of the related certification reports. However, a good indication for the involved work is the number of questions answered when using VCWizard. Even a rather simple device such as an I^2C controller was required to answer nearly 3,800 questions, for the most complex one certified up to now 7,611 questions had to be answered.

Another indicator for the acceptance of the newer certification process is the amount of reusable blocks stored within the IP repository[10]:

[10] Many design organizations indicated that they do not want to submit their blocks to the IP repository because they do not want to spend the effort for the certification. This was a communication issue (the IP repository does not require the certification of a block for its submission), but it is also a good example for the negative effects of a too heavyweight certification.

- In January 2003 there were 84 unique blocks, 3 of them were documents, and 6 of them were software items; 31 of these blocks had 2 up to 16 revisions, in average 3-5. Only one of these items has been certified.
- In August 2004 there were 288 unique blocks, 3 of them were documents, 12 of them were software items; 55 of these blocks had up to 16 revisions, about 50 percent with only a single revision. 22 of them have been certified until now, in less than one year since the creation of VCWizard.

Figure 15.10 illustrates this significant increase of submitted and certified blocks within the IP repository. The complexity of a block is ranging from simple blocks like an universal asynchronous receiver/transmitter (UART), SPI, SCI, to more complex blocks (e.g., FlexCAN), complete processor cores (e.g., Star12, MCore, Coldfire, PowerPC) up to a complete full chip simulator of an airbag device.

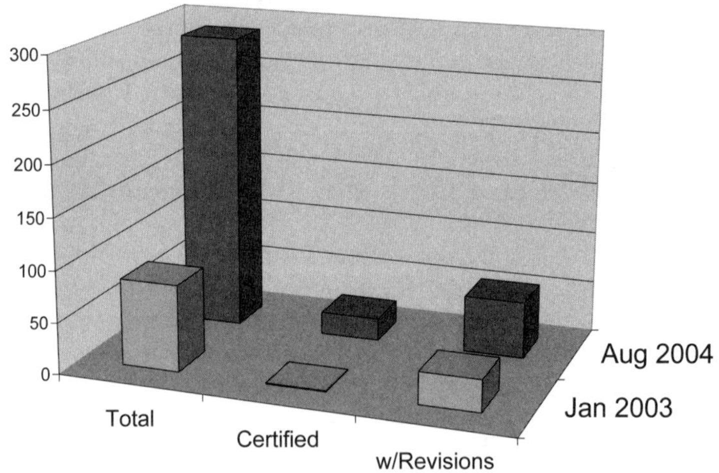

Figure 15.10: *Blocks in the IP Repository*

Some more numbers demonstrate the effectiveness of the pruning process. Until now, 22 items have been certified using the VCWizard toolset. When querying the relational database underlying VCWizard about the involved answers we received the following information (summarized for all items):

Yes	No	Not Applicable	Not Answered [11]	Pruned	Total
34,941	12,725	9,087	1,098	91,333	149,184
23.4 %	8.5 %	6.1 %	0.7 %	61.2 %	100 %

This means that nearly two-thirds of all questions have already been pruned away by the toolset, thus reducing the amount of questions to be answered. However, the relative high number of answers in the "not applicable" category shows that the pruning process can be further refined. Ideally, most of these questions are identified by VCWizard as being not applicable before involving any human being. Although it is impossible to identify all these cases automatically, the previous percentages clearly show the need to revisit the specified pruning rules.

How much effort does certification of an item involve now? Of course, the actual effort varies significantly with the type of IP and its complexity; therefore, it is hard to provide some absolute numbers here. Fortunately, when the VCWizard prototype has been finished, a derivative of the block that has been used for the certification pilot project had to be created. Using this derivative as a pipe-cleaner for the validation of the VCWizard prototype was an obvious choice.

Here is the feedback we received from the reuse officer overseeing this (and the earlier) certification project: "*The data from the original BLOCK_NAME checklists (i.e. the FrameMaker checklists that took the 380hrs to fill out) was subsequently entered into the VCWizard prototype by hand. The data from that system was then imported into the VCWizard that*

[11] The answer 'Not Answered' is used for partial certification and the default value provided by VCWizard. It is useful when not all deliverables are certified (e.g. because the project is not yet finished), or a distributed team is answering the certification questions and one of the team members has not yet certified her/his deliverables. As said in the text above, these numbers were extracted by a snapshot of the internal database underlying VCWizard.

we have today. In the end, we found that only about 75% of the data is complete and accurate. This shows that the checklists[12] *were very ambiguous and error prone. I would estimate that it is about 2-3 man days of effort to complete the remaining certification results (i.e. last 25%) using the VCWizard. From this point, the BLOCK_NAME derivative data can be created and I would estimate that is should only take an additional 1-3 days to get it certified.*"

This encouraging feedback provides some basis not only for effort estimations, but it also makes known the following information:

- The quality of the certification results improves drastically when using the automation features of VCWizard and having less ambiguous assignments between rules and guidelines and deliverables
- Certification of a rather complex block will take a maximum of 8 to 12 man days[13]
- The certification process for a derivative of this complexity can be done in about 1 to 3 man days when the data for the original block has already been entered into VCWizard

When comparing the estimated 8 to 12 man days (64 to 96 hours) with the 380 hours reported for the certification pilot (which just covered 75 percent of the required data), it becomes obvious that the effort savings can be significant. Since the certification effort for a new revision or derivative will be even less, these savings will even increase over time, especially when earlier data is already available within VCWizard.

[12] This refers to the FrameMaker checklists that had to be filled manually in the earlier process.

[13] The actual effort is very likely much smaller. The automated parts of the certification were for sure in the 75 percent that has been covered already, so the 25 percent missing are very likely to be the questions that were harder to answer.

15.14 Incremental Certification

The tight integration of the VCWizard in the CDE enables us to reuse and exchange metadata with other tools within this environment. The basis for this is a common definition of metadata properties and the capability to synchronize any updates.

This is the first step towards a much higher integration and automation of all involved systems. The implementation of MotoSpy on top of SpyGlass is just one example. More checker tools can be implemented or commercial applications (if available) can be integrated into Stingray in a similar manner. Further automation of the certification process is possible by forwarding the run results of those tools into the VCWizard. We are confident that a major portion of the validation work needed for SRS certification can be automated this way. One big hurdle to accomplish this goal was the need to revise the SRS rules to identify the ones suitable for automated verification. Another big hurdle is to convince senior management about the need to provide additional checker scripts to automate the process.

An essential benefit of this approach is that it enables incremental certification. This is a design process where every deliverable can be certified as soon as it has been created or modified. Why is this so important? We already described that there was a tendency not to fix problems found by earlier certification approaches; since the resulting changes might require an additional verification and/or endanger the schedule. As a result, the certification process often just documented reuse obstacles, but did not help to prevent them.

Incremental certification, on the contrary, provides immediate feedback and increases a designer's awareness about reuse issues. Removing standard violations becomes an integral part of the design process when it is possible to easily identify them. This then improves the quality of a VC prior to its completion and consequently avoids any expensive rework. As a result, incremental certification moves the certification process from being reactionary to preventive.

Thus, the early validation of reuse requirements during a design project permits DFR; an aspect of reuse that is as important as DFT for test. Without such a capability, modifying IP for compliance to reuse criteria will always be an activity that is too late to affect the design work itself.

15.15 Rule Characterization and Classification

After some initial successes with VCWizard, the acceptance of certification work—despite enabling incremental certification—still did not reach the expected level. Why was this case? Discussions with people that were involved in the overall process identified several shortcomings, resulting in a much lower acceptance than anticipated:

- The amount of rules to validate is still significant even with rule pruning. There is no problem associated with this, when the certification is automated and there are no rule violations flagged. However, for existing IP, the amount of rule violations that need to be fixed can be an underestimated effort. Equally problematic is that the sheer mass of violations messages can easily mask real errors.
- Less important rules are mixed with important ones, with no distinction when reporting a violation. Again, this is not a problem when there are no violations. However, it is a perfect excuse for any designer not to fix any violation, when there are several thousand violations flagged, and the majority of them are related to naming conventions, placement of comments, or missing information in the file header.
- When the certification process is performed, some rules are more important than others, which depends on the target market, the technology used, and the project phase. For example, rules related to power saving features are considered critical for a wireless device, and less important to other markets. In addition, it is not likely that a designer will want to fix a violated naming convention or commenting style after RTL code freeze.

Rule pruning is clearly very useful, but it is simply not enough. Some additional means are required to differentiate the rules to be certified; and to further influence the certification process. For maximum acceptance, it should be possible to confront a designer only with rules that are considered to improve the work quality of his work.

Rule pruning is the right tool in this case, but it requires some more information about the rules than was available. In a grass-roots effort, some field experts gathered their findings and gave birth to an initiative to improve the RTL quality. This initiative selects criteria for the certification process with the goal to ultimately enhance the quality of the RTL code.

The approach selected by this team is to further characterize the available set of rules according to the following criteria:

1. Rule category—Define a set of categories for the rules (e.g., clocking and reset, DFT, reuse, naming, conventions, etc.), and assign each rule into one or multiple categories.
2. Rule severity—Every rule is associated with a number that specifies the importance of the rule for achieving a high quality end result. The current proposal is to have a numeric scheme ranging from 1 (essential) down to 5 (nice-to-have).
3. Design category—This defines when to check a rule within the design process; especially what is the earliest and the latest project phase to apply a rule. For this work it is of particular help when the design process is standardized, having clearly defined phases, e.g.:
    i. Front-end development—Basic requirements of digital design functions
    ii. Back-end integration—Restrictions for synthesis, flow dependencies
    iii. Sign-off requirements—Including company specific rules applied at this level
4. Dependencies—Every rule is evaluated for dependencies on target markets, the target application, technology features, or other properties that might influence the applicability of a rule. Rules related to safety are more

important when the RTL is intended to be used in an automotive device; e.g. an airbag system. Rules affecting the signal integrity are more important for more advanced technologies. A list is compiled that specifies the criteria identified so far.

This is an ongoing work with limited amount of resources and, as a result, we cannot yet report results based on real design work performed during a real project. However, an initial trial has been done with a set of rules selected to identify and investigate asynchronous inputs on a small, but real design. The results were encouraging. The involved designers indicated that the selected rules were useful, even a potential bug has been found that could not be detected otherwise. However, there was also some feedback that we need to improve the messages and tool support, which requires some tight collaboration with the tool vendors. This might require some more time than we have anticipated so far.

The idea here is to select an applicable subset of rules for a design phase by additional metadata stored with a rule, and to prioritize the rules within this subset. Prioritization is done by the severity level, eventually altered based on information stored within the dependencies. The work packets identified for implementing this approach are:

- Define a classification scheme for all rules, and select the required characterization criteria
- Select severity levels and the related properties
- Review all existing rules and classify them accordingly
- Specify the rule selection process and the algorithm for defining a rule's priority
- Automate the application of this information and assert it is also reflected in the generated logs

Figure 15.11 shows a possible implementation of the corresponding selection process. It starts with a database that contains all rules that have been properly classified. The first step involves removing the rules that do not belong to the selected tiers. As a second step, modifiers are used to change the classification of rules that match a certain property set. Finally, all rules having a lower severity class than selected are blocked from execution by the corresponding checking tool. Basically, the chosen process is similar to the rule pruning process, but it requires a much tighter integration into the corresponding tools and work processes. This is dependent on the tool set used for this task automation and it must only run the selected rules and only provide check results for these rules to avoid cluttering the resulting log files with unwanted information. Further, the steering of the selection process must be made easy and transparent for the user.

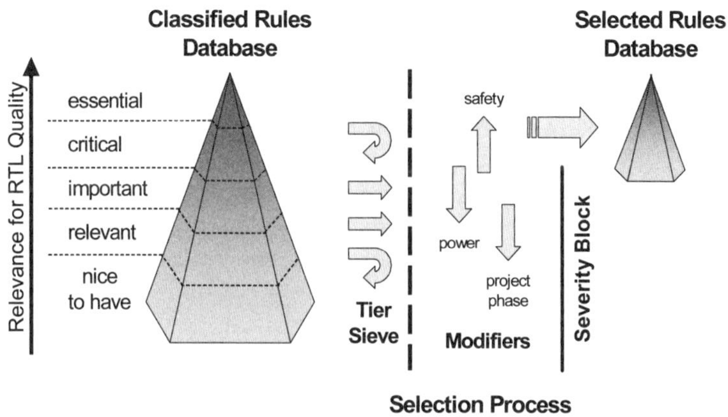

Figure 15.11: *Selection of Relevant Rules to Improve RTL Quality*

There is a much confidence that the definition of this classification scheme will be a great benefit for selecting the most appropriate rules during a project's development cycle. It will even help to identify shortages and holes in the measurement criteria and rule database. A stepwise refinement of the RTL can be achieved by first concentrating on the important rules and subsequently enlarging the rules database with rules having a lower priority. The goal is to select a manageable chunk of meaningful rules at each stage of the design flow. This will also permit an automatic measurement of the related IP quality to be provided when applying these rules.

Management can then request the quality of the IP to be increased over the lifetime of a project or across projects. A formula is needed to map the result of the prioritized rules to a quality indicator. Another part of this work is the proof that this will ultimately result in higher quality of the created RTL. The biggest problem in applying this technology is its likelihood to enhance the tool support to fully automate the selection and prioritization process since this involves changes in third-party tools, which need to be discussed and coordinated with the tool vendors.

15.16 IP Reuse Quality Requires Verification IP Reuse Quality

Verification is a major part of an integrated circuit (IC) design project. This is reflected by many investigations attributing about 60 percent to 70 percent of the design effort to verification [13]. One reason is the complexity of the test benches. Another reason is the sheer size of the code required to cover this complexity; in our design center the testbench code is typically significantly larger than the related design code within a single project. However, comparing VIP with design IP might be misleading since the former is more complex, but also has some other attributes that permit to mitigate the effects caused by its complexity.

Traditional approaches are treating VIP like design IP, which involves the same or similar guidelines on code belonging to the testbench such as in respect to naming conventions, coding style and structure, statement usage, etc. [12] [6]. Although this is a fair approach, since design IP and VIP have many aspects in common, it also has some shortcomings. Some rules do only apply to design IP (usually to assert its acceptance by a synthesis tool [15]; likewise some aspects are only of importance or possible for verification IP. Unfortunately the industry is far less focused on the latter one; until now, there was little effort to identify guidelines to improve the reuse quality of VIP.

An obvious first target is VIP reuse, which is more advantageous than just saving effort within your current project. One reason is the relationship between VIP reuse and its maturity as the more often VIP is used, the higher is its quality and completeness. Holes and shortcomings identified by one test can be addressed for the next user in the same or the next project. Reuse in different contexts (e.g., block level and SoC level) also helps to discover shortcomings that are hard to identify otherwise. This results in a high motivation for the vertical and horizontal reuse of VIP.

Vertical reuse is achieved when a verification component is used for multiple purposes within a project, e.g., in the block level and the SoC level testbench; another example is reuse in a derivative project when the component is again placed in the same level of the testbench. This kind of reuse across projects is called horizontal, and it is of particular importance for device families. The trend towards sharing of peripherals, standard protocols, and common interfaces between devices to increase the connectivity options allows a great amount of horizontal reuse on the verification side. Clearly, both types of reuse provide significant opportunities for reuse of verification IP, but what is needed to enable it?

There is a lot of focus on selecting a verification language; although all of the likely candidates (SystemVerilog [17], *e* [18], OpenVera [20], SystemC [19]) provide a similar feature set, namely the support of randomization, object oriented programming, coverage information, and so forth. Choosing the best language is, of course, beneficial, but this becomes less important when the possible candidates are similar. There are other, important differences between design IP and VIP than can be exploited:

- Verification IP can be coded on a much higher abstraction level than design IP
- It is possible to define requirements for the testbench topology and structure of verification IP

A high abstraction level is beneficial, since IP modeled in a more abstract manner is easier to reuse. More important is the capability to define a common testbench architecture, together with additional rules necessary to permit the reuse of testbench components. This permits a classification of the required testbench objects with common features implemented in a base class, which is well supported by modern verification languages. This is not possible for design IP, since the device architecture is usually defined by the system requirements, leaving little freedom for similar aspects. However, the focus was often on base class definition and their implementation [14] and not on reusable testbench architecture. Unfortunately, the vertical and horizontal reuse of testbench components seemed not to be of highest importance for this work. It is our experience that an appropriate testbench architecture is one, if not the most important prerequisite for any kind of reuse automation for testbenches.

The definition improvement of the various testbench components and their interaction is equally important, so that every item performs its exact tasks. Surprises arising from wrongly written components can become quite costly when someone wants to reuse it in a different environment. However, when the testbench architecture permits implementing most of the code relevant to interactions with other components, then the reuse quality of a component is "built-in."

15.17 Structural and Interface Requirements for VIP

There are conceptually three layers of testbench component in any test bench:

1. The lowest layer contains the drivers, monitors, and responder blocks that are connected to the block or device under verification.
2. The intermediate layer can host multiple levels and layers of different testbench components, e.g.:
    a) Generators receiving orders from the stimulus to create or randomize more complex commands
    b) Hierarchical drivers that are translating a command defined on a higher level of abstraction towards a series of commands defined on a lower level of abstraction for further processing.
    c) Hierarchical monitors that combine simple responses to a more abstract representation.
    d) Response checkers checking expected values calculated on inputs received from various sources.
    e) Reference models for a block, a protocol stack or other functionality that needs to be verified.

3. The layer on the top contains the stimulus driving the test, which connects either to a component in the intermediate layer or directly to a driver, monitor or responder in the lower layer.

Dependent on the complexity of a test bench, the intermediate layer can be empty, but the upper and lower layer is always present. *Figure 15.12* shows two typical SoC level test benches with the corresponding layers, connections, and the VIP reused. The testbench of the project X on the left side is more complex and uses two generators, three response checkers, and a reference model. In contrast, the testbench of the project Y on the right side is rather simple. Here the driver/monitor pair A with its response checker, as well as the monitors B and E and the driver C are used by both testbenches. However, the driver A is driven by generator F in testbench X, but directly driven by the stimulus in testbench Y. Similarly, driver C is directly driven from the stimulus in testbench X, but receives its commands from the generator G in testbench Y. Furthermore, the monitor B connects to the reference model in testbench X, while it connects directly to the stimulus in testbench Y. In both cases, the interfaces have to be compatible, so that we do not have to change driver C when we want to reuse it.

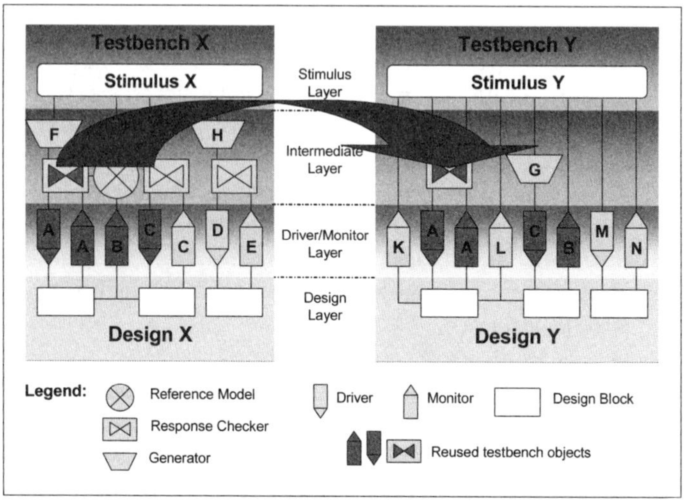

Figure 15.12: *Reuse of Verification IP Across Projects*

Since the topology of a testbench can, and will usually, change during a project or between projects, the capability to insert or remove a component without having to change the connected components is essential. The reuse of testbench components between block level and SoC test benches is a similar problem, resulting in the need for extreme flexibility in the intermediate layer. The wide variety of options for the testbench topology in this layer specifies a first set of requirements:

- Only the testbench components in the lower layer are allowed to interface to the design, and only the components in this layer are using a pin-wiggling interface for the corresponding interfaces.
- All other connections can be done in any combination between the three layers. Consequently, they should have the same interface. A transaction level (TL) interface is used for all connections between the components in the two higher layers and also to connect to the lower layer.
- The TL interfaces used must be compatible between all testbench components, using a standardized, fully defined communication semantic; this permits connecting the stimulus either directly to a driver or to a generator in the intermediate layer, which then generates the commands for the driver.
- The above requirement makes it imperative to separate the communication semantic of the TL interface from the actual commands or data transferred over this interface. It is preferable to enable compile time checks for the actual data transferred to permit only the connection of matching interfaces; a universal serial bus (USB) transaction is simply different from an instruction read on an advanced microcontroller bus architecture (AMBA) high-speed bus (AHB) interface.

Most approaches assume that the usage of a common base class already guarantees the reusability of the components derived from this class. It is our experience that this is not sufficient, several additional rules must be defined to enforce this; the above ones are just the most basic ones. However, the definition of a corresponding set of rules permits the creation of lint-like tools that further automate this step.

Following the structural requirements listed previously, many testbench components become reusable which include:

- Drivers and monitors implementing a common, standard protocol. Such examples include standard protocols as the Joint Test Action Group (JTAG), bus standards like AHB, company internal standards for buses, peripherals and many more. When implementing the full protocol specification, they can be used in any testbench, often multiple times for different blocks.
- Drivers and monitors for a standard protocol, like many communication standards (USB, CAN, universal asynchronous receiver/transmitter [UART], serial peripheral interface [SPI], Flexray, etc). Any block level testbench as well as any SoC testbench of a device employing such an interface is in need of these components.
- Protocol and response checkers, along with reference models, can be reused for SoC, together with the module verification, when the specific block or protocol is used.
- Stacked drivers, or generators can be used any time the implemented command set is used.

Figure 15.13 depicts these testbench components in an example for a typical module level testbench for an USB block.

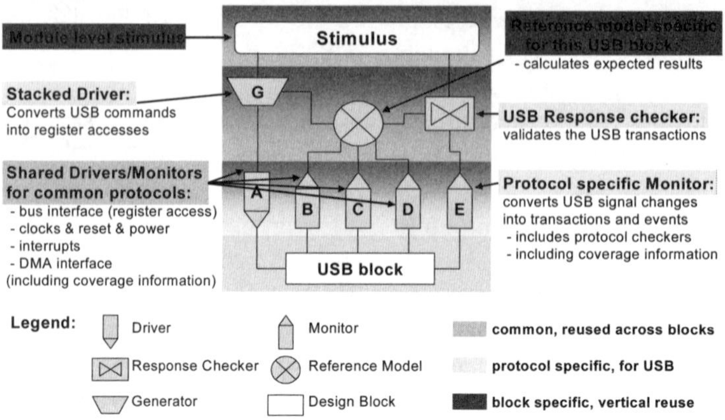

Figure 15.13: *Verification IP Kinds*

Generating good stimulus requires additional rules and guidelines:

- Support the concept of hierarchical stimulus, where each stimulus defines sequences of actions that can be instantiated in other stimuli. This allows the reuse of these sequences, as well as the construction of more complex stimulus.
- Clearly define how stimulus shall be used, as part of a module level testbench or on the SoC level. Sometimes parts of the module level stimulus can be reused to test the integration of the module at the SoC level. Hierarchical stimulus is also beneficial for this purpose.
- Common functionality (like a FINISH command) should be implemented by library functions that provide an application programming interface (API) for their usage. This reduces the creation effort and results in a more common, device independent coding style. SoC or processor specific features can be encapsulated within these functions. As a rule, one should separate the concept and implementation of the test idea from the device configuration.
- Selecting the appropriate language is also important; many tests will use a specific verification language, while directed or processor-driven tests are often written in the C language.
- The capability to generate test documentation from the stimulus code is a further benefit.

Although all of these rules sound simple, common sense, and rather trivial, in actuality, none of the base classes we have evaluated so far are fulfilling all of them. These additional requirements for the ease of reuse will result in rules and guidelines which cannot be ensured by a base class. Rather, the testbench architecture has to be defined and aligned to accomplish for all this. We get a completely new set of rules and guidelines for VIP.

15.18 Verification Flow Automation

To work productively, the verification components not only need to interact in a standardized manner, but they also have to support automation within a regression flow. Hundreds of tests often run daily for a single project. Engineers can not revisit each individual test to identify the status of such a regression environment. Obvious requirements imposed by the need for automation are now state of the art:

- A common message format for relevant information like errors and warnings.
- A standardized definition of a "test pass" criteria, as simple as a PASSED string in the log file.[14]
- Simulation output may need to be compared with reference data in a post processing step.
- Any test shall not influence other tests running within the regression.

The need for automation of a verification flow causes the related verification requirements to be extended into the used environments. This is due to the need for a standardized interaction with a regression system. It is even more important for reuse of an IP block in multiple, different regression systems, when

[14] This requirement sounds extremely simple, which can be misleading. Any testbench component might issue a '… check passed' message at the end of check point, which can indicate a successful indication of this check, but not of the complete test. Determining the correct post-processing of the regression status for complex tests can be rather complicated, and the topics described in this section show only a small portion of the required steps for asserting its correctness.

reusing the block across sites or even companies. Each time automation is necessary, any exception generated by a difference in usage of an IP block becomes costly and should be avoided.

These requirements can be split up into an input and an output side for the regression system. For the input side, the corresponding parameters should be clearly defined and stored in a central place. It is beneficial to manage these parameters by a central component within the test environment. This permits a balanced, common definition of all parameters, which relies upon a clear definition for handling a certain parameter type:

- A common, standardized set of parameters extended with test specific parameters (e.g. the test name, but also special parameters defined by plusargs, defines, or the name of input files).
- Configuration information: e.g., RTL versus gate level, device configurations, and system clock speed
- Tool (simulator) set up like generation of a waveform dump file, timeouts, etc.
- Parameters should not be specific for a certain test bench component unless they are handled separately, so that these parameters can be removed easily, if the component is not present in another environment.

Equally important is an appropriate creation of result data and collateral, in particular:

- All messages must be generated through a common message handler which allows more control by severity levels or other selection criteria to manage the messages
- A clearly defined criteria flagging a successful simulation which is needed for automated retrieval
- Expected results with standardized check steps; files have to be stored in a common, defined place.
- Waivers for expected failures help to avoid duplication of debug efforts when attached to test results; it permits concentrating on problems that are not yet understood.

- Additional information like coverage reports should permit automated pre-processing (e.g., merging coverage reports of multiple tests on the fly) to support the analysis of these reports.

These requirements are essential for post-processing regression results and their development over different regression runs. This kind of automation is needed to enable an effective visualization of the results. *Figure 15.14* shows an exemplary regression system having such capabilities, with a column for each test and the results of a regression run in the columns in a Web page that can be accessed from within the intranet. This is an essential feature when a design team is spread over multiple sites, where some of them have only limited bandwidth available. A quick path for investigating a problem was provided by simply clicking on a FAIL button, which will then display additional information (run result, log files) about the failure with the ERROR messages highlighted.

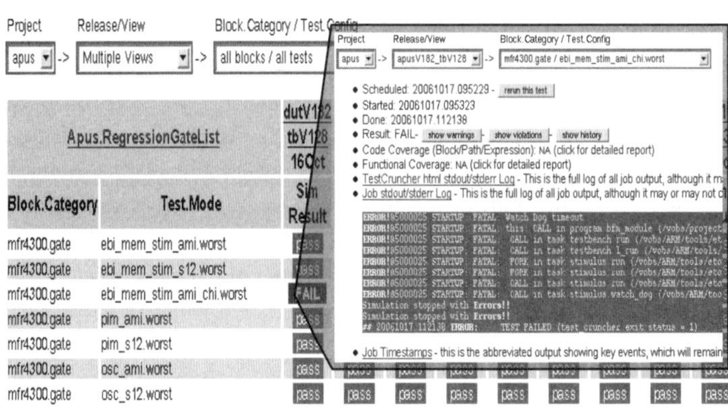

Figure 15.14: Regression Automation

We have described requirements imposed on verification IP by the need to support an automated regression system. It is our experience that this is a part of the automation problem that is often overlooked, although it contributes significantly to the productivity of a design and verification team. To make things worse, many of the corresponding requirements are imposing structural or implementation standards on IP that are hard to fulfill when supporting multiple IP consumers. However, there needs to be a growing awareness about the problems and according standardization on VIP in order to assert its reuse quality.

15.19 Summary

Implementing a reuse infrastructure like the CDE, the associated design flows, training, support, and maintenance work is a huge investment to increase the productivity within a company. Reuse standards like the SRS, design systems, and databases for IP cataloging and searching are basic components of state-of-the-art design environments. Communication of IP availability, incompatible tools or data, incomplete or inconsistent data and/or documentation are needed to overcome the technical challenges involved in sharing IP. Usually each element of such a design environment focuses on solving a specific problem; therefore a tight integration is needed to achieve superior results.

One key issue here is to ensure the fitness of the created IP for reuse. One way to quantify this is to certify the created data against a reuse standard like the SRS. We described enhancements of the corresponding work which would minimize the amount of deliverables and guidelines to be validated. We also explained a significant reduction of the data to be entered and the benefits gained by automated gathering of design specific information. We have investigated the effort savings achieved by this new technology, and detailed its influence on the acceptance of the certification process.

Although these are important achievements, the major improvement will be in the tight and seamless incorporation of all certification work and related activities into the CDE. This is the prerequisite for incremental certification, which enables further automation of related activities, and provides additional capabilities described in this chapter.

It is our experience that certification must happen as early as possible and must become leaner and faster to become fully accepted by designers. Coupling the benefits of incremental certification with a world-class reuse environment will ultimately result in higher quality designs for immediate reuse. Incremental certification is a prerequisite to prevent the creation of IP that can not be reused. We believe that this is the way to instill reuse in a company's design culture. A proper rule classification and prioritization scheme, which permits a project specific customization of the IP validation process, seems to be one of the missing features; also in the current tool landscape. Solving this issue requires a cross-functional collaboration between tool vendors, company internal, and external standardization bodies and the central organization within a company driving the reuse strategy.

We further outlined that the intent to reuse verification IP requires additional methodology and support to ensure effective reusability. Usage of a common base class and additional coding rules is not sufficient. There are structural needs for a good testbench architecture that must be reflected in the components; their general nature makes them an important topic. We found that these requirements are unique for verification IP; the common practice of simply deriving reuse requirements from design IP is a dangerous route for the reuse of verification IP. Lastly, the shape and coding of VIP is strongly linked to the needs of automation in a regression environment. Several classes of rules for easier reusability in this context are of equal importance for the verification engineer and have been touched by the remainder of this chapter.

References

[1] Annual Workshops on Institutionalizing Software Reuse. http://www.umcs.maine.edu/~ftp/wisr/wisr.html

[2] http://www.atrenta.com

[3] Pierre Bricaud and Michael Keating, *Reuse Methodology Manual for Systems-On-A-Chip Designs,* Kluwer Academic Publishers, 1999.

[4] Dan Griffith, *Software Asset Management in the age of Network Licensing.* EC4S 2002, San Jose, October 2002.

[5] http://www.mentor.com

[6] Freescale Semiconductor, Inc. *Semiconductor Reuse Standard, Version 3.2.* Parts of this document can be downloaded from http://www.freescale.com (perform a keyword search for SRSSTANDARDS for the downloadable .zip file)

[7] http://www.openMORE.com (page no longer exists)

[8] Qualis Design Corporation (aquired by Synopsys). *Design Process Re-Engineering.* http://www.qualis.com/method.learning.html (page does no longer exist)

[9] Annette Reutter et al., *Key Issues in Developing a Comprehensive Reuse System for Industrial Usage.* MEDEA Conference on Embedded System Design, Munich, October 2000

[10] Gabriele Saucier, *From Prototyping to Design Platforms.* MEDEA Conference on Embedded System Design, Munich, October 2000.

[11] Peter Schindler et al. *IP Repository, A Web–Based IP Reuse Infrastructure.* CICC 1999, San Diego, May 1999.

[12] Virtual Socket Alliance (http://www.vsi.org): *Quality IP Metric.*

[13] http://www.collett.com

[14] Janick Bergeron, Eduard Cerny, Alan Hunter, and Andrew Nightingale, *Verification Methodology Manual for System Verilog,* Springer, 2006

[15] *IEEE Standard for Verilog® RegisterTransfer Level Synthesis,* IEEE Std 1364.1™-2002, 18th December 2002

[16] http://www.oracle.com

[17] *IEEE Standard for SystemVerilog—Unified Hardware Design, Specification, and Verification Language,* IEEE Std 1800™ - 2005, November 22, 2005

[18] For more information about *e* refer to http://www.cadence.com/verisity/. There is a language reference manual for the *e* language in form of a proposed IEEE standard (P1647)

[19] http://www.open-vera.com

[20] IEEE Standard SystemC® Language Reference Manual, IEEE Std 1666™-2005, March 31, 2006.

Chapter 16

Strategies for IP Evaluation, Selection, and Management

Mobashar Yazdani, ASIC Program Manager and Technologist, Hewlett-Packard

Mike Stahl, Alliance Manager and Design Technology Scientist, Hewlett-Packard

16.1 Introduction

Each new silicon technology node enables increased logic capability for application-specific integrated circuits (ASICs) which leads to larger system-on-chips (SoCs). Solution chip set features are being incorporated into single designs, therefore reducing the number of system components and achieving a reduction in total system cost, power, etc. This has been driving solutions in the handheld and compact device area, which increases the ASIC volume and drives the business. The design complexity has increased, but this does not mean that the time-to-market (TTM) constraints have relaxed. In order to meet this increased design complexity, an intellectual property (IP) validation and incorporation strategy has become an essential part of any SoC design group. Any IP that is inserted into a design must be simulated in order to prove the solution. The customer's expectation is that IP blocks perform flawlessly in the chip without the need for validation or debugging. In order to get working designs right the first time, IP has to be verified and validated all the way through the design implementation phase. IP is verified if the vendors' claims are proved correct. The term "validated," in this case, means that the IP passes all the usage tests so it can be integrated without major issues. Most of the star IP (such as embedded processors) available today is a moving target in regards to its functionality and features set. Star IP is IP which has features that are hard to copy or which are protected by patents. As IP is developed, functional enhancements or standard updates mean that the IP is modified and changed occasionally. The ability of an ASIC provider to model and monitor this goes beyond simply the IP generation capability and has to be carefully looked at if the SoC effort is to succeed.

Audits have become essential to grade and qualify IP for integration in ASICs ranging from custom designs to fully outsourced designs. Un-audited IP increases the risk of achieving successful tapeouts on time. Many facets including IP quality, ease of integration, support and maintenance, and IP generation capability of IP providers have to be looked at in an organized and systematic manner. All this has to be achieved with low cost in mind. This capability development is needed to reduce the cost of designing ASICs, and the industry has been working on standards to allow this to happen more efficiently. However, this capability is still new and not all companies have invested to improve their processes. Ultimately, this results in companies failing to produce successful tapeouts on time. Two of the major areas where progress is essential are standardized IP transfer and management processes, which are later discussed in this chapter.

16.2 The Challenge

Process advances in silicon fabrication technology double the number of gates available for ASIC designs every 18 months. The design process productivity to silicon capability gap has been noticeably widening since 1995. However, design cycle times that end markets dictate have remained the same. This TTM pressure has burdened designers to improve their productivity to the point that cannot be done without drastically newer approaches. To date, the industry has thrown resources at the problem, and this occurs when a design does not perform as expected and specialists are brought in to rectify the problem at a greater than anticipated expense. The added expense of designing in this manner can only be justified with larger market volume. So, in order to minimize risk, companies commonly target multiple products which then makes the SoC even more complicated. The silicon overhead of such chips can easily be at the point where the solution cost is higher than expected. Optimal product cost can only be met by targeting the solution and limiting design to the needed functionality and then integrating it into an SoC. This reduces the silicon area and hence the final part cost. Integration is essential since it also reduces other costs by using fewer suppliers, decreasing supply chain dependencies, and improving the form factor and power consumption.

Improvement in IP selection and validation efforts are driving innovation, and this will be the main driver for product development in the next ten years to come. These methods are prime jewels of companies that employ them since those companies can deliver solutions to market faster than others who do not. However, the lack of IP maturity providers limits third-party IP usage by making it difficult for them to justify merchant IP development. In order to design IP, which will work the first time and integrate seamlessly in designs, it is imperative that a set of standards be developed to obtain and integrate IP economically. It is equally important to be able to access, evaluate, monitor, and partner with companies that can reliably provide robust IP and develop alliances with such companies while the standards take hold. There are many parameters that need to be considered to make sure that the sourced IP comes with the lowest risk. The large number of evaluation parameters, along with the time and effort required to look at them, discourages many companies from doing their own designs, and they thus resort to application-specific standard products (ASSPs). As a result, this decreases in-house IP content and ownership of the differentiation value and hence results in the dilution of their differentiating product features in the market. Target SoC generation then becomes relinquished to part of a cost reduction effort. IP evaluation also takes time, and in some cases, this time may lead designers to conclude that it would have been better to have designed it themselves. In light of this, efforts have been to taken to define the minimum set of parameters to successfully evaluate IP.

Some techniques for judging the quality of an IP and what could be done to reduce the level of effort to use it can be found in this chapter. We will also show why it is important to address this problem by using standards. Working standards ranging from IP evaluation to transfer of an IP will decrease system interface problems, result in ASICs realizing a schedule boost, and it leads to more integration and implementation of innovative customer specific designs.

16.3 The Next Seven Years

As shown in Makimoto's wave in *Figure 16.1*, the industry is currently moving into the cycle of innovation. This translates to the consolidation of functions on ASICs to create new products. IP standards will certainly play a vital role to make this possible. The original ASIC growth came from companies that developed tools to automate the process of silicon implementation, especially through the use of standard cells. In the 1980s, ASIC design was a relatively new field and interfaces developed for netlists and synthesis collapsed to just a few dominant interfaces and data file types supported by major companies. Around these, the computer-aided design (CAD) industry grew up, but nowadays, there are multiple large companies each of which promotes its own tool environment. Future progress in ASIC design will need to come from the standards developed around usage of IP transfer and integration. The users who cannot move data from one system to another without a major effort are the real productivity losers in this current environment. In the last seven years or so, the extra expense of ASIC environments could only be taken on by ASSP providers which flourished by providing general solutions. As we see in Makimoto's wave, the time has now come where standard solutions are running out of steam, and space will be opened for innovative customer specific semiconductor solutions. This means that we are moving towards the time when it will become possible to plug in IPs from different sources and relatively easily implement complete chips. However, the industry presently does not have standards suitable for flexible ASIC IP blocks, and now they require standards that have configurable interfaces that allow the ASIC module behavior to move across interfaces. Newer extensible markup language (XML)–based standards are moving in this direction.

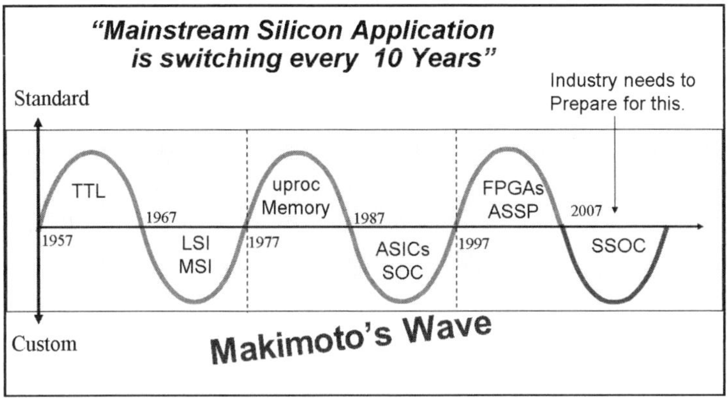

Figure 16.1: *Makimoto's Wave*

16.4 Design or Buy

One of the first challenges addressed in any ASIC design comes at the system design partitioning phase. The design is split into the blocks that will be internally designed and purchased, or designed as IP. Many decision points need to be addressed in order to find the optimum split. If the company will reuse design blocks more than two times in its lifetime, it may be better to develop the blocks internally. It is easier to integrate an internally designed block into the total ASIC because the methodology for design is common to the rest of the ASIC. In addition, if a bug is found, there is internal knowledge to analyze and correct the block. On the other hand, if the block protocol depends on standards, it may be better to negotiate an IP purchase from a supplier who has the resources to track the standard and who can provide better upgrades and maintenance.

16.5 IP Life Cycle

It is very important to consider the full life cycle of the IP, that is to say, whether IP is developed internally or purchased from external sources. This should be done just as a physical product would be, e.g., investigated, sourced, developed, value extracted, and phased out. This approach will give the total cost of ownership of the IP. It is important to decide which design blocks are local to that particular design group and used by the current project team members. Those blocks may be used by others or by future generations of the design. For design blocks local to the particular design, it is acceptable to work with less rigorous documentation since the engineering talent is easily available from within the group. When evaluating other companies' potentials to provide IP, one of the most important things to cover is their full roadmap of IP solutions since IP typically has different versions, all of which need to be tracked.

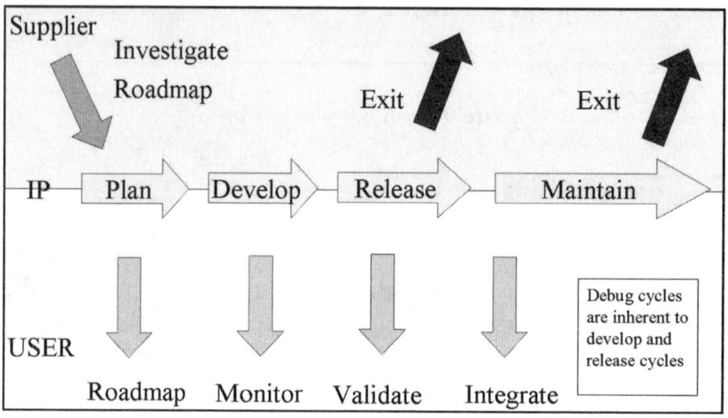

Figure 16.2: *IP Life Cycle*

As expected, IP release is a major point in the IP life cycle where IP development may be terminated without providing further support. If it is important for the user to get support for future implementations, then the user needs to make sure that the IP goes into a maintenance mode at the provider. Ideally, this will have been discussed at the very beginning of the engagement. However, in many instances, the supplier may not be able to continuously extended support especially in a case where the product is both risky and future usage is questionable.

For companies involved in IP alliances, minimum requirements should be the names of the people involved in the design and end-of-life (EoL) documentation filed for future use should the need arise.

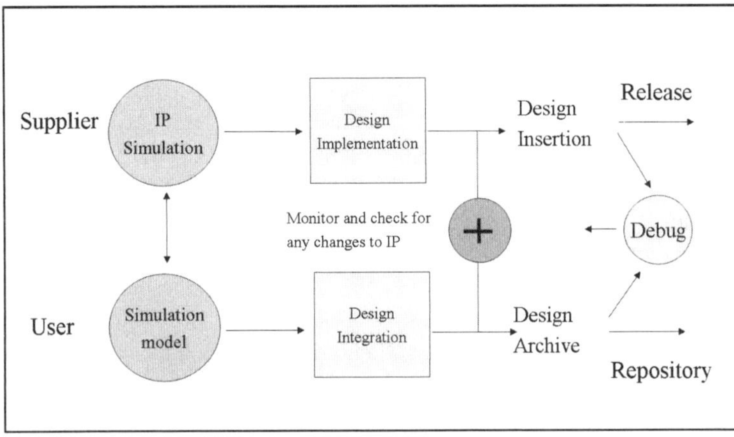

Figure 16.3: *IP Flow to Silicon*

Once an IP has been chosen, the design group requires its simulation model to start the IP interface verification with the rest of the system. At this point, the IP supplier may be engaged in the design implementation. One important requirement is to have the IP supplier to verify that the final version of the model is consistent with the simulation model provided to the integrator. If any changes were made during the development, they need to be brought back to the ASIC integration group to have the design re-verified. Currently, one of the best monitoring mechanisms between the user and the IP provider is weekly meetings. Links that convey the progress between the supplier and the user have generally been developed but they can only be used between companies with significant IP usage and thus justify the implementation expense. We are pursuing standards which allow IP monitoring in development and facilitate instant notification of functional changes implemented in cases where the user needs to be notified.

It is possible that important IP may not be fully developed at the start of the ASIC design. The user has to predict which supplier will be able to deliver the IP as promised. One challenging aspect of the IP evaluation is to develop risk models that correctly indicate which company will be capable of delivering the IP on time without any issues. IP suppliers will always put their best foot forward to get an ASIC contract. The hope is that once they have the contract, any deficiencies in the IP will be rectified later with additional resources. In order to assess the risk in using an IP with a high degree of confidence, the user should rely on ASIC IP auditors. The IP provider, as well as the IP itself, are equally evaluated. Can the IP be provided as required and will it be integrate-able into the user's design environment? Furthermore, the job of the ASIC design auditors is to assure that critical components such as technology, IP, and verification capability are not going to become obstacles. More and more time is now being spent on efforts to validate the IP claims of suppliers.

16.6 IP Categorization

Before even developing, validating, or auditing, one has to categorize the IP from the user's perspective. IP vendors, suppliers, integrated device manufacturers (IDMs), and foundries all have different IP categorization strategies. Users should adopt or select the one which suits their development environment. This means that having a document that clearly lists how the IP can be categorized. IP is classified on its design and encapsulation (packaging) and in which business category it is sold. From the business perspective, the first categorization is by IP type which is based on the market valuation. Each of these can then be distributed in either: type or form (e.g., soft, firm, or hard) and will be described further in the next section.

Finally, the IP is structured and put in a catalog listing, based on different criteria, called IP taxonomy. Users prefer design block functional listing while suppliers prefer a market-based product-like IP listing.

IP Types

Type classification is important to define the value of an IP. IP design takes time and can involve the incorporation of innovative algorithms to implement the design function. There are many other factors which are important in terms of the value of the IP. This leads to a definition on the basis of IP types and their values. Suppliers may have additional types not defined here, but they would generally fit in one of these categories.

Generally IP types are defined as: star, foundation, input/output (I/O), general, or standard IP. Star IP, as the name suggests, is IP which gives a clear advantage over the others by virtue of its significant IP value. This would by either performance or functionality which is commonly protected by patents or IP which has become very popular and is available from only a few providers. Foundation IP (or, cell library) is concerned with the IP provided to the chip designer that uses a foundry's process technology, and it is mostly evaluated by looking at the design methodology. This IP consists of gates, basic memory cells, and I/O cells. Specific technology-dependent I/O library blocks, such as pad drivers, are considered a different category and have an extensive set of constraints that include looking at test outputs. General IP is functional blocks which can be easily designed, but it needs a supporting environment such as test vectors. The user saves time by using general IP. Standard IP is IP which is standardized in consortiums and provided by IP suppliers.

IP Forms

Each IP type can have multiple forms. Since an IP structure could be soft, firm, hard, or module, each type of IP has its own criteria of evaluation. IP is traditionally defined based on its construction type, such as hard (graphic data system [GDS] II), soft (netlist, program), firm (soft with constraints), or module (generator). Soft IP is popular since it facilitates implementation at the point of synthesis and can be silicon process independent. However, soft IP can result in as much as 20 percent performance degradation over hardened IP. Soft IP can also be routed as part of a larger standard cell block to allow ports to be moved around the design. Nowadays, many designers for standard performance SoC ASICs prefer soft IP if it is available at the desired performance point.

IP Taxonomy

The IP taxonomy shows IP in a portfolio, and it can be driven by IP providers or users. IP providers create the taxonomy that allows them to market their IP to maximize the value of their portfolio. IP users define the taxonomy that allows them to design their SoCs efficiently. Further, IP providers can design target IP from scratch or extract IP from their existing projects for sale. Each case comes with different kinds of risks and so each has a different auditing strategy. The first step for an IP user is to define how a company structures their own IP. One method is to look at IP from the designer's point of view which allows for better design synthesis and design tradeoffs. This has dependencies on their previous projects and their future needs.

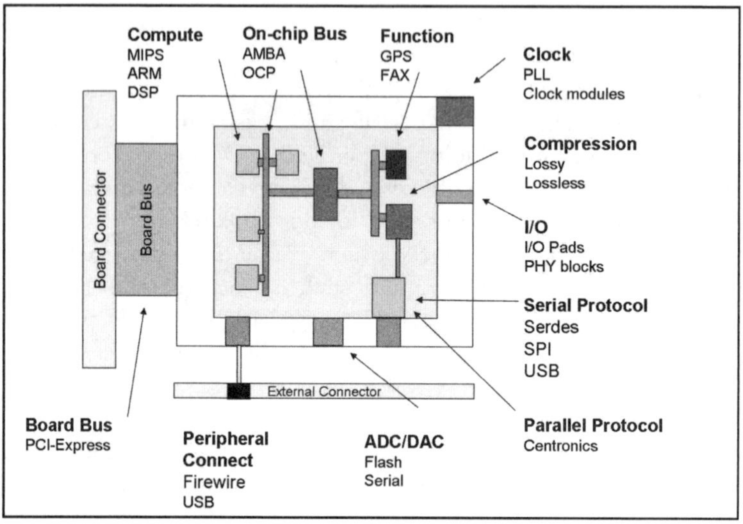

Figure 16.4: *IP Taxonomy and Category Sectioning*

IP providers may sell IP which have other IP blocks embedded in them. These IP blocks frequently have to be extracted to allow for design instantiation. This may end up in extra work and may negate any benefit of IP sourcing for the users in the first place. In general, a user wants discrete IP blocks, provided with all the necessary wrappings, ready for integration, and not extracted IP. The design–based taxonomy does not consider IP as valid IP blocks if it is embedded in a design but rather regards them as design blocks. If a company can provide its own taxonomy to its partners allowing them to see how IP is structured, it simplifies the IP generation process for the supplier. Designers insist that suppliers need to be encouraged to package IP in this manner to be efficiently used in chip design. Roughly ten categories ranging from embedded processors to compression blocks can represent the full IP portfolio typically used across a company with multiple SoCs.

Each category can have from a few to dozens of IP to be evaluated. A portfolio which is well-structured and robust in quality allows designers to readily choose a solution which will meet their functional, performance and cost goals.

16.7 Point IP or Portfolio Assessment

The first evaluation aspect is to decide if a customer company will look at specific IP or companies with an IP portfolio. Individually developed IPs may differ greatly in their quality even if the IP is developed by the same company. Robust audit and analysis methods are needed to validate such IP. Strong contracts need to be written to make sure that the IP can be modified if the need arises (which it almost always does). In the case of IP portfolio companies, a user needs to look at the overall IP generation capability and commitment to specific IP in the providers' roadmap. The best combination seems to be when a design company qualifies an IP provider for the bulk of their IP offering and generates methods to check specific IP.

Point IP Evaluation

Understanding the Effort

Point IP evaluation mostly occurs on a project where the design group needs to procure a particular IP due to lack of internal resources or specific skills. IP evaluation and integration may not be a regular activity of such a group and research and development (R&D) engineers may be involved in validating the IP provider's claims. This activity needs to be done in a streamlined fashion in order to avoid any unexpected outcomes. Many R&D groups underestimate the work involved in doing this evaluation. The basic guidelines are to get the following information about the IP:

- Datasheet
- Lifecycle roadmap
- Target foundry
- Manufacturability
- Cost, design efficiency
- Design practices
- Tools environment

- Test strategy
- Integrating guidelines
- Design repository
- Maintenance and support

Quality IP (QIP) standard developed by the Virtual Socket Interface Alliance (VSIA), with ongoing support and enhancement available from the Institute of Electrical and Electronics Engineers (IEEE) provides an excellent way to get started in this. This spreadsheet consists of questions which when answered give a quality number which can be used to grade IP.

Evaluating the IP Portfolio

Today, IP is available from many different sources such as IP vendors, IDMs, design houses, or foundries. Getting to a metric that represents a particular company becomes difficult. A metric has to gauge the benefits available at all the IP providers in order to come up with a metric that could be applied across multiple companies. For example, a design house will have plenty of knowledge about an IP, but it may not have hardened it in silicon. On the other hand, a foundry will have it hardened and maybe also provide fault analysis, a key knowledge element to determine if the IP works. A grade in the form of a number composed of six parts can typically represent IP from current sources. A rating of 6 would indicate that the IP provider has ownership of the full life cycle and has built it in silicon. One would need a 3 or 4 rating to feel comfortable using the IP knowing that additional effort to makeup for missing abilities will be required. This particular grading system allows the users to get IP from the source which best matches their capability reducing overlap and being a better cost fit.

Each IP category is given points which are normalized to the same number. The portfolio evaluation gives a very good idea of a particular company's IP match to an end user company needs. A great amount of time is saved because engineers can zoom to the IP vendor which has the closest match. The engineer can then engage in a second level evaluation to check the details.

In *Figure 16.5*, a portfolio number is given to each IP. The number will be low in cases where the user does not care for the breadth of the company's IP. However, specific IPs may still be a match and thus it is important to look at individual IP as well. Such a spreadsheet covering the IP needed and the potential supplier's grade gives a good indication of which provider can provide the IP. The specific details used for the grading is a detailed discussion and is commonly generated by consultation of engineers using the IP. The spreadsheet use used by the author has over 100 IP blocks and much detailed reasoning for giving points to IP providers. Each supplier has a separate sheet, and a filter is used to target the IP needed and compare the solution from multiple companies.

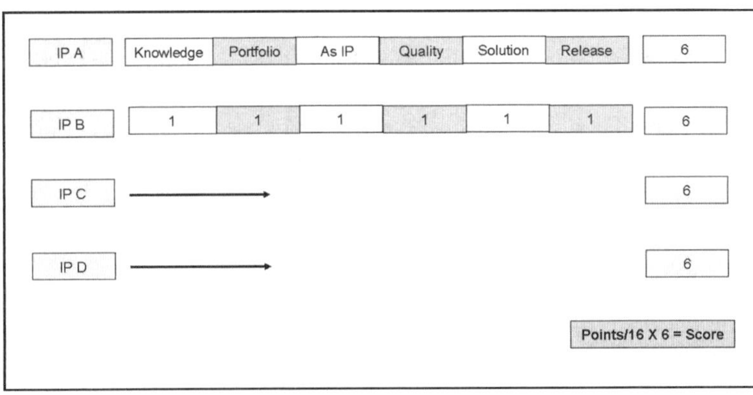

Figure 16.5: *IP Portfolio Grading Measures*

IP Portfolio Attributes

Category attributes such as knowledge, portfolio, as IP (meaning the IP is packaged for IP distribution), and solution can be very specific to the user and thus are difficult to measure. Hence, the user needs to clearly list the expectations and grade the IP provider against it, and others like quality and release are more quantifiable as discussed in the following section.

IP Quality

IP quality is a very misunderstood attribute, and it has been generally used to indicate if an IP functions as desired. However, IP quality pertains more to the quality of construction of IP and its generic interface-ability to other IP. One may even regard IP as being sub-standard on the fact that it is not capable of interfacing with other IP. This could be incorrect for the IP being considered. Let us consider an IP which uses level sensitive scan design (LSSD) test structures and is very robust. Its test strategy will not allow it to work with edge-triggered test structures. From an interface requirement perspective relative to a system, the IP would not be a match even though its quality is high. Hence, the quality measure defining the quality of multiple IP being used together will be different than that of individual IP.

Companies designing IP need to define such a quality measure, along with a measure which can be used to gauge the company's progress on its IP development. A checklist created for the purpose of grading IP quality numbers is one big step in getting to that point. A generic quality checklist will have many items which will not be representative of any randomly chosen IP. Based on the evaluated IP, a user will select the attributes which are rightly applicable.

The quality measure can cover the high level design portion or the mapped netlist and silicon. Silicon quality measures relate to the margin implemented in the design and design rule check/electrical rule check (DRC/ERC) requirements. Typically, the performance or compaction needed is what drives deviation from best practices and results in a sensitive design. This again calls for design audits to be conducted to check the design implementation.

IP Release

IP release can be defined to be at different points of the development or life cycle. A user should make a careful assessment of when he can take the IP and use it, and also at what points he or she starts to engage for validation of the solution. Best case scenario is where the milestones in user design methodologies match exactly those of the IP provider. This is generally not the case which forces the user to understand the provider's milestones and design flow and relates them to his or her own milestones and design flow. This, along with assessing the supplier's ability to meet the milestones on time, will give an indication of the least risky supplier alternative.

16.8 IP Vendor/Supplier Evaluation Areas

It is equally important to qualify both your supplier and the IP. In any engagement, there will most likely be challenges along the way, and you should make sure you have selected someone you can work with. It is best to audit and qualify your partner up front. In other words, it is better to have this done before time pressures are upon you. One important reason to look at the IP lifecycle is to get all the parties who will be involved in the IP development involved in the selection process. This way, the risk can be calculated over the life of the IP. The following areas called attributes are initially considered in this process:

- Alliances
- Quality
- Design infrastructure
- Maintenance and support
- Program management

These attributes are scored by giving the goodness number for the company. These scores can then be assessed to get the main factors which indicate whether a company should be pursued for an alliance or as a partner. Each of these attributes has many parameters which have to be checked in order to satisfy the attribute.

A major portion of IP evaluation is related to reducing the risk of getting IP from external sources. The largest risk is in not being misled into thinking that the IP does what it is not capable of doing. Then, there is the risk of needing to massage the IP to get the right testability, interface structure, or even functionality. Hence, well-defined and robust IP are required here, and such factors as QIP and IP alliances are good indicators of the capability of a supplier company to provide this level of IP. The factors which show a company is capable of delivering the solution depend on their IP infrastructure and support. The link joining the consumer and provider is the full view of IP development program management.

Alliances

As product design companies move towards higher levels of integration, IP available at a single company becomes unable to satisfy all the needs of the ASIC designer. Our experience when we, as IP users, talk with suppliers for an SoC IP which they do not presently have, the general answer we get is that they can get it. However, in practice, it is not possible to quickly obtain and integrate IP from companies that have not been looked at or used before. There are always issues because of design tools and methodologies, and some IP blocks actually need considerable work before the IP can be incorporated into the ASIC design. The IP alliance factor checks to see if the fabless provider or a larger IP provider has an established interface with IP providers. Checklist items list IP validity, progress, interface development, etc. are reviewed to get a sense of the maturity of the supplier and IP provider relationship. One of the strongest indicators of a close working relationship is if money has been committed or spent on the engagement. Some of the fab alliances would seem strong from this point of view because they address the usage of IP across their alliances. However, a user should investigate the ability to share IP because most of these efforts are still in their early stages. We find that many will claim to have over ten IP partners in various stages of qualification.

Alliance Development Factors

A company may have a few point IPs, but the real test is when a company can engage at an alliance level. This requires commitment to publish its IP process and improve it. To look for companies that one could potentially have an alliance with, many more capabilities need to be looked at. Most of these capabilities can be summed into four major factors (as shown in *Figure 16.6*). Even when our regular evaluation is done for IP, it is structured in a way which makes it easy to classify the company into one which is a good candidate for alliances. These major factors or categories are:

- Practice
- Infrastructure
- Development
- Service

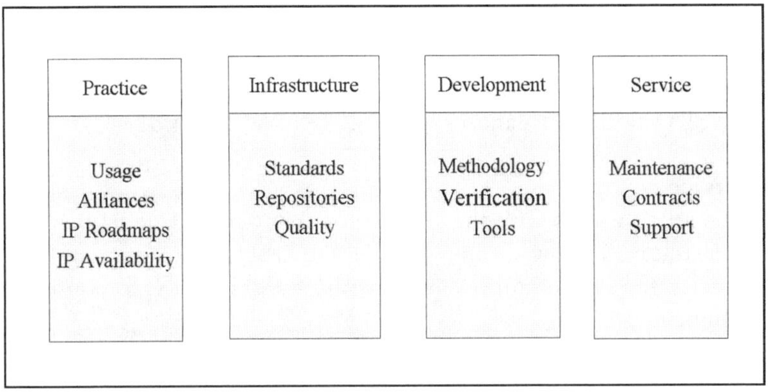

Figure 16.6: *Factors for IP Alliance Capability*

Practice

This category covers a company's strategy from IP generation to implementation. It describes how a company decides what IP to work on, its success rate, its business models, its dedicated IP resources, the IP portfolio examination, and whether the roadmap is verified. IP alliances are thoroughly examined, and the company is given a score for the maturity or strength of the alliances.

Infrastructure

IP infrastructure is very important since this is the core of any IP development or fabless company. The company should have much more than just a design database containing IP. It should be able to go from the design database to storage or archive, and checklists should be used to make sure that the IP passes minimum quality standards. The company should likewise be able to create a package containing the desired IP and its version. In addition, when IP is downloaded the system, it should keep track of who has downloaded it and be able to notify the users if there is a bug fix, update, or any other significant change associated with the IP.

Development

IP development is perhaps the most important capability of an IP provider. One needs to check if the company has enough tools and resources to create the desired IP. Verification scripts and consistent system environment are essential for the leverage of design information inside the company. The company should have a methodology focusing on major aspects of design such as performance, low power, high integration etc. Proper reviews, although time-consuming, should accompany design milestones and all critical information should be well documented.

Service

This section deals with the company's ability to service IP that has been released. How do they archive the IP blocks? What are their policies for end of life? What are the rates to provide continual support?

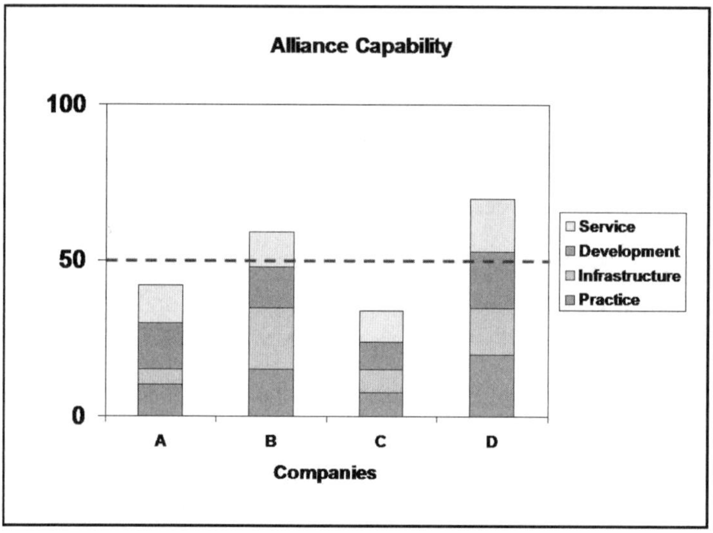

Figure 16.7: *Selecting IP Alliance Partners*

The final rank provides a number that grades the effectiveness of a particular company or the IP it produces. The way the point system should be selected should give a clear indication of where the line should be drawn, i.e. when could a company be approved as an alliance partner. If this is when only 50 percent of the requirements are met then the approval line should run horizontally at the 50 point mark.

Efficient and effective IP evaluation is a long process involving many criteria and extensive auditing by experts. We have tried to encapsulate the activities as much as possible. Active development is needed in this area to make the process fast, reliable, efficient and less costly. This can be further improved by creating standards around the IP assessment process that include IP quality measures.

Quality
This is a detailed evaluation of the vendor quality strategy process. It looks at the criteria used by the supplier to establish its quality strategy. Quantifiable measures are required that indicate the goodness factors for design, integration, implementation and support.

Design Infrastructure
Typical design means setting up an environment based on the tools used. However, if the end goal is to create IP for internal or external use, one has to invest in tools which allow IP generation, distribution, analysis, etc. The ASIC tool industry has been extremely slow to provide such tools. This is again mostly due to a lack of established standards. Each major tool vendor is busy improving his design environment to keep its users tied to its system and to make it difficult for the users to move to a different environment. There are standards like graphic design system (GDS) II and SPF, but they emerged to allow the design process to work efficiently and not to enable the exchange of data between IP users. We look at the IP providers design infrastructure to see how they exchange data internally. Efficient design mechanisms are a big plus, and they are an indication of successful methodologies.

Maintenance and Support
The main difference between a design reuse block and IP is that the latter is created and supported in a way that can be used externally. As the varied standards based on IP are developed, they are incorporated into the IP and test vectors are created for design verification. However, since maintenance takes resources, IP providers will always want to work on the IPs that are being actively used. Therefore, it is important to get the criteria the supplier uses for maintenance and an attribute defined for the level of maintenance. One easy way to check if the IP is supported is to check the details on the IP's Website, where one should look for bug reports, new versions notes, and any tool dependencies.

Program Management

Since leading IP is typically not fully designed and implementable at project initiation, one needs to monitor the progress of the IP's design. The usual method to mange this problem is to have weekly meetings to assess the state and progress of the IP. An improvement is to use e-room, share point, or similar tools to monitor the progress. A definition of the attributes is needed which will be exchanged and shown on a dashboard. These could be attributes which define the various stages of IP implementation and the progress within the stages. Before doing this, one has to be sure to define the deliverables that will be completed in each of the stages. These deliverables then have progress attributes. Generally, at a high level, the progress attributes have statuses of not started, in progress, or done. The deliverable attributes that are called milestones are defined in the kickoff meeting. One should note that this progress interface should be driven from the design environment as well as the IP repository at the IP provider. The IP transfer mechanism defines attributes that provide the status of IP. This information can be exchanged by transferring the IP package with the status information but without the detailed data. In short, the program manages everything, and there is no such thing as an "off the shelf" purchase in regards to ASIC IP.

16.9 IP Acquisition

Once a user decides to use IP from a company he needs to acquire it, but this endeavor can be quite costly. Methods are now being developed to ease this process, but they are in their infancy.

In general, a user can buy point IP in specific areas or enter into an alliance to obtain a broader set of IP from IP provider companies. An IP supplier company may dedicate itself to work in point solution areas like memory interfaces, wireless applications, storage, etc. In this case, the company usually wants to be at the leading edge in developing such applications. However, since each IP would be obtained in an independent manner, it is important to develop system interfaces to facilitate monitoring of IP status and IP transfer. Point IP purchases involve lots of overhead to validate and obtain the IP and one reason to work with IP portfolio companies is to decrease this effort. Currently, the industry is trying to focus on alliances to keep the level of overhead in IP acquisition low. As standards (tools, SPIRIT, VSIA etc.) are developed, the overhead of IP acquisition will decrease. This will help the alliances by lowering their cost on one hand and allowing new players to enter the business on the other.

Establishing an alliance with a company means that strategic intent is exchanged for at least the next three years. In the case of IP, this means sharing roadmap information to allow partners to know what is needed and what will be available. Today, TTM is extremely important. If the provider misses milestones on its roadmap, the customer can go to another alliance partner. Typically, this is what the IP provider fears most as their investment in the IP may be lost. However the alliance mechanism allows other IP to be purchased from within the alliance guaranteeing revenue for the alliance partner. (see *Figure 16.8.*)

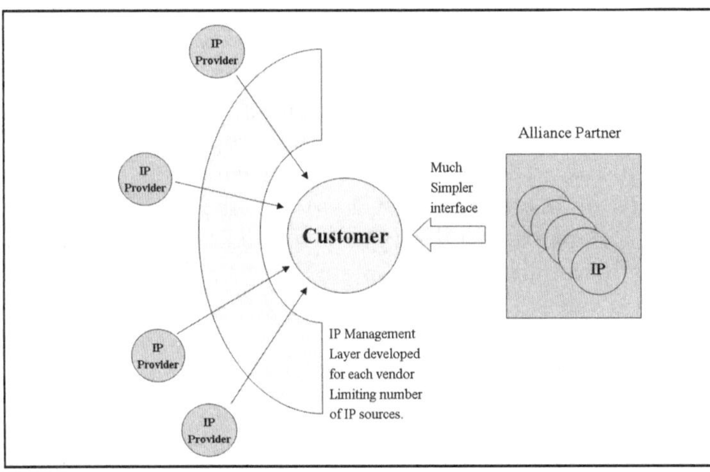

Figure 16.8: *IP Provider Links for Individual IP and IP Alliance Partner*

Currently, multiple interfaces need to be developed between all the alliance partners. There is a strong desire to have standard interfaces to allow IP transfer without unnecessary overhead. Standards will allow customers to interface individually with IP providers and reduce their expenses of interfacing with multiple vendors. It will also allow larger IP providers to work on the actual

IP instead of investing valuable resources on multiple interfaces. Larger companies will be able to roadmap and deliver more IP producing greater value in this scenario as shown in *Figure 16.9.*

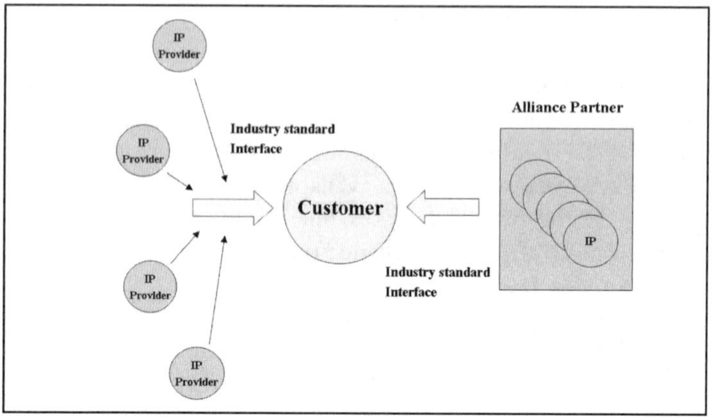

Figure 16.9: *Standardization Reduces Overhead of IP Providers*

These interfaces also extend into program management and monitoring of IP development. Alliances are important nowadays because it takes a significant amount of time and resources to validate IP to determine if it will work in the user's system. We have been trying to ease this process to obtain IP from individual providers by working to develop a mechanism for transferring IP from IP providers.

16.10 Transferring IP

Today, most IP is transferred through .tar or .zip files, and these are excellent file packing programs. However, they do not have the content specifications and dependencies as part of the package. Considerable time is spent in retrieving the IP and loading it on a system in order to evaluate or design with it. An easy way to transfer the IP with the knowledge of its files and design details is required. There are many design environments which allow such transfers, but they are within their own domains. Since the design transfer is within the domain many of the system environmental conditions and system parameters are assumed known or easily accessible, they are not included in the transfer stream. In order to transfer IP, this information has to be encapsulated and inserted in the package with known attributes. There have been efforts to capture the system and design information in attributes which could be standardized in an XML environment. These are typically specific to the systems which generate them. A successful transfer mechanism will have just the right number of attributes to allow transfer of IP from one system to another and still meet the goals of design implementation. *Figure 16.10* shows an example on an IP encapsulated package.

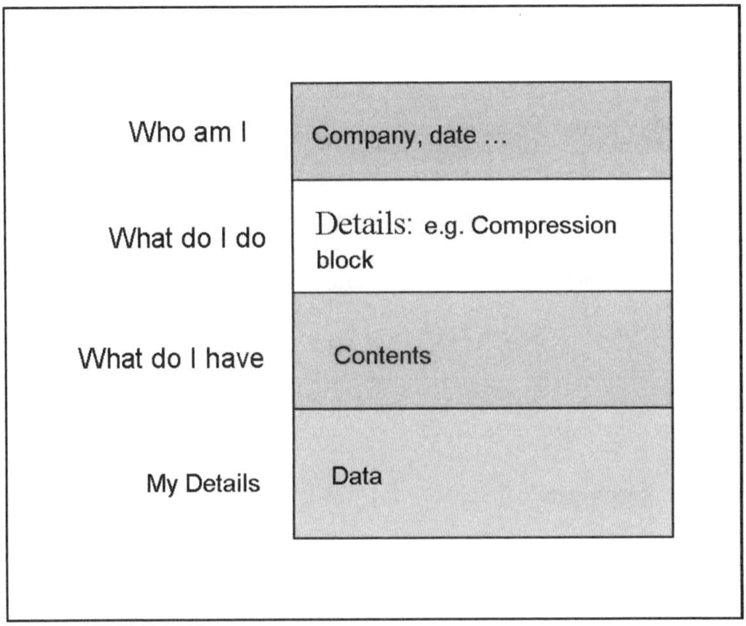

Figure 16.10: *An IP Encapsulated Package*

Evaluation of IP using a standard IP package will allow faster evaluation and give more design options to designers. The package inherently reduces the risk of receiving incorrect or missing information. In addition, simple scripts can be written to allow package creation from design repositories or design systems.

Ideally when an IP is encapsulated for transfer, the design environment is saved with it. It is challenging to develop a general structure which will allow IP to be transferred between multiple systems in a manner that each system environment will be able to recognize and use it. This can be done to an extent by using an XML structure with a directory structure defining the IP attributes as well as the design system attributes.

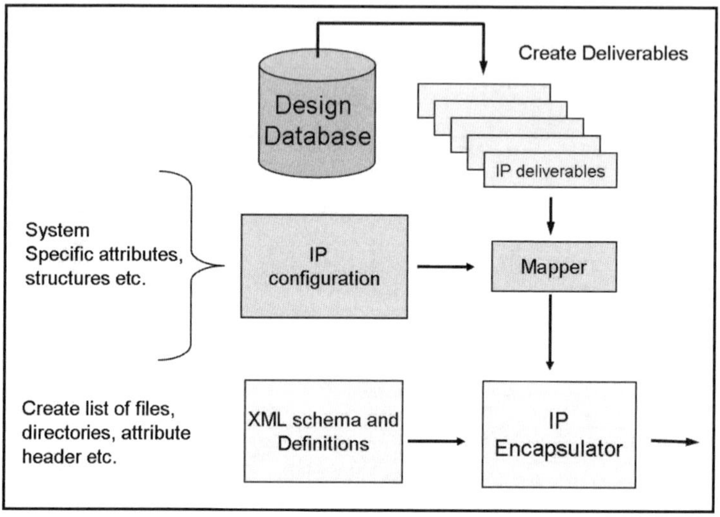

Figure 16.11: *IP Encapsulation for Transfer*

16.11 Summary

Successful design today and in the future depends on the ability of a design company to source IP externally. This involves IP selection, transfer, and integration. An ASIC project's success depends on deciding where to get the IP and how to go about using it. In this chapter, we showed the difficulties associated with evaluating and obtaining IP from different sources. One method was presented which results in the formation of alliances with IP partners. Improvements in IP transfer will greatly improve the IP that can be obtained to allow end users to widen their IP sources. In order to come up with potential alliance partners, IP attributes can be categorized in a manner to indicate the overall health of a company. These attributes assess the risk associated with a particular company and the issues that need to be resolved before engaging with them.

References

[1] K. Werner, T. Anderson, "Specs Eye Functional Verification Quality" EEtimes, July 12, 2004 issue.

[2] M. Yazdani, T. Halbac, "Evaluation and Procurement of IP for SoCs" FSA IP Forum 2003.

[3] M. Yazdani, "Component Repository and Supplier Connect," DesignCon 2001.

[4] A. Walker., "IP Quality: In the Eye of the Beholder?" IP 2000, Santa Clara, CA Mar. 2000.

[5] T. Zimmermann, "Providing Reusable IP through a Repository and Librarian," Portable Design, Mar. 2000.

[6] M. Birnbaum, H. Sachs, "How VSIA Answers the SoC Dilemma," Computer, June 1999.

[7] www.vsi.org

Chapter 17

Considerations for ASIC Physical Design

Ramesh Banda, Distinguished Member of Technical Staff,
LSI Logic Corporation

17.1 Introduction

Agere Systems, since acquired by LSI Logic, has developed well-balanced and aggressive register transfer language (RTL) hand-off flows, audits, and considerations originating mainly from various customers, electronic design automation (EDA) tool vendors' evaluations, interactions, and discussions. This chapter will enlighten developers, managers, and marketing teams in order to eliminate risks and re-spins. It also describes various kinds of key flows, audits, checks, and considerations that have to be systematically reviewed during front-end development to generate a robust functional net-list that have been explained in the form of design and test flow diagram, along with audits in the form of tools, deliverables, RTL, design for testability (DFT), and synthesis.

Today's design complexity, process technology, time-to-market (TTM), and many other factors are of major concern and worrisome in a multi-dimensional way at every stage of the application-specific integrated circuit (ASIC) process which includes design architecture, implementation, verification, mask order, processing and production. Agere Systems RTL hand-off flow, audits, and considerations minimize both the design effort and design cycle during back-end place and route (P&R) activities and timing closure. Agere Systems provides full flows to enable predictable ASIC design environment to help meet the challenge of today's multiple systems on a chip and ever-increasing ASIC complexity using the advanced nanometer technologies. The standard ASIC flow and methodologies provide customers with predictability in schedules, design productivity, and cost for today's multi million gate designs.

This chapter addresses key components of the flows, audits, and checks that have to be kept track of during front-end design cycle for first time design success. The list here details all of the sections that are following by front-end flow audits checklist and conclusions.

- Front-end design and test flow diagram—This section in general details architecture, design, and test flow.
- Tools and associated versions audits—This section describes the usage of various tools and associated versions in design flow, such as design analysis, logic optimization and design verification.
- Deliverables audits—This section covers the audits related to database, source code, verification, software scripts and proprietary tools, synthesis and documentation required for the project development or to provide to an ASIC vendor.
- RTL audits—This section describes the audits for communicating design issues, specifications, bus definitions, latency/backward compatibility requirements, testability, IP's/memory integration, area, speed, hardware description language (HDL) description, Lint, RTL partition, design nature and design reviews.
- DFT/verification audits—This section describes various audits for functional tests and formal verification of the design.
- Synthesis, timing and constraints audits—This section highlights various audits related to designs associated with synthesis, timing and constraints.

17.2 Front-End Design and Test Flow Diagram

Overview

The front-end design and test flow diagram are depicted in *Figure 17.1*. The diagram shows partition between front-end (Check Point4) and back-end (Check Point3, Check Point2, Check Point1, and Check Point0) flows. Each of these check points are designed to meet goals at every stage of the milestone.

The diagram has two components for Check Point4 flow where one section describes the design aspects, and the other section describes testing. The arrows in these two components point to each other and go back and forth depending on results and the stage of the components during the design cycle.

There are five components in design aspect namely architectural specification, micro-architectural specification, RTL development, verification development, synthesis, and gate-level net-list development.

Verification development section, as depicted in *Figure 17.1*, has four blocks to test the design, namely high-level model, cycle–based cycle accurate model, verification test, and formal verification used to verify RTL versus gates.

The high-level model will be developed and used to analyze architecture, bus interfaces, software code development, initial floor-planning, feasibility of performance, area, power, clocking, and test management at the architectural specification stage. Cycle–based accurate models will be developed and used at micro-architecture specifications stage, verification tests will be developed and used at RTL, gate-level net-list development stage, and finally formal verification is part of testing RTL versus net-list.

RTL audits corresponds to architectural and micro-architectural specifications and RTL development, verification development section corresponds to DFT/verification audits and synthesis, timing and constraints audits belong to RTL synthesis and gate-level net-list development.

Flow Diagram

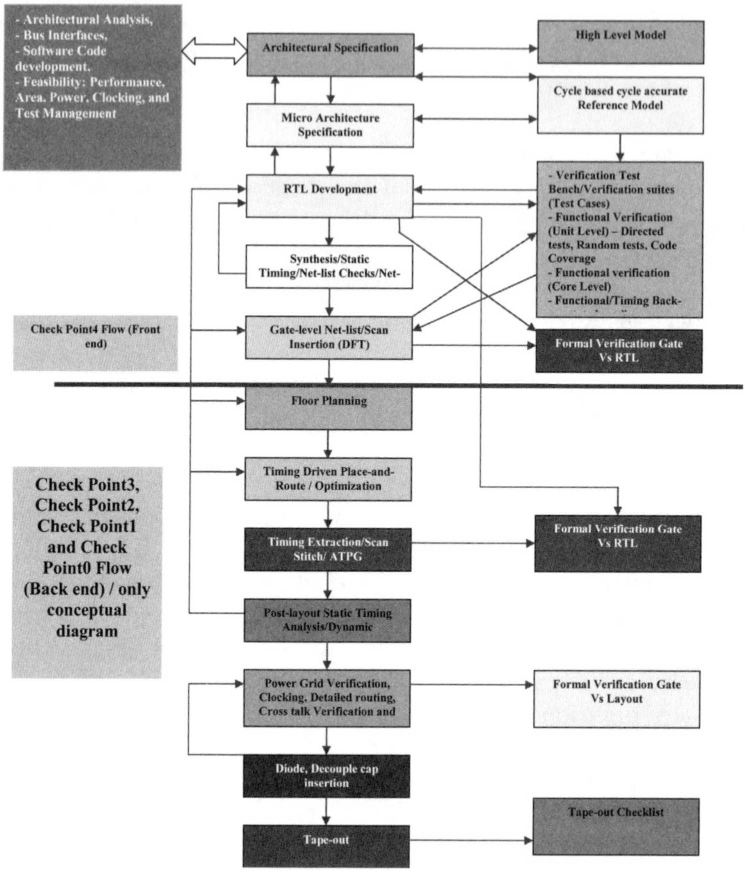

Figure 17.1: *Front-End Design and Test Flow*

17.3 Tools and Associated Versions Audits

Overview

The design flow and associated tools used at every stage of the design flow are depicted in *Table 17.1*. Most of these tools belong to a third party except in very few cases. Agere uses internal tools for development or debugging purposes. The front-end design-flow consists of the following:

- Design analysis—Verilog/VHDL simulators, power analysis, RTL, RTL DFT, synthesis constraints and net-list audits, design structure and design analysis), logic optimization part includes (synthesis, test synthesis, floor planning) and design verification covers (test/code coverage tools, formal verification, static timing analysis and gate-level simulations).

Design Flow and CAD Tools

Design Flow	Task	CAD Tools
Design Analysis	Verilog/VHDL Simulation	NC Verilog, NC VHDL, NC-Sim (Cadence) ModelSim (Mentor Graphics) VCS (Synopsys)
	Power Analysis	Design Power (Synopsys) Watt Watcher-OLA (Sequence)
	RTL, RTL DFT, Synthesis constraints and Net-list Audits	Spyglass (Atrenta) Magnify(Agere)
	Design structure and design analysis	Debussy (Novas)
Logic Optimization	Synthesis	Design Compiler, DesignWare, Power Compiler, Physical Compiler (Synopsys) Envisia Ambit Synthesis PKS,

		(Cadence)
---------?	Test Synthesis	ICBIST (LogicVision)
--------?	Floor Planning (RTL , Gates)	First Encounter (Cadence)
Design Verification		
	Test Bench Automation	Specman (Verisity)
	Formal Verification	Formality (Synopsys) Conformal-LEC (Cadence)
	Static Timing Analysis	Prime Time (Synopsys)
	Gate-level Simulations	NC Verilog, NC VHDL, NC-Sim (Cadence) ModelSim, (Mentor Graphics) VCS (Synopsys)
Design Verification	Test Tools /	DFT Advisor, Fastscan, FlexTest (Mentor) Test Compiler, Tetramax (ATPG) (Synopsys) Covermeter (Cadence)
DFT	Test Tools /	DFT Advisor, Fastscan, FlexTest (Mentor) Test Compiler,

		Tetramax (ATPG) (Synopsys) Logic Vision, ICBIST, MEMBIST, Boundary Scan

Table 17.1: *Design Flow and CAD Tools*

Audits

- Design analysis tools
  - Verilog/VHDL behavioral simulation tool(s) and version used
  - Power analysis tool(s) and version used
  - RTL, RTL DFT, synthesis constraints and net-list audits tool(s) and version used
  - Design structure and design analysis tool(s) and version used

- Logic optimization tools
  - Logic synthesis tool(s) and version used
  - Test synthesis tool(s) and version used
  - Floor planning tool(s) and version used

- Design verification tools
  - DFT tool(s) and version used
  - Test code coverage tool(s) and version used
  - Formal verification tool(s) and version used
  - Static timing analysis tool(s) and version used
  - Verilog/VHDL gate-level simulation tool(s) and version used

17.4 RTL Deliverables Audits

Overview

The following audits are helpful for either handing off the database to ASIC vendors or for internal implementation teams.

Audits

- Database
    - Installation scripts, associated design release version, and operating systems used with instructions for installing the database

- Source code
    - All design related RTL source code is compiled and is cleaned of RTL audits, DFT/verification audits and synthesis, timing, and constraints audits

- Verification
    - Cycle–based simulation or emulation models
    - Bus functional, memory models, and monitors used in testbench
    - Design simulation tests, testbenches, run scripts, and simulation results log files
        - Functional verification tests, testbenches, run scripts, and simulation results log files
        - Random verification tests, testbenches, run scripts and simulation results log files
        - Regression verification tests, testbenches, run scripts and simulation results log files
        - Instructions verification tests, testbenches, run scripts and simulation results log files
        - Gate-level verification tests, testbenches, run scripts and simulation results log files
        - Code coverage verification tests, testbenches, run scripts and simulation results log files
        - Manufacturing verification tests, testbenches, run scripts and simulation results log files
        - Software and applications verification tests, testbenches, run scripts and simulation results logs files
        - Formal verification run scripts and simulation results logs files

- Software scripts and proprietary software tools
    - Software scripts, if any, to generate source code, tests, and synthesis scripts for any or particular mode of chip operation
    - Proprietary internal tools, if any, that are required for the database

- Synthesis
    - All the final synthesis scripts used in database to generate the final net-list along with log files if any available

- Documentation
    - Soft copy of chip architectural specification documentation
    - Soft copy of chip functional/micro-architecture specification documentation
    - Soft copy of verification specification documentation.
    - Soft copy of bus functional, memory models, and monitors specification documentation
    - Soft copy of all IP's specification documentation
    - Methodology documentation and directory organization specific to the design.
    - Soft copy of flow diagram of the clocks and resets in the design.
    - Soft copy of floor plan diagram of the design.

      or
    - Soft copy all the above specification documents consolidated into one.

17.5 RTL Audits

Overview

There are many design issues that need to be resolved before the design is coded. If any of the following items in the checklist are not properly executed, there is a greater risk for the functionality, speed, area, productivity, and consequently the RTL description of the chip development. When RTL is coded, one needs to know what it implies both logically and physically. Depending on back-end solutions to solve timing and other issues is not a recommended practice. One has to have a good micro-architecture implementation and test specifications and has to code RTL correctly in the beginning with some preparation, knowledge, and understanding. The following RTL audits are the most common requirements, which need to be considered during RTL implementation.

Audits

- Communicating design issues
    - Following common methodology guidelines of naming conventions for hierarchical blocks, signals and designating active (high/low) states for signals
    - Type of revision control used
    - Directory trees organization
- Specifications
    - Architectural specification
        - Architectural simulations for proof of concept, performance evaluation and preliminary power estimations have been performed to derive functional/micro-architecture and verification specifications
    - Functional/micro-architecture specification
        - Definition of ASIC functional/micro architecture specifications has information regarding on-chip buses (PCI, PCI-X, In-fin band, PCI-Express, VCI, AHB/APB etc.,), block level interfaces, register implementation, clocking and synchronizing schemes, timing of each functional block, I/O definitions (pin definitions), functional/physical partitioning, and integration of soft/hard IP blocks.
    - Verification/test specification
        - Definition of ASIC verification/test specification has verification plan (test cases) for creating high level models (used

for floor planning to discover top level inter connect and as well as first pass constraints/timing budgets) and performing block-level, multi-block module level, ASIC level, system level and as well as hardware-software co-verification.

- Bus definitions (bi-directional, uni-directional)
    - Bi-directional bus interconnects have been brought out from core and/or module level logic of hierarchy
    - Bi-directional buses are always driven or have bus holders
- Latency/backward compatibility requirements
    - Tester specifications and behavior of the chip is available for legacy chips.
- Testability (e.g., scan, memory built-in self-test [MBIST], Joint Test Action Group [JTAG], and pad-ring)
    - Test for manufacturing defects have been added to design (scan insertion, MBIST, JTAG and pad-ring].
- IPs/memory integration
    - List of all IP's and number of each IP instantiated in the design (E.g., memories, phase-locked loops [PLLs], cores, buses, first in first outs [FIFOs], etc.) are available.
    - Functional and timing accuracy needs to be verified in the design at all IP, memory and custom IP interfaces with rest of the logic.
- Pin-limited or gate-limited design
    - Package and die size availability is determined based on the design requirements
- Speed is barrier for technology
    - Pipelining, register retiming (partition of data paths/logic) and duplicate logic cones have been considered to meet the speed of the chip for targeted technology.
- HDL description
    - Use standard header at beginning of each

file

  - Create block level drawing before coding
  - Break the design into manageable RTL code
  - Write comments in the code for every assignment or block
  - All top-level modules should be used just
  - for connectivity only without any logic
  - Keep same signal name through different
  - hierarchies
  - Do not mix active low and active high
  - logic in the design
  - At the top level there should be four blocks I/O pads, clock generator, JTAG logic and the core design (hierarchy)
  - Use separate always at block process for sequential logic and combinatorial logic
  - Specify all branches of all conditional statements
  - Initialize the output of conditional statements prior to defining the statements
  - For mutually exclusive conditions, use case statements
  - For multiple conditions happening at same time, use "if" statements and prioritize the conditions using "else if" for each subsequent conditions
  - Use high level constructs (case, if, always@) as much as possible
  - Use non-blocking assignment in synchronous process and blocking assignment in asynchronous process
  - Use parameters instead of definitions
  - Use one common file for all definitions that cannot be changed into parameters

- Lint
  - The RTL is clean with some of the most common lint warnings
    - When all possible conditions are not given in Verilog case statement

        - When a net is driven by multiple drivers but is not declared to be a tri-state
        - Any operand to logic binary operation has more than one bit
        - When a self-assignment is present in the code, since it will create a latch
        - Constant in the sensitivity list because it has no effect
        - If a loop index variable is modified inside the body of the loop
        - When a variable is referenced in a not blocking assignment that was previously assigned but not present in the sensitivity list of the "always" block
        - If a variable is assigned in both blocking and non-blocking assignments
        - If different bits of variable are assigned values in multiple "always" blocks.
        - When a function is being assigned using the "assign" construct rather than "="
        - When a complex conditional expressions are used that could be misinterpreted due to precedence rules
        - When blocking or non-blocking assignments have delays associated with them
        - When RTL constructs are mixed inside gate-level designs.
        - When a variable is set but not used.
        - Compiler constructs "ifdef" will be warned.
        - When input or an inout is declared as a registered variable

- RTL partition
    - Make the partition of the design such that the entire design can be used as an IP (design reuse), which can be easily portable and used in new designs
    - Avoid logical sneak paths that cross multiple hierarchical boundaries

- Related combinatorial logic should be kept in the same module
- Partition boundaries have been placed immediately to registers.
- Prioritize the partitioning of PLL's, clock, reset, IP, memory and custom IP modules in the design
- Break down the logic structures associated with huge muxes/de-muxes/priority – encoders/decoders to avoid congestion during back-end place and route activity.
- Pay special attention to intensive arithmetic data path logic circuits such as instruction fetch, decode, execute and write-back cycles [encoders, decoders, muxes/demuxes, adders, multipliers and shifters (e.g., arithmetic logic units [ALUs]).

- Design nature
    - Completely synchronous design. Avoid asynchronous logic.
    - Finalize reset strategy used (synchronous versus asynchronous). At reset time all bi-directional ports should be in input state.
    - Avoid internally generated clocks and conditional resets. Exception to this unless it is low power design or have any prior knowledge of implementing clock gating and consequences for testing and verification. In this case global clock gating logic must be kept in a separate module outside the functional module.
    - Turn off sections of the design whenever they are not in use for low power designs.
    - Design with minimum number of interconnecting wires on core interfaces. Do not make buses wider than necessary.
    - Signals that cross-different clock domains should be double sampled (synchronizer flop) after crossing domains.
    - Do not use clocks or resets as data or as enables unless they read as status signals for debug purposes.
    - Use minimum number of clock domains per

design.
    - Avoid using flops with negative edge clocks. Analyze duty cycle if used.
    - Latches should be avoided in the design wherever possible.
    - Sensitivity list should be as complete as possible.
    - Avoid multiple driven nets.
    - Avoid multi-cycle paths for efficient timing analysis. If used should be specified completely from source register to destination register.
    - Avoid delay elements.
    - Avoid asynchronous/combinational feedback loops.
    - Avoid high fan-out nets for most of the design and especially timing critical nets.
    - All inputs and outputs are registered.
    - No direct combinational paths from inputs to outputs.
    - All internal interfaces are sampled.
    - Do not use statements that assign initial values to signals and variables.
    - Resynchronization of all asynchronous set/reset/clear signals.
    - Synchronized read/write at memory boundary.
    - Synchronization for on-chip buses.
    - Use synchronous single/double port generic memory blocks.
    - Registered block inputs/outputs.
    - Avoid using internal tri-state signal/buses.
    - Do not instantiate gates unless it is required. Make code technology independent.
- Design reviews
    - Design reviews for RTL audits have been conducted.

17.6 DFT/Verification Audits

Overview

Testability is the amount of control and observation of data in the scan chain. The following audits are helpful for device's DFT and verification.

Audits

- Internally gating the reset signal
    - Design has internally controlled gated reset signals, unless there is an external bypass signal available to override the internal controls during scan chain activation.
- Clocks—Internally generated gated, muxing clocks with data or other clocks
    - All gated clocks have an external control capability as well as no clocks are being used as data inputs to registers.
    - Test plan exists for multiple clock domains.
    - Design with multiple clock domains, have clock muxing to a single clock source during testing. Mux control is provided so that all clocks can be switched over during scan.
- Tri-state drivers
    - Multiple tri-state drivers on a bus do not overlap (mutually exclusive) while driving the bus.
    - Bus hold cells are used in design, which will hold state when all three state drivers are disabled.
- JTAG
    - JTAG logic is kept at separate level of hierarchy.
- Random access memory (RAM) and BIST
    - Design has RAMS.
    - Test plan for RAMS exists.
    - Plan to test external ports, ring of registers, self contained BIST.
- Functional tests
    - Verified the performance of the chip as per specifications.
    - Verified internal pipeline structure.
    - Protocol and timing verification for on-chip and internal buses.
    - Generated test cases for module/chip functional verification.

  - All the device features have been tested.
  - Unit level simulations are performed.
    - All the directed tests pass the simulations.
    - All the random tests pass the simulations.
    - All the regression tests pass the simulations.
    - Any applications have been run on the RTL.
  - Gate-level simulation done (functional, timing, protocol)
    - All the directed tests pass the simulations.
    - All the random tests pass the simulations.
    - All the regression tests pass the simulations.
    - Any applications have been run on the RTL.
  - Code coverage [toggle and node coverage] is done
  - Fault grading and simulation done
  - Fault coverage
  - SDF back-annotated timing simulations done
  - DFT rules (such as scan rules) check is done
- Formal verification
  - RTL versus gates equivalence is done.
  - Gates versus transistor equivalence is done.
- Design reviews
  - Design reviews for DFT/verification have been conducted.

17.7 Synthesis, Timing, and Constraints Audits

Overview

The following audits cover the designs associated with synthesis, timing and constraints.

Audits

- Synthesis
  - Technology library—It has at least four power levels for each cell in the library including registers. Logic cells have bubbled inputs.

- False and multi-cycle paths—Identify all the false and multi-cycle paths in the design.
- Critical paths—These are identified in the design.
- Design constraints
    - Constrains for I/O's and as well complete design is done.
    - Design has minimum constraints: set_load, set_driving_cell, create_clock, set_clock_latency, set_clock_uncertainity, set_input_delay, set_output_delay, set_max_transition, set_max_capacitance, set_false_path, set_multi_cycle_path, and set_wire_load_model.
    - Design is over constrained.
- Wire load models—Timing analysis is performed using standard wire load models.
- Default compile—Used default compile (e.g., compile-map_effort medium –scan) and met the timings.
- DW implementation
    - What are the DW implementations selected. "set_implementation" command is used to specify which DW component is used. "set_dont_use" is used for particular implementations to use for a DW part. Are there any ripple carry adders?
    - DW components are ungrouped.

- don't_touch attributes—These are present in the design.
- Design Hierarchy—There are hierarchical, flatten or mixed (hierarchical and flatten) types of synthesis.
- Clock skew and buffering.
    - Clock skew within specified margin.
    - Buffered clocks at the block boundary.
- Design rule violations.
    - Design rule violations are checked in the design.

- report_constraints –all_violators is used to verify where the design rule violations occur in the design.
- "check_design" is used to verify RTL net-list problems with design. It reports unconnected ports, multiple drivers driving a signal, mismatched port sizes between blocks.
- "check_timing" is used to check all timing paths in the design path are constrained.
- Pay attention to clock pins driven by multiple clocks, un-clocked registers, un-constrained endpoints, no input delays, combinatorial feed back loops and generated clocks.

    - Net list
        - File name should contain the device part number (or name) and revision code (or date code). E.g., 123_rev1.v, 123_rev1.sdc.
        - Do the net-list have any behavioral constructs?
        - Do the net-list have any 1'b0/1'b1?
        - TIELO (VLO) and TIEHI (VHI) cells should be properly addressed.

    - Log files
        - Synthesis generated log files are fully read and warnings were justified.

- Timing
    - Negative slack amount—If this is greater than 100 percent of the clock period (continue analysis), 100 percent to 19 percent of the clock period (continue analysis), 20 percent to 1 percent (compile with map_effort high –incremental).
    - Number of paths—How many paths violate

timing and by how much. 50 percent of paths are violating timing with large negative slack there is a fundamental problem in architecture. Focus can be made to improve these paths with a "critical_range".

- Sharing start and end points—Paths share any start points or endpoints with each other. Constraints are properly implemented.
- Large single item delays—Is there any large single item delay in the critical path? Fixing this path may fix all the problems.
- DW components—Are there too many DW components in the critical path?
- Type of critical path—Is the critical path a control signal or a data-path signal?
- Multiple levels of hierarchy—Critical path cross multiple levels of hierarchy.
- Rearranging data paths—This eliminates the critical path.
- Restructuring the design—This eliminates the critical path.
- Room for optimization—Design can be further optimized.
- Levels of logic—Number of logic levels in critical path.
- Fan-out—The number along the critical path.

- SDC
    - Does SDC have all the parameters that are essential for the design? SDC file is recommended to have (Specifying clock information [frequency, uncertainty, latency, transition etc], describing a clock network, specifying clock-gating setup and hold checks, creating internally generated clocks, setting port input delay, setting input delay on clocks and setting port output delay along with timing exceptions [false paths, setting maximum and minimum path delays, wire load model and setting multi cycle paths], disabled timing arcs, output load capacitance and output loads). For

timing driven layout SDC file is recommended to have (source capacitance, input delay, output delay, driving cells, false paths, multi-cycle paths, case analysis).

- Design reviews—These have been conducted for design for synthesis, timing, and constraints.

17.8 Front-End Flow Audits Checklist

Design Documents Audits Checklist

Audit Types	System Architecture Document	Micro Architecture Document	Test Architecture Document	Software/Firmware/Board Specifications/ Tester Document
Block Level	-	Completed – Y/N	Completed – Y/N	-
Chip Level	Completed – Y/N	Completed – Y/N	Completed – Y/N	Completed – Y/N

Table 17.2: *Design Documents Audits and Checklist*

Tools and Deliverables Audits Checklist

Audit Types	Tools Audit	Deliverables Audit
Block Level	-	-
Chip Level	Completed – Y/N	Completed – Y/N

Table 17.3: *Tools and Deliverables Audits Checklist*

RTL Audits Checklist

Audit Types	RTL Audits	RTL DFT Audits	RTL Design Reviews
Block Level	Completed – Y/N	Completed – Y/N	Completed – Y/N
Chip Level	Completed –	Completed –	Completed –

	Y/N	Y/N	Y/N

Table 17.4: *RTL Audits Checklist*

Synthesis, Timing, Constraints, Formal Verification, and STA Audit Checklist

Audit Types	Synthesis Audits	Synthesis Timing Audits	Synthesis Constraints Audits	Formal Verification Of RTL versus Synthesized Net-list	Static Timing Analysis	Synthesis, Timing, Constraint Audit Design Reviews
Block Level	Completed – Y/N	Completed – Y/N	Completed – Y/N	Completed – Y/N	Completed – Y/N	Completed – Y/N
Chip Level	Completed – Y/N	Completed – Y/N	Completed – Y/N	Completed – Y/N	Completed – Y/N	Completed – Y/N

Table 17.5: *Synthesis, Timing, Constraints, Formal verification and STA Audits Checklist*

Functional Tests and Verification Audits Checklist

Audit Types	Functional Testing	Functional Timing Tests	Code Coverage Analysis	Test Design Reviews
Block Level	Completed – Y/N	Completed – Y/N	Completed – Y/N	Completed – Y/N
Chip Level	Completed – Y/N	Completed – Y/N	Completed – Y/N	Completed – Y/N

Table 17.6: *Functional Tests and Verification Audits Checklist*

Gate-Level Simulations Audits Checklist

Audit Types	Gate-level Simulations	Back-annotated Gate-level Simulations (Set Of Selected Tests)
Block Level	Completed – Y/N	-
Chip Level	Completed – Y/N	Completed – Y/N

Table 17.7: *Gate-Level Simulations Audits Checklist*

17.9 Unique Features in Design Flow

Automated RTL Audits

The very nature of the RTL interface, while beneficial to the customer, removes direct responsibility for complete correctness in RTL design quality, testability, and timing constraint specification. Without a comprehensive audit prior to handover, problems in these areas become the responsibility of the ASIC vendor, and they can lead to contentious and expensive iteration with the customer.

The purpose of automated audits is to provide a painless interface for the customer to ensure minimum requirements are met prior to handover in a manner, which fits naturally into their environment, with a minimum of new learning. Field experience has shown that this can dramatically reduce iterations and generally increases customer satisfaction overall.

Atrenta has worked closely with Agere to develop a standard set of SpyGlass®audits around Agere RTL handover requirements. These audits check a range of requirements, which are detailed in this paper. We expect that with further experience the rich set of checks that are covered as part of Agere RTL handover requirements will be developed and extended in the SpyGlass product.

17.10 Conclusion

The front-end flow is a milestone from initial freezing of the specification (Architecture) through RTL coding (implementation), verification (functional, timing, formal) till net-list (synthesis, constraints, timing) hand-off. The front-end hand-off flow, audits and considerations eliminates the need to worry about tools, and issues in later part of back-end design cycle typically done at an ASIC vendor. The audits are prepared to address complex high speed and low power, SoC, MCM, ASSP, and IP designs that involves both control and data-path (e.g., processors, digital signal processors [DSPs], etc) designs. The flow and audits addresses projects related to both cost reduction (or, legacy) as well as new developments.

Chapter 18

Case Study, Multiformat Video Decoder, IP Design Based on Configurable and Extensible Processor IP

Gulbin A. Ezer, Hardware Design Manager, Tensilica, Inc.

Dennis Moolenaar, Applications Engineer, Tensilica, Inc.

18.1 Introduction

This chapter describes a case study for creating a star IP by using building blocks as a special kind of star IP, namely, configurable and extensible processors. We have designed a fully programmable multiformat video decoder IP based on Xtensa LX [1] [2] configurable and extensible processor technology, and we have achieved low frequency and low-cost technology goals by both architecting a multi-processor solution and accelerating the application through the design of custom instructions. Our video decoder is a combined hardware and software IP package designed to provide a complete solution for video decompression and is targeted for the standard definition television (SDTV) space, for potential market segments such as digital camcorders, personal video recorders, personal video players, satellite and Internet protocol (IP) TV. It supports all resolutions from quarter common intermediate format (QCIF) up to and including D1 (720 x 480 @ 30 fps NTSC or 720 x 576 @ 25 fps PAL). The video decoder supports all the four major video decoding standards shown in the following and, since it is programmable and accelerates the block-based video compression, it can easily be reprogrammed to support new emerging standards such as the audio-video standard (AVS) from China.

- H.264 main profile @ level 3

- Moving Picture Experts Group (MPEG)-4 advanced simple profile @ level 5
- MPEG-2 main profile @ main level
- Windows Media Video (WMV) 9/ video codec (VC) 1 main profile

The goals of the project in order of decreasing priority were: (1) short schedule with limited engineering resources, (2) low cost process technology: 130nm G process, (3) low frequency: < 200MHz, (4) small area/power: < 10 mm^2/80 mW, (5) low system memory bandwidth: < 20 paper of 16-bit double data rate (DDR) @ 200 MHz. To meet all these goals, our design methodology had to rely heavily on automation as well as hardware/software (HW/SW) co-development. In the following sections, we will take you on our design journey for the video decoder, talk about the challenges and trade-offs we faced and the solutions we devised to reach our goals.

18.2 Block-Based Video Decompression Algorithms

Here is a brief summary of the block based video decompression algorithm that is common to H.264 AVC, MPEG-4, MPEG-2 and VC1 video standards [3] [4]. A macroblock, corresponding to a 16x16 pixel region of a frame, is the basic unit for these compression algorithms. *Figure 18.1* shows the block diagram of the decoder data flow.

The compressed bitstream consists of header information, coded motion vector parameters, and coded residual coefficients for each macroblock. The decoder receives a compressed bitstream, and the entropy decodes the data elements to extract quantized coefficients, motion vectors, and header information for each macroblock. Run-level coding and re-ordering are reversed to produce quantized, transformed macroblock residual coefficients. The coefficients are then both inverse quantized and inverse transformed to give the residual coefficients shown as D. Using the header information decoded from the bitstream, the decoder creates a prediction block. If the block mode is intra-predict, it uses previously decoded neighboring pixel information to create the prediction block. If the block mode is inter-predict, it uses the motion vector parameters, together with one or more previously decoded reference frames, and performs motion compensation to create the prediction block. The final block is reconstructed by adding the residual coefficients to this predicted block. Then, there may be an optional deblocking filter step before the block is written to memory to be used as a reference frame for further decoding, as well as output to the video display.

As one can see, the video decoder algorithms common to these multiple standards pose big computational and data movement challenges and have complex control with many modes and options. The next section summarizes the architectural trade-offs we evaluated and the steps we followed during the IP selection process to implement the video decoder engine.

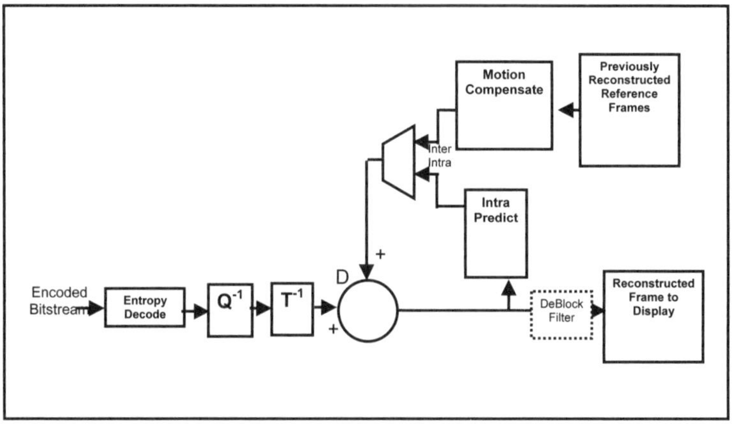

Figure 18.1: *Decoder Data Flow Diagram*

18.3 Architectural Evaluation Phase and IP Selection

In order to meet the real-time requirements of video decoding, the most compute intensive functions (also called hot spots) of the application need hardware supported acceleration. In order to provide multiformat decoding, hardware acceleration must support a superset of such functions in all the formats. However, such hardwired register transfer language (RTL) is very specific to the function and cannot be reused. Furthermore, it has communication overhead to transfer its inputs and outputs to the control processor's registers or memory. Most importantly, hardwired RTL is a big schedule risk, because bugs cannot be fixed post-silicon and they need to be redesigned in response to changes in the still evolving standards. On the other hand, a general purpose processor with a fixed instruction set requires expensive silicon resources and high frequencies to emulate the hardware with fixed data types and computational resources. So, our solution was to use configurable processor technology with a base instruction set architecture (ISA) and accelerate the application by designing custom instructions. Configurable and extensible processors with high compute power and high input/output (I/O) bandwidth can reduce the risk of complex state-machine design by replacing hard-to-design, hard-to-verify state-machine logic blocks with pre-designed, pre-verified processor cores and application firmware. The processor interfaces can be customized to allow data transfer from the external system resources directly into the execution units. This eliminates the I/O bottlenecks of traditional microcontrollers or digital signal processors (DSPs). Configurable processors provide most of the advantages of hardwired logic, yet at the same time maintain the time-to-market and programmability advantages of general purpose processors. As a result, we opted to use the Xtensa LX (Tensilica) processor technology and designed all hardware acceleration with Tensilica instruction extensions (TIE) and made them as generic as possible to enable reuse by many functions across different formats. In this methodology, both instructions and state are reusable, the state machines are soft and more malleable to morph into a different state machine for a different function usage. Schedule risk is also considerably lowered, since bugs may be fixed post-silicon and standard changes can be implemented by modifying the software package. With the decision of the basic building block IP of our design behind us, we moved to the architectural evaluation phase of the video decoder.

First, we developed the C-code for the applications and verified them against the reference decoders for each format using conformance bitstreams downloaded from the standard committee's Web sites [5] [6]. We also purchased multi-format bitstreams from commercial companies such as Tektronix [7] that provide a diverse set of short video clips designed to test the video decoders to the limits of their abilities. The reference models served as the golden model against which code running on various models of the system was compared and checked for bit exactness of decoded data. The reference code was run on the native machine (Windows or Linux PC) and the decoder software was run on the cycle accurate instruction set simulator (ISS) of Xtensa LX processors. Then, using the Web-based Xtensa processor generator graphical user interface (XPG GUI), we chose the configuration options that we needed. We used the same GUI and the processor development tools that are available to all Tensilica licensees. The evaluation license from Tensilica enables the user to configure the processor IP on the XPG GUI and download the software tools including the ISS model of the customized processor, compiler, assembler, debugger, profiler, operating system (OS) kit and an field-programmable gate array (FPGA)-based emulation bitstream. The user can then evaluate their application on the Xtensa processor before deciding to purchase the tools and the processor license. The processor generation takes about one hour and the synthesizable hardware design and the software environments are ready for performance evaluation and architectural iterations.

For the video application, we selected multimedia enhancing optional features of the processor such as multiply-accumulate, 16-bit multiply, zero-overhead loop, min, max, clamp, sign-extend, synchronization instructions, local memory sizes, direct memory access (DMA) ports, on-chip-debug, and so forth. The estimator slidebar on the configuration GUI gave us a first estimate on area and frequency interactively as each feature was selected. The processor generation itself took one hour and the synthesizable hardware design and the software environment were ready for performance evaluation. We then simulated the most challenging decoder formats, H.264 VC 1, on the Xtensa ISS model with a variety of bitstreams and gathered profiles on the execution of the code. A one second video of D1 resolution stream (30 frames in NTSC or 25 frames in PAL) ran on the ISS model in one to two hours. The profiles tabulated various aspects of the program executions, such as how much time is spent on each function, how often each function is called, branch overhead, cache misses, pipeline interlocks and the memory access patterns. Without any acceleration, the H.264 application required 1.5 billion cycles (> 1.5

GHz) to decode hard streams. We identified the most compute intensive functions which are the "hot spots" and marked them for hardware acceleration. Even with the estimated cycle reduction, we ended up with around 500 million cycles which would mean greater than 500 MHz clock speed for the processor. We then decided to choose a multiprocessor implementation in order to reduce the clock frequency so that our IP could be implemented on a low cost process technology.

Next, we evaluated alternatives for algorithm partitioning: data plane partition or functional partition? A homogeneous dual core system could work on multiple data sets in parallel, would be easier to program, and yet would need duplication of all acceleration hardware and instruction and data memory. Then, we evaluated a heterogeneous dual core partition, where each core is optimized for the specific functions it needs to perform. This partition resulted in a smaller memory footprint and lower hardware cost. We decided to design a DMA engine to handle the data movement between the processors. A block diagram of the final architecture of the video decoder is shown in *Figure 18.2*.

The video decoder subsystem consists of two heterogeneous processors: a stream processor and a pixel processor both enhanced by application-specific instructions; a point-to-point interconnect, and a multichannel DMA controller. The stream processor is accelerated for serial processing such as bitstream parsing, entropy decoding, and control functions. The pixel processor works on the data plane and performs parallel computations on pixel data using a single instruction multiple data (SIMD) instruction architecture. As shown in the block diagram, both processors have different local memory and data width configurations as required by their functional partition. The Xtensa LX processor allows direct interfaces to be created in other RTL blocks through the design of custom instructions. The transpose block shown in the diagram is an array of latch-based storage used to perform matrix transpose functions. This block is directly attached to the pixel processor through a customized TIE queue interface and is controlled by the new instructions we designed. Processors can also communicate seamlessly with each other through this custom I/O. Most of the data movement between the processors and memories is performed by the DMA engine. It can be setup to perform chaining DMA with circular descriptor buffers. It has five channels; each one can queue up multiple DMA operations and has an intelligent arbitration scheme based on programmable bandwidth allocation per channel. The processor interface (PIF) interconnect block is basically a high bandwidth, low latency crossbar capable of transferring data at 3.2GB per second connecting all the video decoder components and the system ports together. The video decoder subsystem connects to the system memory controller through two ports with identical PIF and communicates with the system controller through a shared memory application programming interface (API).

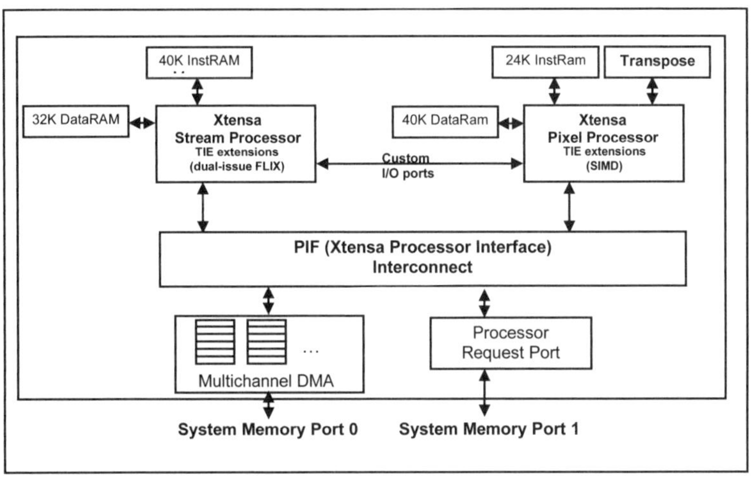

Figure 18.2: *Video Decoder IP Block Diagram*

In the following sections, we detail the methodology to arrive at the final processor configurations, implementation, and refinement of the design through quick iterations which depends heavily on the processor generator and system builder toolset. The tools for creating extensible processors allow the automatic generation of new processor hardware design and software tools from one high-level description. Automating hardware and software generation guarantees consistency in all the representations of the processor definition, namely, the hardware design, testbenches, simulation models, compilers, assemblers, debuggers, and real-time operating systems.

18.4 Design Phase

As soon as we finalized on the coarse system architecture, we planned the major design tasks, allocated resources, and started a parallel hardware and software co-design flow to refine the micro-architecture, design the components, integrate the full system and validate the application with conformance suites. As seen in *Figure* 18.3, we were able to parallelize the majority of the tasks and integrated them at various synchronization points after each subflow met its design goals. This methodology, at the heart of which lies the automated processor generation and system build tools, significantly reduced our development schedule.

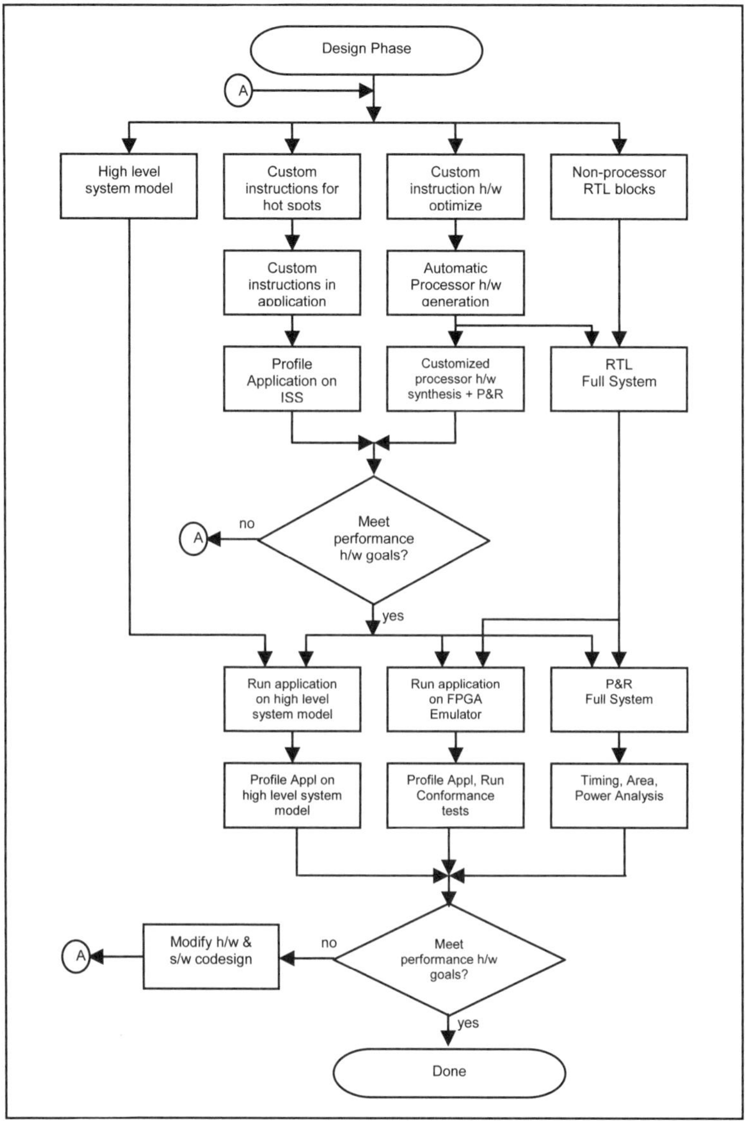

Figure 18.3: *Design Methodology Flow Chart*

Parallel and pipelined design flows are summarized in the following bullets:

1. Custom instruction design and verification
    - Functional design and verification
    - Hardware optimization with feedback from physical design
2. RTL design and verification of non-processor blocks
3. Automated system level netlist creation of the video decoder
4. Xtensa modeling protocol (XTMP) based electronic system level (ESL) model design for multi-processor video decoder
5. Multi-processor partition of the application and port to XTMP-based ESL model of the video decoder
6. Video decoder emulation on hardware prototyping platform
7. Physical implementation for ASIC flow and area/timing/power analysis on routed netlist

The design phase can be loosely split up into component design phase and system integration phase. The first four tasks aforementioned here belong to the component design phase and will be described in detail in this section. Tasks 5 through 7 belong to the system integration phase and will be detailed in the following sections. In the component design phase, the first task was the custom instruction design for each processor using the high level Tensilica instruction extensions (TIE) language to accelerate the hot spots. In parallel, we also started optimizing the new instruction hardware. The second flow is the RTL design of non-processor blocks such as the DMA controller in the Verilog language. In the third flow, we used the system build tools that we had designed to automatically configure and generate top level netlists and a testbench that hook up the fully synthesizable video decoder IP and the test components to create the full system RTL netlist and verification environment. The fourth flow is design and test of the MP system modeled in C, using the ISS API which is called XTMP. This model is used by the software engineers to verify the functionality and performance of the application optimized with customized instructions and is split between the two processors in a full system environment.

In the system integration phase, where the hardware and software implementation is finalized, we started three major design tasks in parallel. In the fifth flow we ported the application onto multiple cores, and evaluated the performance on the cycle-accurate XTMP model. The sixth flow is to port the RTL netlist on the ZeBu-XL system emulator from EVE [8] and running the application on the real hardware. We used the emulator mainly to perform conformance testing and code validation for a variety of bitstreams and test suites obtained from video standards committees. The last flow is the generation of EDA scripts for an ASIC reference flow; we performed the place and route of the full video decoder and evaluation of the final implementation by running gate level simulations, timing and signal integrity checks and performing power analysis.

Throughout the design phase, we compared the results from parallel flows and made decisions as to whether to iterate on modifying or refining the design or if the trade-offs between performance and hardware cost were acceptable. For example, some TIE instructions were redesigned or processor configurations were modified (e.g., memory/system interface width) if they had a major frequency impact or caused significant routing congestion or did not meet the real time performance requirements. The processor generation automation for both hardware design and software tools was the key pain reliever to enable us to converge on the goals since these iterations took hours instead of weeks as would have been in traditional system design flows.

Extending Processor Instruction Set and Interfaces

We designed the custom instruction extensions using a high level instruction extension language which is called the TIE language. Similar to hardware design languages (HDLs), TIE uses Verilog syntax to describe instruction semantics for logical, arithmetic and load/store operations. Unlike HDLs, TIE has constructs to express key software information about instruction names, operands, pipelining, and hints for compiler inference; this information exposes the new instructions in the high level programming environment. Some TIE instructions with certain properties can be directly inferred by the Xtensa C/C++ compiler and all are available to the software developers as assembly mnemonics or intrinsic C/C++ functions. TIE constructs include instruction formats and encodings, customized datapaths, interfaces to external logic, state registers, register files, variable instruction lengths, etc. Using these TIE files, the Xtensa processor generator automatically produces customized versions of the processor hardware, RTL, testbench, verification suite, physical design scripts (e.g., synthesis, place and route, timing, signal integrity check, etc.). In addition, the generator creates a customized software environment that includes a vectorizing C/C++ compiler with intrinsic support for user extensions, an ISS, debuggers, linker scripts, a basic runtime and handlers (XTOS), commercial real-time operating system (RTOS) support, and various other software development tools. The generation time for the full hardware and software environment was around one hour. However, since most of the time we only required the customized software model and tools to experiment with the new TIE instructions we designed, we used the TIE development kit (TDK) which is a local software development environment that is compiled in a matter of minutes.

We then instantiated the TIE instructions as intrinsics in the application C code, ran it on the ISS model extended with TIE libraries, analyzed the profile results to see if the cycle count reduction goals were achieved and then looked for new hot spots. In parallel, we also optimized the hardware for the new TIE instructions. This was done by breaking critical paths into multiple pipe stages and minimizing area by sharing hardware resources, which can all be expressed easily in TIE language. Once again, we utilized the processor generator to obtain the customized processor RTL and physical scripts and performed synthesis and place and route iterations to keep track of our hardware goals.

The most important point to be made here is that 95 percent of the design was verified in the high-level ESL environment using the actual application. This was possible because of hardware design automation and software tool chain generation from a single source which is the TIE description file. Since the generated processor RTL is correct by construction, we did not have to do any simulation runs for the new instructions with the slow RTL verification environment. This level of automation has enabled us to quickly perform many design iterations. We designed a mixture of 24- and 32-bit wide instructions, where the latter was used for multiple instruction issue. In total, we designed around 200 TIE instructions, which resulted in increasing the hot spot function performance from anywhere between 10 percent and 700 percent. For example, the H.264 entropy decoder for main profile requires a context adaptive binary arithmetic coding (CABAC) scheme which needs 800M cycles on a 200MHz reduced instruction set computer (RISC) processor for a midrange complexity bitstream and is usually targeted for external hardware accelerator implementation. We implemented the CABAC state machine using 12 TIE instructions, two of which were used in wide instruction format and issued in parallel, reducing the cycle count from 710M to 12M and achieving 55 times performance improvement for an area cost of around 20K gates.

We utilized three techniques while designing the custom instructions:

1. Vector operations—Single instruction on multiple data (SIMD)
2. Flexible instruction extensions (FLIX)—Multiple operations on multiple data with variable instruction length; e.g., very-long instruction word (VLIW) without code bloat
3. Fusion—Multiple operations combined in a single instruction, e.g., load + add

We will now show four examples of using TIE during the whole design process. The first example shows how easily you can improve the performance using FLIX. The second example shows an interesting use of the Fusion technique. The last two examples show how the system can be accelerated by adding custom ports on the processor using TIE.

TIE Example 1:

Xtensa LX supports two sets of three different instruction lengths (16/24/32-bit or 16/24/64 bit) which can be mixed seamlessly without any mode bits. FLIX instructions may be composed of multiple operations of variable size encodings. It is very easy to create a static superscalar architecture by replicating Xtensa base ISA instructions in different slots. Since these instructions are recognized by the compiler, they are automatically inferred and need not be instantiated in the application source code as intrinsics. We configured the stream processor to have 64-bit instruction width and created a two-way static superscalar architecture using the following three TIE instructions:

format far64 64 {slotar0, slotar1}
slot_opcodes slotar0 { L16SI, L16UI, L32I, L32I.N, L32R, L8UI, S16I, S32I, S32I.N, S8I, ABS, ADD, ADD.N, ADDI, ADDI.N, ADDMI, ADDX2, ADDX4, ADDX8, NEG, SUB, SUBX2, SUBX4, SUBX8, AND, OR, XOR, ALL4, ALL8, ANDB, ANDBC, ANY4, ANY8, ORB, ORBC, XORB, SLL, SLLI, SRA, SRAI, SRC, SRL, SRLI, SSA8B, SSA8L, SSAI, SSL, BALL, BANY, BBC, BBCI, BBS, BBSI, BEQ, BEQI, BEQZ, BEQZ.N, BF, BGE, BGEI, BGEU, BGEUI, BGEZ, BLT, BLTI, BLTU, BLTUI, BLTZ, BNALL, BNE, BNEI, BNEZ, BNEZ.N, BNONE, BT, J, JX, MOV.N, MOVEQZ, MOVF, MOVGEZ, MOVI, MOVI.N, MOVLTZ, MOVNEZ, MOVT, MUL16S, MUL16U, EXTUI
}
slot_opcodes slotar1 { ABS, ADD, ADD.N, ADDI, ADDI.N, ADDMI, ADDX2, ADDX4, ADDX8, NEG, SUB, SUBX2, SUBX4, SUBX8, AND, OR, XOR, SLL, SLLI, SRA, SRAI, SRC, SRL, SRLI, SSA8B, SSA8L, SSAI, SSL, MOV.N, MOVEQZ, MOVF, MOVGEZ, MOVI, MOVI.N, MOVLTZ, MOVNEZ, MOVT, EXTUI }

We defined a new 64-bit instruction format composed of two slots, where a slot is a subset of bits of a FLIX instruction that encodes one operation. We used a subset of Xtensa ISA instructions in both slots in order to take advantage of the Xtensa compiler optimizations. Since these instructions are predefined, we did not have to design them. This new format is capable of performing two Xtensa LX base ISA instructions in parallel. The generation of the parallel functional units and the additional read/write ports to the general purpose register file is all automated. Furthermore, the Xtensa compiler automatically bundles the operations from each slot without requiring the use of intrinsics. We achieved 30 percent performance gain for an area increase of 25K gates and a frequency loss of 15MHz by using FLIX.

The performance gain was, however, not high enough so we tried other TIE extensions.

TIE Example 2:

To achieve higher performance gain TIE instructions can be designed that combine multiple core instruction operations into one TIE instruction. An example is a load with address increment or an *add* and *shift* instruction. Yet, fusion of operations is not limited to solely combining calculations. A more interesting example is a saturate instruction that implements the following C code:

```
Int A;
If (A<0)
   A = 0;
Else if (A>255)
   A = 255;
```

This is an operation that often occurs in video since the main data in most video coding algorithm is a byte size color component. The whole sequence shown above can be replaced by the following TIE instruction.

```
operation sat8u {out AR Result, in AR Data}{}{
   wire min       = Data[31]; // sign bit indicates negative number
   wire max       = (Data[30:8]!=0);
   assign Result  = min ? 0 : max ? 255 : Data;
}
```

This is the TIE description for a saturate instruction called "sat8u." It generates an AR register output which is the Xtensa core register file, and it takes an AR register as an input. The logic of the instruction follows the C description shown above. Two intermediate values are calculated that indicate if it is a negative number (min) or if the value is larger than 255 (max). Then, the result is assigned with 0.255 or the input value depending on the intermediate values. This instruction can be used as:

```
Int R, A;
R = sat8u(A);
```

All TIE instructions on the stream processor use mainly the fusion technique to accelerate it. The SAT8u instruction is actually located on the pixel processor as a SIMD instruction where 8 or 16 parallel saturates can be performed on 12 or 24 bit data.

TIE Example 3:

The ability to add custom I/O ports to the processor with the TIE instructions was very handy in designing seamless communication channels between the Stream processor and the Pixel processor. Here is an example where one processor can post an interrupt to the other processor through a direct wire connection between their I/O ports. The following TIE code adds two state registers and two custom I/O ports that are connected to these registers. This concise description also automatically adds new instructions to the processor ISA to read/write/swap the contents of the state with the processor's general purpose register file. These state output ports from the sending processor are simply connected to the interrupt input ports of the receiving processor.

```
// Used to generate interrupts if an error occurs, or a system interrupt is needed.
state CodecInterrupt 1 1'b0 add_read_write export
```

The new ISA instructions to manipulate the state registers are: write user register (WUR), read user register (RUR), exchange contents of user register (XSR). The C-code example to generate an interrupt simply toggles the output signal values from 0 to 1 in order to generate an edge-triggered interrupt as shown here:

```
WUR_CodecInterrupt(0);
WUR_CodecInterrupt(1);
```

TIE Example 4:

The previous example showed a very simple example of custom ports for generating interrupts. A more complex example of custom ports is used by the Pixel processor. The pixel processor contains a SIMD engine that can operate on 8 data elements of 24 bits in parallel. The SIMD engine works well when operations are performed between the same elements of different vectors, but none of the SIMD instructions operates on elements within the same vector. This creates a problem for algorithms like the transform block (T) as show in *Figure 18.1*. A transform first performs calculations on the elements of each row and then on the elements of each column of a two-dimensional array (8x8 elements).

```
For (i=0;i<8;i++){
        Perform Transform on one row
}
For (i=0;i<8;i++) {
        Perform Transform on one column
}
```

The computations on the columns can easily be implemented using the SIMD engine, but calculating the row transforms can not efficiently be implemented using the SIMD instructions. To solve this, we can generate a whole set of instructions that operate on vector elements, but a more efficient solution would be to transpose the two-dimensional array. After the transpose, the operations on the row would be operations on columns. The flow of the TIE accelerated transform using SIMD TIEs would now become:

```
// Begin of row transforms
Transpose 8x8 block
Perform all column transforms using SIMD TIE
Transpose 8x8 block
// End of row transforms
Perform all column transforms using SIMD TIE
```

The cost of using the SIMD instructions on the row transforms is the two transpose functions required. Ideally, you do not want to spend any cycles on performing the transpose operation. One way to achieve this is to have a transpose memory connected to the pixel processor by custom ports. The TIE that implements the ports is shown here:

```
regfile SR 128 8 sr

queue TQ128_out          128   out
queue TQ128_in           128   in
```

```
queue TQWrAddr_out        4      out
queue TQRdAddr_out        4      out
// State export to control the external Transpose logic block
state TQMode_out               7      3'b0 add_read_write export

// Instruction semantics to move data between Transpose logic and the vector
register file
operation RdTQ128u {out SR Result}{in TQ128_in}{
      assign Result = {64'h0, TQ128_in};
}
operation WrTQ128 {in SR Data}{out TQ128_out}{
      assign TQ128_out = Data[127:0];
}
```

We create a vector register file that can operate on 16 elements of 8-bits. For our application, a depth of 8 was sufficient. The queue structures add two output ports (128-bit data and 4-bit address) and 2 inputs ports to send/receive the matrix data to be transposed. *TQMode_out* state is exported which also creates a 7-bit output port out of the processor and is used to control the transpose block modes. Depending on the control encoding, the transpose block can provide simple row-to-column transpose or butterfly transpose of 16 by 8-bit element or 8 by 16-bit element rows. The operations describe the semantics of queue read/write operations. In a read operation, the data from the queue input port is assigned to *SR* which is defined as an entry of the vector register file. In a write operation, the data from the *SR* vector register file is assigned to the queue output port. The push/pop control I/O is automatically generated for the processor logic from these definitions.

To hide the cycles needed for performing a transpose, we can put the data in the transpose memory at the end the inverse quantize block (Q) as shown in *Figure 18.1*. This function would normally write the results to the local memory of the processor with a SIMD store instruction and now that store instruction will be replaced by the WrTQ128 instruction. Now the transform function uses the RdTQ128 instruction instead of the SIMD load instruction to load the data. This allows us to have a transpose that does not cost us any cycles.

Table 18.1 shows the performance improvement obtained with TIE instruction acceleration in H.264. The profile results are given for both options of entropy decoders: context adaptive binary arithmetic and variable length decoders for the Rugby video stream at 5.4 Mbps bitrate. The TIE acceleration has brought the processor clock frequency requirement from 1.3 GHz down to less than 200 MHz, saving significant area and power. Even without the benefit of hardwired RTL acceleration, we have achieved significant performance gain through the design of custom instructions and configuring the processors options to fit the application requirements. The video decoder IP is implementable in low cost 130nm G-process with a modest clock frequency goal of less than 200 MHz.

Baseline Xtensa Rugby CABAC 5.4 mbps	Baseline Xtensa Rugby CAVLC 5.4 mpbs	TIE accelerated Xtensa Rugby CABAC 5.4 mpbs	TIE accelerated Xtensa Rugby CAVLC 5.4 mpbs
1300 MCPS	701 MCPS	172 MCPS	168 MCPS

Table 18.1: *Performance Data for H.264 Bitsreams Before and After Custom Instructions*

RTL Block Design and Verification

The RTL blocks of the video decoder, which consisted of the DMA controller, interconnect network and the transpose block, were designed in Verilog HDL language and verified in standalone unit level testbenches. The unit level verification consisted of directed and randomly generated self-checking diagnostics and assertion checkers. RTL design is the most time consuming mainly because of the low level verification effort. Fortunately, since we implemented all the complex video functions as instruction extensions and verified them at the application level, this slow RTL design cycle was kept to a minimum and was never in our critical path.

Automated System RTL and Testbench Generation

For the hardware system integration, we designed an automated flow to generate the full system netlist and testbench. This flow is based on an internal system description and configuration language, called system architecture and integration (SARI) and the database compiled from the description. We have designed an API to access the database and various client tools which use this API to automate system integration and verification process. The SARI language is developed to configure systems, describe the connectivity of their components, and to specify potential constraints on the system components. The two major functional component categories used in constructing systems are: (1) leaf nodes, such as processors, memories, registers, DMA controllers, and I/O devices (memory mapped or instruction mapped as in the case of TIE custom interfaces), (2) interconnects, such as routers (e.g., crossbars, etc.) and bus bridges.

We wrote a SARI description of the video decoder, which details each component's interface widths, connectivity to other components and its address map. The SARI compiler created a system database from this description. Next, the system build tools that use the database API were invoked. The netlister tool generated a top-level synthesizable RTL netlist and the system-level testbench generator tool created a Verilog-only testbench for the full system which instantiated the automatically generated video decoder top-level module and connected its two ports to a shared memory model. This automation enabled us to start top level physical implementation and floor-planning early in the design phase even when the component interfaces and netlist hierarchies were changing as we experimented with different bus widths and connection topologies for the multi-processors and the DMA engine. The system level testbench was used for two major purposes: a quick test of system validation before the netlist was ported to the emulation environment, and gate-level and power simulations post-synthesis and post-route during physical implementation flow. The system level testplan on the RTL testbench targeted the following areas:

- The top-level netlist and the interconnects of the different components of the video decoder as well as all the I/O of the video decoder IP
- The system reset and the boot sequence of the processors
- All types of system interrupts and system errors

There was no need to conduct extensive verification since the RTL netlist was emulated with the decoder applications and conformance bitstreams on the hardware emulator.

The system testbench was also used for gate-level netlist verification and power simulations. We performed this by replacing the RTL netlist for the video decoder with the routed gate-level netlist and used the SDF delay files and associated physical libraries. We ran all the decoder applications with a representative sample of the bitstreams that exercised all the major functions and options within each format to get an accurate estimate of the power. We measured both core power as well as the memory power for the local processor memories and tabulated the external dynamic random access memory (DRAM) bandwidth usage to estimate the impact of the decoder IP on the final system power.

ESL Model for the Video Decoder and the System Testbench

We wrote a high level simulator program for the full system in C++ using the ISS API for the Xtensa LX processors which is referred to as the XTMP. This is a multi-threaded transaction-level simulation environment in which each Xtensa core runs in a separate thread. One can also create user threads for modeling time-based devices. We created two more user threads, one for the DMA device and the interconnect arbiter and another for the external memory controller which will not be part of our IP product but which is necessary to simulate the system testbench. We used a combination of the XTMP simulation components provided by the Xtensa tools and our custom device models to create the system model. The system components include the processors, each instantiated with its parameter file that describes its unique configuration including the TIE extension ISS library, memories (both external and local), connectors with address translation to model PIF interconnect, TIE queue transactor device function, and the DMA controller device model. The XTMP memory model, API, provides features to model delays on single/block transactions for both reads and writes which allowed us to design a realistic model of the latency and bandwidth of a system memory controller. The transactor block was modeled as a callback function corresponding to the TIE queue interface identified with a port name in XTMP API. The DMA controller device was modeled as a memory mapped XTMP device using the transaction-oriented interfaces. When an Xtensa core issues a load/store to this device, its callback function is invoked, which receives the information about the transaction, processes the information and sends back a response to the core with appropriate delay modeling. This model was sufficient for simulating the video decoder applications that run on the video decoder but not the API functions for communicating with the system control processor that would be the master of the video decoder. For this purpose, we instantiated another Xtensa processor with a default controller configuration and designed an example C-program on it that showed the Video Decoder API calls just like it would be used in the final SOC.

18.5 MP Partition of Decoder Applications and Validation on ESL Models

As the XTMP model of the decoder system was being designed, we partitioned the application by restructuring the code to separate the functions of each core and simulated DMA with memcopy functions and validated it on a single ISS model which had the combined TIE instructions for both cores. Then, we ported the application onto the XTMP model, enabled the DMA channels for data prefetching and optimized the C-code for efficient software pipelining and load balancing between the two processors and the DMA controller. We measured the effects of memory latency and arbitration for shared resources by utilizing the detailed, nonintrusive profiling results obtained from the XTMP simulator. The functional correctness of the application code is verified by comparing the XTMP generated decoded YUV files against the golden reference model running on the native machine for bit exactness.

To debug the application on the XTMP model, we used the Xtensa gnu debugger (XT-GDB), which is customized to understand the TIE instructions and state we added to each processors. A debugger was connected to each core which communicated independently with the debugger; so when a core was being debugged, the rest of the system could continue to run, which mimics what happens in real hardware systems. We could also debug in synchronized mode, in which all processors run in lock-step, cycle-by-cycle and can be stopped and started synchronously. The printf features of the XTMP environment was another powerful debugging tool that enabled us to quickly debug the failing conformance test cases from the regression runs on the hardware emulation platform which had less visibility.

We basically used the XTMP model for the detailed non-intrusive profiling of the decoder functions during the software (application code) and hardware (TIE refinement) optimizations. The XTMP model was fairly fast and cycle accurate at the processor level and transaction accurate at system level. It could run one second of D1 resolution video stream (30 frames) in 2.5 to 3.5 hours, which was little overhead to create profiles on interesting streams that would then take the engineers a few days or a week to analyze and make decisions on code optimization or new instruction design based on the analysis. In contrast, a simple 8-frame video stream would take seven hours to simulate on the RTL testbench. Being able to perform most of our validation and performance analysis on the ESL environment greatly shortened our development schedule.

18.6 System Emulation on FPGA Platform

We used the ZeBu-XL Emulation system from EVE [8] for accelerated system level testing of the video decoder in two capacities: (1) full system RTL verification using the application code, (2) conformance testing of the multiple formats. Video decoder conformance testing requires running multiple bitstreams with multiple bitrates on different complexity video frames. The emulation system running at 5 MHz provided us with the performance required to decode many seconds of video frames as well as the cycle accuracy of final RTL for real time performance validation. We used the same system netlist as instantiated in the Verilog testbench and synthesized it for Xilinx FPGA (V2-8000) libraries. The system partition resulted in the placement of the two Xtensas and their local memories in two FPGAs, and the rest of the system (DMA controller, interconnect, system memory controller) into a third one. We used ZeBu's SDRAM models as the external system memory of our video system testbench. The C code for each Xtensa was compiled using basic runtime (XTensa operating system [XTOS]) and manually customized linker script packages, LSPs. The object files were loaded into the SDRAM by means of JTAG controlled on-chip debug (OCD) port of one of the Xtensa processors with the XT-GDB tool (GDB customized by the Xtensa processor generator to understand the new state additions). We designed an additional bus transactor interface port to the SDRAM controller and performed the download/upload of the encoded bitstream and decoded frame data files from the host PC at the start and end of the application runs. We automated the conformance runs with regression scripts and were able to validate each format within a couple of days. We also measured the performance across various bitstreams in terms of cycles taken to decode one second of video and compared against the XTMP model to find that we were within 1-2 MHz of accuracy out of 200 MHz.

18.7 Physical Design of Full System

The reference physical flow for individual system components, such as the DMA controller and the Xtensa processors, as well as for the full system was fully automated to allow quick iterations. We extended the physical flow scripts provided by the Xtensa processor generator tools to work on the multicore top-level netlist. The sequence of physical flow steps we followed for the video decoder is listed here:

- Synthesis—A hierarchical synthesis methodology is followed. The processors with their local memories are individually synthesized and then the whole design is flattened and synthesized at the top-level.
- Physical synthesis and scan chain insertion
- Clock tree synthesis and route
- Resistance/capacitance (RC) extraction
- Timing verification and signal integrity checks
- Power analysis in-gate level testbench using application code with a representative bitstream set to test all functions and options of each format
- Automatic test pattern generation and fault coverage analysis

The feedback from physical results were used in the customized instructions hardware refinement, interconnect reduction, DMA channel buffer storage optimizations, local memory size reduction tradeoffs against software complexity.

One of the major challenges in providing IP is to validate its implementation across the most widely-used commercial EDA tool chains. Although our first product release provides automated scripts for the reference flow mentioned previously, we usually expand our product offering with other EDA flows or mix-and-match of tools from different companies, such as RTL Compiler from Cadence or BlastFusion from Magma, etc.

18.8 Packing Video Decoder IP

The video decoder IP will be delivered as an integrated HW/SW IP package that can be downloaded from our FTP site and will include the following deliverables:

- Software package
    - Standard Xtensa LX software tools
    - Software evaluation testbench—Cycle accurate ESL model and the system testbench
    - Bootcode that initializes each processor and jumps to main
    - Codec binaries for each format
    - Video decoder API documentation and example source code that calls the API and decodes example bitstreams for each format
    - Documentation and data sheets

- Hardware package
    - Synthesizable RTL model in Verilog language
    - Sample testbench with C-source test programs and run scripts for out-of-the-box testing and for gate-level model verification for all major EDA simulation tools
    - Synthesis, floorplan and place-and-route scripts for reference design flows from major EDA vendors

- System bus development kit for designing interfaces to the video decoder IP interface protocol
- Documentation and data sheets

To protect our IP we employ various techniques such as watermarking, encryption, copyright tagging, and customer ID tagging of our deliverables.

Future releases of the video decoder package will also include support for an FPGA–based emulation platform or encrypted FPGA netlists for major FPGA vendor tools that can be imported to the user's board platform.

Use Model

The user unpacks and installs the hardware and software packages.

- Software evaluation testbench usage—The user points to the Xtensa tools, compiles the XTMP model, and runs the example program on the system control processor ISS model that calls the video decoder API functions with the example bitstreams. The testbench uses the actual codec binaries. The user can provide their own bitstreams for decoding and evaluate the results. The user may choose to use the XTMP environment for their ESL modeling and extend the system model to add other devices or other processors. If the user chooses an Xtensa processor as the video control processor, then the user downloads the full Xtensa software/hardware environment customized for the control processor configuration and develops the full audio/video system application for the target market. If the user chooses a different processor architecture for the video controller, then the user may use the SystemC thread package and the SystemC interface of the XTMP model to import video ESL model.
- Hardware package usage—The user points to the Xtensa tools, compiles the C-diags and runs them on the RTL Sample testbench as out-of-the-box sanity check. The user then performs synthesis, place and route (P&R), automatic test pattern generation (ATPG), signal integrity checks, etc.

The user sanity checks the physical implementation by running the gate-level RTL on the sample test-bench and does further verification using formal tools such as equivalence checkers between RTL and gate-level netlists.

- SoC Integration—The video decoder is instantiated in the SoC with its two PIF ports connected to a memory controller (of SDRAM) shared with the video control processor. Video controller and the video decoder communicates through shared memory-based API and interrupts in both directions. The interrupts from the video controller are tied to the stream processor only which handles the control functions and API implementation. The pixel processor functions in slave capacity to the stream processor during decoding. The video processors communicate with each other using interrupts or through direct read/write to each other's local DataRam with a consumer/producer model. Both cores ResetVectors are configured to point to the same location in SDRAM. Upon power-up, the video controller keeps the video decoder engine asleep by using either reset or stall-after-reset signals until it loads the codec binaries and boot code into the SDRAM. After the reset/stall inputs to video decoder are deasserted, both cores start fetching instructions from the same reset vector. The reset handler initializes each core based on its processor ID. The video decoder puts out the physical addresses as specified by its memory map and the system designer needs to translate these addresses appropriately with external logic to direct it to the SDRAM. There is a 256Byte region in memory that video controller initializes to tell the video decoder how to offset to various parts of the code (reset handler, permanent code, overlay codes and data section). After the cores are reset, they jump to the API code and wait for the video controller to tell them what to do.

18.9 Conclusion

In summary, any multiprocessor SoC (MPSoC) hardware/software co-design benefits greatly from a well formulated parallel design methodology as we have outlined in this chapter. Our MPSoC flow offers innovations to previous design flows in these major areas:

1. We use extensible processor generator technology rather than using standard component IP; our heterogenous processors share the basic core Xtensa ISA and use the same development tools (compiler/RTOS) and the customized instructions are integrated into the hardware and software tools automatically. Furthermore, the communication network is simplified because both processors use the same hardware protocol and software MP synchronization API.
2. Our stream processor can function as an accelerator engine (uses VLIW to speed up arithmetic decoding) as well as a control processor thus making more efficient partitioning of the video algorithm across both processors.
3. No fixed function coprocessors or hardwired engines are used; application-specific functions are designed with new state and instructions that operate on them.
4. RTL and the cycle accurate C-model for the new functionality is generated automatically from a common source expressed in a high level language (TIE) and verified directly by running the application code where they are instantiated as intrinsics; since the RTL is correct by construction, additional verification of the new instruction hardware in a low level RTL testbench is not necessary significantly reducing our verification schedule.
5. Automated RTL generation allows quick feedback from synthesis and P&R analysis to validate the cost/performance tradeoffs during design.
6. Our Sari-based configurable system build tools (patent pending) automate the creation of top-level netlists when processor attributes (e.g., I/O and memory requirements) change and physical implementation of the full system can start earlier in the design phase.

We have proven the methodology on a real application: the multiformat video decoder IP, which is based on a configurable and extensible processor design technology. Full programmability has the flexibility of adapting to standard changes and post-silicon bug fixes. Even without the benefit of hardwired RTL acceleration, we have achieved significant performance gain through the design of custom instructions and configuring the processors options to fit the application requirements as depicted in *Table 18.1*. The video decoder IP is implementable in low cost 130nm G-process with a modest clock frequency goal of less than 200 MHz. Furthermore, this automated design methodology considerably reduced our verification schedule. We verified the customized instructions in the C-source testbenches and directly in the application code where they were instantiated as intrinsics and since the RTL is correct by construction, additional verification of the new instruction hardware in a low level RTL testbench was not necessary. We have used split hardware and software development flows: application code was verified and optimized with successive refinements of custom instructions and with feedback from profiling on the high level system model. In parallel, system hardware optimizations were performed based on feedback from physical synthesis and place and route results. This automated MPSoC design methodology considerably speeds up development schedules by allowing refinement of the design with short iteration cycles. We have achieved all our design goals within 18 months with the following team: one engineer responsible for the TIE design and optimization, two engineers for the RTL, verification, emulation and ESL model design and six engineers to develop four decoder applications.

Acknowledgements

This project is a joint development with our partner, Ittiam Systems [9] who designed the application for the multi-format codecs and our team designed the multi- processor subsystem and the application specific instruction extensions.

References

[1] Chris Rowen, *Engineering the Complex SoC,* Prentice Hall, 2004

[2] Steve Leibson, *Designing SoCs with Configured Cores,* Morgan Kaufmann, 2006

[3] Iain E.G. Richardson, *H.264 and MPEG-4 Video Compression,* John Wiley & Sons Ltd., 2003
[4] Peter Symes, *Video Compression Demystified,* McGraw-Hill, 2001
[5] http://www.itu.int/ITU-T/studygroups/com16/jvt
[6] http://smpte-vc1.org/
[7] http://www.tek.com/products/video_test/vclips/index.html
[8] http://www.eve-team.com
[9] http://www.ittiam.com

Author Biographies

Ramesh Banda is a distinguished member of the technical staff at LSI Logic Corporation. He has more than 13 years of experience in the industry, having worked at Sandmicro Electronics, Phoenix Technologies, inSilicon Corporation, Synopsys, Cradle Technologies, and Agere Systems. He has focused on various front-end and back-end development stages of ASIC and FPGA design for SoCs, ASSPs, and IPs. Mr. Banda has served as a DesignCon technical program committee member and track organizer for chip-level architectural and functional design since 2005. He received an M.S. degree in electrical engineering from the University of New Orleans and a B.E. degree in electronics and communications engineering from Osmania University in Hyderabad, India.

Behfar Bastani is a registered U.S. patent agent. He provides patent prosecution and strategy for Silicon Valley start-ups in software, algorithms, hardware, semiconductors, and mechanical and medical devices. Mr. Bastani has also served as in-house intellectual property manager for a software start-up and has developed software for high-performance clustering and distributed file systems. He holds an M.A. degree in computer science from the University of Chicago, where he studied theoretical computer science and served as lecturer, and a B.A. degree in computer science from the University of California, Berkeley.

Bill Beale is an application engineer for the Hillsboro PHY group of Synopsys. He worked for five years on electronic packaging, specifically the electrical modeling aspects, and later worked for Ansoft as an application engineer supporting signal integrity tools. In 2000, Mr. Beale joined Accelerant Networks and continued signal integrity work on high-speed signals of serial interfaces and moved into application support. He joined Synopsys, Inc., through an acquisition in 2004 and is currently the point of support for both internal and external customers. He earned a B.S.E.E. from Santa Clara University in 1992.

Henry Chang is vice president and co-founder of Designer's Guide Consulting, Inc. He is an expert in analog and mixed-signal design methodologies, next-generation tools and flows, and relevant standards in SoC design. During his career, Dr. Chang has designed and implemented several analog integrated circuits, written EDA software, developed SoC design methodologies, and spent extensive time working with designers to understand their short- and long-term needs and provide solutions. From 1995 to 2005, Dr. Chang worked at Cadence Design Systems, Inc., in research and development, methodology services, product marketing, corporate strategy, and the office of the chief technology officer. He is the author of three books: *A Top-Down, Constraint-Driven Design Methodology for Analog Integrated Circuits* (Springer, 1996), *Surviving the SoC Revolution: A Guide to Platform Based Design* (Springer, 1999), and *Winning the SoC Revolution: Experiences in Real Design in 2003* (Kluwer, 2003). Dr. Chang received his Ph.D. and M.S. in electrical engineering from the University of California, Berkeley in 1994 and 1992, respectively. He received his Sc.B. degree in electrical engineering from Brown University in 1989.

Pallab Chatterjee is chief technological officer at SiliconMap, LLC, a design and training services firm that started in 2000. Since 1985, and prior to SiliconMap, he served as president and chief technology officer of P&D Engineering Consultants, Inc., an independent consulting firm specializing in mixed-signal designs and CAD flows. Before working as an independent consultant, Mr. Chatterjee held R&D positions at Philips and Telmos Corp. He participates on the technical advisory board of several semiconductor, EDA, and IT/security companies. Further, Mr. Chatterjee holds and has held positions on the board of directors of the University of California at Berkeley Engineer Alumni Society, as the founder/conference chair for the Magma User Group, the IC section chair for the Mentor User Group, and on the organizing committee of the ISQED conference. He holds an M.S.E.E. from San Jose State University—where, along with at Silicon Valley Technical Institute, he has taught graduate coursework—and a B.S.E.E. from University of California, Berkeley.

Larry Cooke is an engineering consultant providing SoC design methodology, technical marketing, and intellectual property services. He was previously executive director for the VSI Alliance and later vice president of technical marketing at eASIC. Mr. Cooke helped start and held executive positions at Vertex, Crosspoint Solutions, Chameleon Systems, and On-Chip Technologies.

Ahmed Dabbagh is the design-for-reuse team leader at Freescale Semiconductor. He is one of the architects of the S12X micro-controller family and is also working at the New Product Development Center. Some of his many side activities include leading the "quality of RTL" effort and participating in external standardization activities such as the VSI Alliance. Before joining Freescale Semiconductor, Mr. Dabbagh worked for several years in the domain of micro-controllers and a wide range of microprocessor architectures ranging from 16 bits to super-scalar architectures employing 64-bit wide instruction words. He is the inventor of several patents in this important field. Mr. Dabbagh holds a Ph.D. in telecommunications and an M.S. (DEA) in images and signal processing from the University of Rennes in France.

Gulbin A. Ezer is a hardware design manager at Tensilica, Inc., working on configurable and extensible processor technologies. She has extensive industry experience in the architecture and design of dataflow supercomputers (Cydrome, Inc.), workstations (MIPS Inc.), general purpose, and multimedia processors (SGI, Inc.). Ms. Ezer also holds three patents in multimedia processor design. She has an M.S. degree in electrical and computer engineering from University of California at Santa Barbara and a B.S. degree in electrical engineering from Yale University.

Joachim Fader is a principal design engineer at Freescale Semiconductor. He works on design and hardware verification in the New Product Development Center. Currently, he works as a verification lead for products for the automotive industry. In earlier assignments, he was responsible for a broad range of products targeted toward the wireless and consumer market. His main interests are in the functional verification area, where he is spearheading the development of new methodologies, testbench architectures, verification IP, and other related topics. Mr. Fader holds a Dipl.-Ing. (FH) from Fachhochschule Furtwangen.

Saverio Fazzari is an associate at Booz Allen Hamilton. He has more than 15 years of electronics industry experience and has held a variety of roles in engineering, marketing, and field management. Mr. Fazzari has experience in using and supporting digital verification technology. He spent eight years at Cadence Design Systems, where he focused on IP technology and third-party support, including a recent role as the technical marketing director responsible for compliance testing in the Cadence OpenChoice IP program. Mr. Fazzari holds an M.S.E.E. from the University of Pittsburgh and a B.S.E.E. from Johns Hopkins University.

Dennis Fernández is managing partner for Fernández & Associates, LLC. He has more than 20 years of experience working in Silicon Valley in the high-tech industry as a patent prosecutor and intellectual property litigator, a venture capitalist, and an engineering manager. Mr. Fernández specializes in developing offensive and defensive patent strategies for start-up electronics, software, and biotech companies and their investors. He serves as strategic advisor to leading venture capital firms, including Sevin Rosen, Venrock, Charles River Ventures, and Walden International working with such clients as Marvell Technology, SiRF Technology, Ayala Corporation, Stanford University, and Northwestern University, as well as various start-up companies acquired by Cisco, Broadcom, Ciena, and Cadence Design Systems. Mr. Fernández also serves on the editorial board of the *Nanotechnology Law & Business Journal* and the board of directors of the Association of Patent Law Firms and the Science and Technology Advisory Council. He is also an inventor of several U.S. and international patents in digital television, sensor networks, and bioinformatics. Mr. Fernández has a law degree from Suffolk University Law School and an electrical engineering degree from Northwestern University.

Susan Fintz is an attorney at Wilson, Sossini, Goodrich, and Rosati. She provides contract revision and negotiation services for technology companies in Silicon Valley. Ms. Fintz has also advised individuals and companies on intellectual property strategy and the formation of legal entities. She is a parliamentarian and speaker for The Greater San Jose Hispanic Chamber of Commerce, and a member of the Silicon Valley Association for Start-up Entrepreneurs and the American Bar Association. Ms. Fintz holds J.D. degrees from the University of Miami and Mackenzie University in Brazil.

Jeff Freeman is a member of the technical staff in the transportation and product standard group, Microcontroller Division's Core Platform organization at Freescale Semiconductor in Austin, Texas. Prior to Freescale, he was with Motorola for 18 years. Mr. Freeman has written articles on test and synthesis methodology, microprocessors, and platform digital design. He has spoken at conferences in the United States as well as given presentations in Europe, the Middle East, and Japan. Mr. Freeman holds a B.S.E.E. from Texas A&M University.

Jack Horgan is managing editor of EDACafe, an on-line portal for the EDA, IP, PCB, and semiconductor industries, where he publishes weekly commentaries and quarterly financial revenues. He has dedicated almost his entire professional career to the CAX arena in various industries. Dr. Horgan first led a team for a large A&E firm that developed programming for various engineering disciplines. He later joined Applicon in the PCB area and then moved to the mechanical sector, where he managed software development during a period of explosive revenue growth. Dr. Horgan left Applicon to co-found Aires Technology, a pioneer company in desktop CAE, where he was vice president of product development and managed both hardware and software development. Afterwards, he moved to the West Coast and joined CADAM, Inc., a Lockheed subsidiary, as a technical advisor to the president where he ended up holding numerous marketing positions. Dr. Horgan then became senior vice president of marketing, sales, finance, and operations for MICROCADAM, a subsidiary of Tokyo-based CSC. He has advanced degrees in physics and computer science from the University of Massachusetts and Boston University.

Robert B. Likovich, Jr., has been an IBM employee for nine years and has worked extensively in the areas of design and verification.

Recently, he has been involved in I/O interface design and verification, specifically in the area of elastic interface designs.

Grant Martin is a chief scientist at Tensilica, Inc., and senior member of the IEEE. He previously worked for Burroughs in Scotland for six years, Nortel/BNR in Canada for 10 years, and Cadence Design Systems for nine years, eventually becoming a Cadence fellow in their labs. Mr. Martin is a co-author and co-editor for nine books dealing with SoC design, SystemC, UML, modeling, EDA for integrated circuits, and system-level design, including the first book on SoC design published in Russian. He has also presented many papers, talks, and tutorials and has participated in panels at major conferences. Mr. Martin was co-chair of the DAC technical program committee for methods from 2005 to 2006. His particular areas of interest include system-level design, IP–based design of SoC, platform-based design, and embedded software. Mr. Martin received his bachelor's and master's degrees in mathematics, with a specialization in combinatorics and optimization, from the University of Waterloo in Canada in 1977 and 1978, respectively.

Joseph D. Mendenhall has been an IBM employee for five years, primarily focusing on network hardware emulation and processor verification. He has worked most recently on developing a verification environment for elastic interface designs.

Dennis Moolenaar is an applications engineer at Tensilica, Inc. He previously worked at IMEC on low-power CAD tools and designed a custom processor for a GSM/DECT phone. Mr. Moolenaar later worked at National Semiconductor as a systems architect. He received his degree in computer architecture from the electrical engineering faculty of the Technical University in Delft, Netherlands.

Michael Rohleder is principal staff engineer, now working in design and hardware verification at Freescale's New Product Development Center in Munich, Germany. His current focus is on verification methodology, especially for the Joint Development Program between Freescale Semiconductor, Inc., and ST Microelectronics. Mr. Rohleder worked for about eight years as a software technologist on system-level design and full chip simulation. He also led process improvement activities, worked on semi-custom design systems, and was one of the architects of Motorola's Unified Design System, the predecessor of Stingray. Mr. Rohleder is a member of the SRS technical committee and is also active in external standardization activities such as the SystemVerilog development, SPIRIT, and the QiP Pillar at VSIA. He holds a Dipl.-Inform. from the Technical University of Munich.

Ross Segelken is a design engineer at nVidia. Prior to his current position, he was at Synopsys, Inc., where he led all digital design activities in the Hillsboro PHY group. He worked for five years at Intel Corporation on micro-architecture and circuit design for the Pentium 4 and next-generation IA32 processors. Mr. Segelken joined Accelerant Networks in 2002 and has worked on architecture and digital design for multiple generations of SerDes and PHY implementations. He obtained a bachelor's degree in electrical engineering from the California Institute of Technology in 1997.

Leena Singh is a senior staff manager and engineer at Qualcomme, Inc. She has more than 15 years of experience in ASIC design and verification and until recently was working for Rambus, Inc., as a verification architect responsible for verification IPs, models, and methodologies. Ms. Singh has worked as a methodology specialist in Cadence Designs Systems, Inc., for SystemVerilog, SystemC–based flows across design abstractions and platforms. Prior to that, she was a verification manager and architect for networking and multimedia start-ups, where she managed the entire verification cycle from setting up the verification teams to tapeouts. Ms. Singh has written several papers and books on verification, including *System-on-a-Chip Verification – Methodology and Techniques* (Kluwer, 2000) and *Advanced Verification Techniques: A SystemC–Based Approach for Successful Tapeout* (Springer, 2004).

Sean Smith is the chief verification architect at Denali Software. Prior to his current role, he was a verification engineer at Cisco Systems, where he led verification efforts on numerous systems, ASICs, and SoCs using a wide variety of verification techniques and technologies, which included directed random testing, assertion-based verification, and coverage-based verification. In addition to his work at Cisco, Mr. Smith was also active in industry standards bodies, including the IEEE and Accellera, and he is the moderator of the Specman mailing list.

Mike Stahl is an alliance manager and design technology scientist in the hardware systems and technology division at Hewlett-Packard in Fort Collins, Colorado. He has worked at Hewlett-Packard for 27 years in VLSI capacities of design engineer, product engineer, R&D project manager, manufacturing test and assembly product line manager, and technology alliance manager. Mr. Stahl holds an M.S. in electrical engineering from Colorado State University, an M.E. in engineering management from the University of Colorado, and a B.S. in electrical engineering from Kansas State University.

Neal Stollon is principal engineer at HDL Dynamics. Prior to his current position, he was a systems engineer and director of technical marketing with the first silicon solutions division of MIPS Technologies, Inc. Mr. Stollon has more than 25 years of experience in digital design and processor development, having worked at Texas Instruments, LSI Logic, Alcatel, and other leading companies. He has written more than 30 technical papers and holds seven patents. Mr. Stollon has a Ph.D. in electrical engineering from Southern Methodist University.

John T. Stonick is an application engineer with Synopsys Professional Services. His interests include system architecture, clock and data recovery, and use of adaptive digital techniques to compensate for analog circuit imperfections in wireless and wireline transceivers. Mr. Stonick is a member of the IEEE and a member of the ISSCC 2006 technical program committee (wireline). Starting in 2000, he was a principal design engineer with Accelerant Networks until they were acquired by Synopsys in 2004. From 1997 to 2000, Mr. Stonick was an assistant professor with the electrical and computer engineering department at Oregon State University and a co-director for the NSF Center for the Design of Analog-Digital Integrated Circuits (CDADIC). From 1993 to 1997, he held a postdoctoral research position in the electrical and computer engineering department at Carnegie Mellon University. He received his Ph.D. degree in electrical and computer engineering from North Carolina State University in 1992.

Adam Traidman is president of Chip Estimate Corporation, an EDA company focused on early chip-planning solutions. Prior to joining Chip Estimate, Mr. Traidman ran North America West sales at Hier Design, a FPGA design automation company acquired by Xilinx in 2004. His industry experience includes various management and technical roles at Adaptec, Monterey Design Systems, Texas Instruments, and the NASA Jet Propulsion Laboratory. He obtained a B.S. degree in computer and systems engineering from Rensselaer Polytechnic Institute.

Ken Umino is a consultant with Synopsys Professional Services. He has been involved with numerous chip design projects as an ASIC, RTL, and circuit designer for more than 15 years. For the past six years, he has been employed with Synopsys, working on various aspects of timing closure, physical design, and synthesis. Mr. Umino earned a B.S.E.E. and an M.S.E.E. from the University of Washington at Seattle.

Jason Upton recently joined the Synopsys Hillsboro PHY group to perform RTL to GDS implementations of high-speed SERDES PHYs. He has previously worked for Hewlett-Packard, Agilent Technologies, and Avant! Mr. Upton was an independent consultant for RTL to GDS services for five years before joining Synopsys in September 2005. He obtained a B.S.E.E.T. from Oregon Institute of Technology in 1993 and has completed many post-graduate courses at Oregon State University.

Richard Vlamynck is a verification engineer at Mentor Graphics, specializing in the integration of hardware and software. He previously operated an independent consulting firm and has worked for Ready Systems, Apple Computers, Microtec Research, and Mentor Graphics. His consulting clients include Raytheon Missiles and Space, Silicon Optix, Mentor Graphics, Mediaworks-ISI, Hughes, Hyperchip, and Nortel.

Chad E. Winemiller works at IBM, where he primarily focuses on ASIC designs, verification, and static timing analysis. Some of his past job responsibilities include network processor design and verification. Most recently, he has been involved in the verification of a high-speed elastic interface design utilizing Verisity's Specman.

Mobashar Yazdani is an ASIC program manager and technologist at Hewlett-Packard. As part of the HP Labs VLSI Group, he has worked on the development of Hewlett-Packard word architecture, and he has also developed the flow for design of large high-performance ASICs. Mr. Yazdani managed the ASIC/ASSP technology center in HP supply-chain operations for many years, and he is currently involved in global engineering services that support ASICs development for HP worldwide. He manages the central ASIC supplier assessment process, including IP. Mr. Yazdani has worked extensively on industry IP standardization efforts over the last eight years. His current interests include new models of ASIC development, IP standards, and ASIC outsourcing. Mr. Yazdani holds B.S.E.E. and M.S.E.E. degrees from the University of Texas at Austin and the University of Kentucky, respectively.

Acronym Guide

AFE	analog front end
AI	artificial intelligence
ASIC	application-specific integrated circuit
ATE	automated test equipment
BER	bit-error rate
BERT	bit-error rate tester
BGA	ball grid array
BIST	built-in self test
BU	bounded uncorrected
BUJ	bounded-uncorrelated jitter
CDF	cumulative distribution function
CDR	clock and data recovery
CDRC	clock and data recovery circuit
CJTPAT	compliant jitter tolerance pattern
CML	current-mode logic
CMOS	complementary metal-oxide semiconductor
CMRR	common mode rejection ratio
CMU	clock multiplication unit
CR	clock recovery
DAC	digital-to-analog converter
DCD	duty cycle distortion
DDJ	data-dependent jitter
DDN	data-dependent noise
DFE	decision-feedback equalizer
DFT	design for testability

DJ	deterministic jitter
DN	deterministic noise
DRSL	differential Rambus signaling level
DUT	device under test
EMC	electromagnetic compatibility
EMI	electromagnetic interference
EPA	edge placement accuracy
EPD	eye-pattern diagram
ESL	equivalent series inductance
ESR	equivalent series resistance
FC	Fibre Channel
FCC	Federal Communications Commission
FDTD	finite difference time domain
FEC	forward error correction
FEXT	far-end crosstalk
FFT	fast Fourier transformation
FIR	finite impulse response
FPGA	field programmable gate array
FT	Fourier transformation
GbE	Gigabit Ethernet
GJ	Gaussian jitter
GTL	Gunning transistor logic
HPJ	high-probability jitter
HSS	high-speed serial
HSSL	high-speed serial link
HVM	high-volume manufacture
IC	integrated circuit

IFFT	inverse fast Fourier transform
IIR	infinite impulsive response
ISI	intersymbol interference
ITRS	International Technology Roadmap for Semiconductors
LICA	low-inductance capacitor array
LTI	linear time invariant
MG	multiple Gaussian
MoM	method of moment
MTS	mega transfers per second
NEXT	near-end crosstalk
NF	noise figure
NRZ	non-return to zero
ODR	octal data rate
OIF	Optical Internetworking Forum
OS	over-sampling
PAM	pulse amplitude modulation
PBGA	plastic ball grid array
PC	personal computer
PCB	printed circuit board
PDF	probability density function
PDN	power distribution network
PEEC	partial element equivalent circuit
PHY	physical layer
PI	phase interpolator
PJ	periodic jitter
PLI	programming language interface
PLL	phase-locked loop

PMD	polarization mode dispersion
PRBS	pseudo-random bit stream
PRF	parallel resonant frequency
PSD	power spectrum density
PSRR	power supply noise rejection ratio
PTH	plated through hole
PVT	process, voltage, and temperature
PWB	printed wiring board
RC	resistive and capacitive
RJ	random jitter
RN	random noise
RSL	Rambus signaling level
Rx	receiver
SCA	static crosstalk analysis
SerDes	serializer/deserializer
SG	single Gaussian
SI	signal integrity
SoC	system on chip
SRF	self resonant frequency
SSC	spread spectrum clock
SSN	simultaneous switching noise
SSO	simultaneous switching output
STA	static timing analysis
TDR	time-domain reflectometer
TDT	time-domain transmission
TIA	time-interval analyzer
TJ	total jitter

TMM	transmission matrix method
TN	total noise
Tx	transmitter
UDSM	ultra-deep sub-micron
UI	unit interval
VCS	Verilog compiler simulator
VNA	vector network analyzer
VRM	voltage regulator module